The Facts On File

DICTIONARY of INORGANIC CHEMISTRY

The Facts On File DICTIONARY of INORGANIC CHEMISTRY

Edited by
John Daintith

Facts On File, Inc.

The Facts On File Dictionary of Inorganic Chemistry

Facts On File, Inc.
132 West 31st Street
New York NY 10001

Library of Congress Cataloging-in-Publication Data

The Facts On File dictionary of inorganic chemistry / edited by John Daintith.
p. cm.
Includes bibliographical references.
ISBN 0-8160-4926-2 (alk. paper).
1. Chemistry, Inorganic—Dictionaries. I. Title: Dictionary of inorganic chemistry. II. Daintith, John.

QD5.F34 2003
546'.03—dc22 2003049490

Facts On File books are available at special discounts when purchased in bulk quantities for businesses, associations, institutions, or sales promotions. Please call our Special Sales Department in New York at (212) 967-8800 or (800) 322-8755.

You can find Facts On File on the World Wide Web at http://www.factsonfile.com

Compiled and typeset by Market House Books Ltd, Aylesbury, UK

Printed in the United States of America

MP 10 9 8 7 6 5 4 3 2 1

This book is printed on acid-free paper

CONTENTS

Preface	vii
Entries A to Z	1
Appendixes	
I. The Periodic Table	244
II. The Chemical Elements	245
III. The Greek Alphabet	247
IV. Fundamental Constants	247
V. Webpages	248
Bibliography	248

PREFACE

This dictionary is one of a series covering the terminology and concepts used in important branches of science. *The Facts On File Dictionary of Inorganic Chemistry* has been designed as an additional source of information for students taking Advanced Placement (AP) Science courses in high schools. It will also be helpful to older students taking introductory college courses.

This volume covers inorganic chemistry and includes basic concepts in physical chemistry, classes of compound, reaction mechanisms, and many important named inorganic compounds. The entries on individual chemical elements are designed to give a basic survey of the chemistry of the element. The definitions are intended to be clear and informative and, where possible, we have illustrations of chemical structures. The book also has a selection of short biographical entries for people who have made important contributions to the field. There are appendixes providing a list of all the chemical elements and a periodic table. A short list of useful webpages and a bibliography are also included.

The book will be a helpful additional source of information for anyone studying the AP Chemistry course, especially the section on Descriptive Chemistry. It will also be useful to students of metallurgy and other related fields.

ACKNOWLEDGMENTS

Contributors

John O. E. Clark B.Sc.
Richard Rennie B.Sc., Ph.D.
Tom Shields B.Sc.

A

AAS *See* atomic absorption spectroscopy.

absolute temperature Symbol: T A temperature defined by the relationship:

$$T = \theta + 273.15$$

where θ is the Celsius temperature. The absolute scale of temperature was a fundamental scale based on Charles' law applied to an ideal gas:

$$V = V_0(1 + \alpha\theta)$$

where V is the volume at temperature θ, V_0 the volume at 0, and α the thermal expansivity of the gas. At low pressures, when real gases show ideal behavior, α has the value 1/273.15. Therefore, at $\theta = -273.15$ the volume of the gas theoretically becomes zero. In practice, of course, substances become solids at these temperatures. Nevertheless, the extrapolation can be used to create a scale of temperature on which –273.15 degrees Celsius (°C) corresponds to zero (0°). This scale was also known as the *ideal-gas scale*; on it temperature interval units were called *degrees absolute* (°A) or *degrees Kelvin* (°K), and were equal in size to the Celsius degree. It can be shown that the absolute temperature scale is identical to the THERMODYNAMIC TEMPERATURE scale, on which the temperature interval unit is the kelvin.

absolute zero The zero value of thermodynamic temperature; 0 kelvin or –273.15 degrees Celsius.

absorption A process in which a gas is taken up by a liquid or solid, or in which a liquid is taken up by a solid. In absorption, the substance absorbed goes into the bulk of the absorbing material. Solids that absorb gases or liquids often have a porous structure. The absorption of gases in solids is sometimes called *sorption*. *Compare* adsorption.

absorption indicator (adsorption indicator) An indicator used for titrations that involve a precipitation reaction. The method depends upon the fact that at the equivalence point there is a change in the nature of the ions absorbed by the precipitate particles. Fluorescein – a fluorescent compound – is commonly used. For example, in the titration of sodium chloride solution with added silver nitrate, silver chloride is precipitated. Sodium ions and chloride ions are absorbed in the precipitate. At the end point, silver ions and nitrate ions are in slight excess and silver ions are then absorbed. If fluorescein is present, negative fluorescein ions absorb in preference to nitrate ions, producing a pink complex.

absorption spectrum *See* spectrum.

abundance **1.** The relative amount of a given element among others; for example, the abundance of oxygen in the Earth's crust is approximately 50% by mass.
2. The amount of a nuclide (stable or radioactive) relative to other nuclides of the same element in a given sample. The *natural abundance* is the abundance of a nuclide as it occurs in nature. For instance, chlorine has two stable isotopes of masses 35 and 37. The abundance of ^{35}Cl is 75.5% and that of ^{37}Cl is 24.5%. For some elements the abundance of a particular nuclide depends on the source.

acac Abbreviation for the bidentate

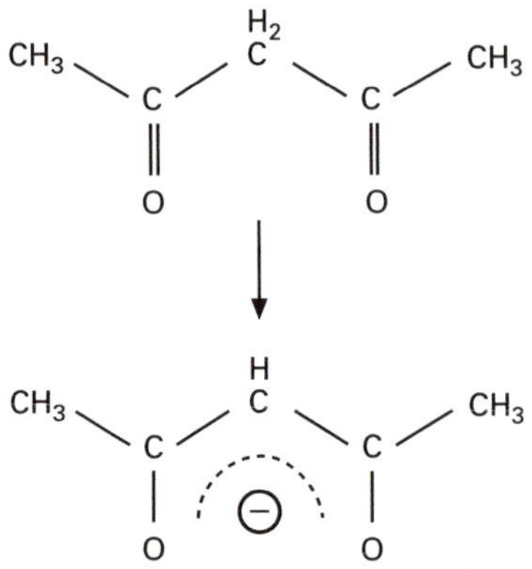

Acac: the bidentate acetylacetonato ligand formed from a diketone

acetyacetonato ligand, derived from acetylacetone ($CH_3COCH_2COCH_3$).

accelerator A CATALYST added to increase the rate at which a chemical reaction occurs.

acceptor The atom or group to which a pair of electrons is donated in a coordinate bond. Pi-acceptors are compounds or groups that accept electrons into pi, p or d orbitals.

accumulator (secondary cell; storage battery) An electric cell or battery that can be charged by passing an electric current through it. Because the chemical reaction in the cell is reversible, current passed through it in the opposite direction to which it supplies current will convert the reaction products back into their original forms. The most common example is the lead-acid battery used in automobiles and other vehicles powered by internal combustion engines.

acetate *See* ethanoate.

acetic acid *See* ethanoic acid.

acetyacetonato *See* acac.

acetylene *See* ethyne.

Acheson process *See* carbon.

achiral Describing a molecule that does not exhibit optical activity. *See* chirality.

acid A substance than contains hydrogen and dissociates in solution to give hydrogen ions:

$$HA \rightleftharpoons H^+ + A^-$$

More accurately, the hydrogen ion is solvated (a hydroxonium ion):

$$HA + H_2O \rightleftharpoons H_3O^+ + A^-$$

Strong acids are completely dissociated in water. Examples are sulfuric acid and tricholoroethanoic acid. *Weak acids* are only partially dissociated. Most organic carboxylic acids are weak acids. In distinction to an acid, a *base* is a compound that produces hydroxide ions in water. Bases are either ionic hydroxides (e.g. NaOH) or compounds that form hydroxide ions in water. These may be metal oxides, for example:

$$Na_2O + H_2O \rightarrow 2Na^+ + 2OH^-$$

Ammonia, amines, and other nitrogenous compounds can also form OH^- ions in water:

$$NH_3 + H_2O \rightleftharpoons NH_4^+ + OH^-$$

As with acids, *strong bases* are completely dissociated; *weak bases* are partially dissociated.

This idea of acids and bases is known as the *Arrhenius theory* (named for the Swedish physical chemist Svante August Arrhenius (1859–1927).

In 1923 the Arrhenius idea of acids and bases was extended by the British chemist Thomas Martin Lowry (1874–1936) and, independently, by the Danish physical chemist Johannes Nicolaus Brønsted (1879–1947). In the *Lowry–Brønsted theory* an acid is a compound that can donate a proton and a base is a compound that can accept a proton. Proton donators are called *Brønsted acids* (or *protic acids*) and proton acceptors are called *Brønsted bases*. For example, in the reaction:

$$CH_3COOH + H_2O \rightleftharpoons CH_3COO^- + H_3O^+$$

the CH_3COOH is the acid, donating a proton H^+ to the water molecule. The water is the base because it accepts the proton. In the reverse reaction, the H_3O^+ ion is the acid, donating a proton to the base

CH_3COO^-. If two species are related by loss or gain or a proton they are described as *conjugate*. So, in this example, CH_3COO^- is the *conjugate base* of the acid CH_3COOH and CH_3COOH is the *conjugate acid* of the base CH_3COO^-.

In a reaction of an amine in water, for example:

$$R_3N + H_2O \rightleftharpoons R_3NH^+ + OH^-$$

The amine R_3N accepts a proton from water and is therefore acting as a base. R_3NH^+ is its conjugate acid. Water donates the proton to the R_3N and, in this case, water is acting as an acid (H_3O^+ is its conjugate base). Note that water can act as both an acid and a base depending on the circumstances. It can accept a proton (from CH_3COOH) and donate a proton (to R_3N). Compounds of this type are described as *amphiprotic*.

One important aspect of the Lowry–Brønsted theory is that, because it involves proton transfers, it does not necessarily have to involve water. It is possible to describe reactions in nonaqueous solvents, such as liquid ammonia, in terms of acid–base reactions.

A further generalization of the idea of acids and bases was the *Lewis theory* put forward, also in 1923, by the US physical chemist Gilbert Newton Lewis (1875–1946). In this, an acid (a *Lewis acid*) is a compound that can accept a pair of electrons and a base (a *Lewis base*) is one that donates a pair of electrons.

acid-base indicator An indicator that is either a weak base or a weak acid and whose dissociated and undissociated forms differ markedly in color. The color change must occur within a narrow pH range. Examples are METHYL ORANGE and PHENOLPHTHALEIN.

acidic Having a tendency to release a proton or to accept an electron pair from a donor. In aqueous solutions the pH is a measure of the acidity, i.e. an acidic solution is one in which the concentration of H_3O^+ exceeds that in pure water at the same temperature; i.e. the pH is lower than 7. A pH of 7 is regarded as being neutral.

acidic hydrogen A hydrogen atom in a molecule that enters into a dissociation equilibrium when the molecule is dissolved in a solvent. For example, in ethanoic acid (CH_3COOH) the acidic hydrogen is the one on the carboxyl group, –COOH.

acidic oxide An oxide of a nonmetal that reacts with water to produce an acid or with a base to produce a salt and water. For example, sulfur(VI) oxide (sulfur trioxide) reacts with water to form sulfuric acid:

$$SO_3 + H_2O \rightarrow H_2SO_4$$

and with sodium hydroxide to produce sodium sulfate and water:

$$SO_3 + NaOH \rightarrow Na_2SO_4 + H_2O$$

See also amphoteric; basic oxide.

acidic salt *See* acid salt.

acidimetry A volumetric analysis or acid-base titration in which a standard solution of an acid is gradually added to the unknown (base) solution containing an indicator. In the converse procedure, *alkalimetry*, the standard solution is of a base and the unknown solution is acidic.

acidity constant *See* dissociation constant.

acid rain *See* pollution.

acid salt (acidic salt) A salt in which there is only partial replacement of the acidic hydrogen of an acid by metal or other cations. For polybasic acids the formulae for such salts are of the type $NaHSO_4$ (sodium hydrogensulfate) and $Na_3H(CO_3)_2.2H_2O$ (sodium sesquicarbonate). For monobasic acids such as HF the acid salts are of the form KHF_2 (potassium hydrogen difluoride). Although monobasic acid salts were at one time formulated as normal salts plus excess acid (i.e. KF.HF), it is preferable to treat them as hydrogen-bonded systems of the type $K^+(F–H–F)^-$.

actinic radiation Radiation that can cause a chemical reaction; for example, ultraviolet radiation is actinic. *See also* photochemistry.

actinides *See* actinoids.

actinium A soft, silvery-white, highly radioactive metallic element of group 3 (formerly IIIB) of the periodic table. It is usually considered to be the first member of the ACTINOID series. It occurs in minute quantities in uranium ores as a result of the natural radioactive decay of ^{235}U. The metal can be obtained by reducing AcF_3 with lithium or it can be produced by bombarding radium with neutrons. It is used as a source of alpha particles and has also been used to generate thermoelectric power. The metal glows in the dark; it reacts with water to produce hydrogen.

Symbol: Ac; m.p. 1050±50°C; b.p. 3200±300°C; r.d. 10.06 (20°C); p.n. 89; most stable isotope ^{227}Ac (half-life 21.77 years); other isotopes have very short half-lives.

actinoid contraction The decrease in the atomic or ionic radius that occurs in the actinoids as the atomic number increases from actinium through nobelium. The increase in atomic number in the actinoids is associated with the filling of the inner 5f subshell. It is similar to the LANTHANOID CONTRACTION.

actinoids (actinides) A group of 15 radioactive elements whose electronic configurations display filling of the 5f level. As with the lanthanoids, the first member, actinium, has no f electrons (Ac [Rn]$6d^17s^2$) but other members also show deviations from the smooth trend of f-electron filling expected from simple considerations, e.g. thorium Th [Rn]$6d^27s^2$, berkelium Bk [Rn]$5f^86d^17s^2$. The actinoids are all radioactive and their chemistry is often extremely difficult to study. The first eight, actinium, thorium, protactinium, uranium, neptunium, plutonium, americium, and curium occur naturally, although with the exception of thorium and uranium only in trace amounts. The others are generated by artificial methods using high-energy bombardment. *See also* transuranic elements.

activated charcoal *See* charcoal.

activated complex *See* transition state.

activated complex theory A theory of chemical reactions in which the rate at which chemical reactions take place is related to the rate at which the *transition state* (activated complex) is converted into products. Activated complex theory is sometimes known as TRANSITION STATE THEORY. It was put forward by Henry Eyring in 1935.

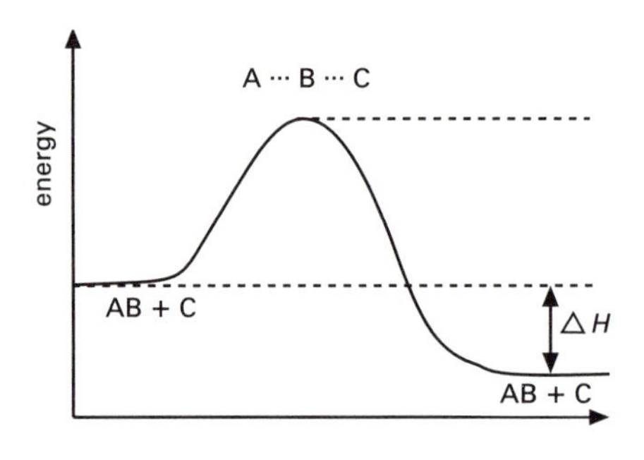

Activation energy

activation energy Symbol: E_a The minimum energy that a particle, molecule, ion, etc. must acquire before it can react; i.e. the energy required to initiate a reaction regardless of whether the reaction is exothermic or endothermic. Activation energy is often represented as an energy barrier that must be overcome if a reaction is to take place. *See* Arrhenius equation.

activator *See* promoter.

active mass *See* mass action.

activity **1.** Symbol: a A corrective concentration or pressure factor introduced into equations that describe real solvated systems. Certain thermodynamic properties of a solvated substance are dependent on its concentration (e.g. its tendency to react with other substances). Real substances show departures from ideal behavior and thus require such correction factors.

2. Symbol: A The average number of atoms disintegrating per unit time in a radioactive substance.

activity coefficient Symbol: f A measure of the degree of deviation from ideality of a solvated substance, defined as:

$$a = fc$$

where a is the activity and c the concentration. For an ideal solute $f = 1$; for real systems f can be less or greater than unity.

acyclic Describing a compound that is not cyclic (i.e. a compound that does not contain a ring in its molecules).

addition reaction A reaction in which additional atoms or groups of atoms are introduced into an unsaturated organic compound, such as an alkene or ketone. A simple example is the addition of bromine across the double bond in ethene:

$$H_2C{:}CH_2 + Br_2 \rightarrow BrH_2CCH_2Br$$

Addition reactions can be induced either by *electrophiles*, which are ions or molecules that are electron deficient and can therefore accept electrons, or by *nucleophiles*, which are ions or molecules that can donate electrons.

adduct *See* coordinate bond.

adiabatic change A change during which no heat enters or leaves the system.

In an adiabatic expansion of a gas, mechanical work is done by the gas as its volume (V) increases, its pressure (p) decreases, and its temperature (T) falls. For an ideal gas undergoing a reversible adiabatic change it can be shown that

$$pV^{\gamma} = K_1$$

$$T^{\gamma}p^{1-\gamma} = K_2$$

$$\text{and } TV^{\gamma-1} = K_3$$

where K_1, K_2, and K_3 are constants and γ is the ratio of the principal specific heat capacities. *Compare* isothermal change.

adsorbate A substance that is adsorbed on a surface. *See* adsorption.

adsorbent The substance on whose surface adsorption takes place. *See* adsorption.

adsorption A process in which a layer of atoms or molecules of one substance forms on the surface of a solid or liquid. All solid surfaces take up layers of gas from the surrounding atmosphere. The adsorbed layer may be held by chemical bonds (*chemisorption*) or by weaker van der Waals forces (*physisorption*).

Compare absorption.

adsorption indicator *See* absorption indicator.

aerosol *See* sol.

AES *See* atomic emission spectroscopy.

affinity The extent to which one substance is attracted to or reacts with another.

afterdamp *See* firedamp.

agate A hard microcrystalline form of the mineral chalcedony, which is a variety of quartz. Typically agate has greenish or brownish bands of coloration, and is used for making jewelry and ornaments. *Moss agate* is not banded, but has mosslike patterns resulting from the presence of iron and manganese oxides. Agate is also used in instrument bearings, because of its resistance to wear.

air The mixture of gases that surrounds the Earth. At sea level the composition of dry air, by volume, is nitrogen 78.08%, oxygen 20.95%, argon 0.93%, carbon dioxide 0.03%, neon 0.0018%, helium 0.0005%, krypton 0.0001%, and xenon 0.00001%.

Air also contains a variable amount of water vapor, as well as particulate matter (e.g. dust and pollen) and small amounts of other gases.

air gas *See* producer gas.

alabaster A dense, translucent, fine-grained mineral form of dihydrate CALCIUM SULFATE ($CaSO_4.2H_2O$). Due to its softness it is often carved and polished to make ornaments or works of art.

alchemy An ancient pseudoscience that was the precursor of chemistry, dating from early Christian times until the 17th century. It combined mysticism and experimental techniques. Many ancient alchemists searched for the *philosopher's stone* – a substance that could transmute base metals into gold and produce the *elixir of life*, a universal remedy for all ills.

alcohol A type of organic compound of the general formula ROH, where R is a hydrocarbon group. Examples of simple alcohols are methanol (CH_3OH) and ETHANOL (C_2H_5OH).

By definition alcohols have one or more –OH groups attached to a carbon atom that is not part of an aromatic ring. Thus, C_6H_5OH, in which the –OH group is attached to the ring, is a phenol whereas phenylmethanol ($C_6H_5CH_2OH$) is an alcohol.

Common alcohols are used as solvents, denaturing agents, chemical feedstocks, and in antifreeze preparations. Ethanol is the intoxicating ingredient in alcoholic beverages. Propane-1,2,3,-triol (glycerol) is used to make polymers, cosmetic emollients, and sweeteners.

alkali A water-soluble strong base. Strictly the term refers to the hydroxides of the alkali metals (group 1, formerly subgroup IA) only, but in common usage it refers to *any* soluble base. Thus borax solution may be described as mildly alkaline.

alkali metals (group 1 elements) A group of soft reactive metals, each representing the start of a new period in the periodic table and having an electronic configuration consisting of a rare-gas structure plus one outer electron. The alkali metals are lithium (Li), sodium (Na), potassium (K), rubidium (Rb), cesium (Cs), and francium (Fr). They formerly were classified in subgroup IA of the periodic table.

The elements all easily form positive ions M^+ and consequently are highly reactive (particularly with any substrate that is oxidizing). As the group is descended there is a gradual decrease in ionization potential and an increase in the size of the atoms; the group shows several smooth trends that follow from these facts. For example, lithium reacts in a fairly controlled way with water, sodium ignites, and potassium explodes. There is also a general decrease in the following: melting points, heats of sublimation, lattice energy of salts, hydration energy of M^+, ease of decomposition of nitrates and carbonates, and heat of formation of the '-ide' compounds (fluoride, hydride, oxide, carbide, chloride).

Lithium has the smallest ion and therefore has the highest charge/size ratio and is polarizing with a tendency towards covalent character in its bonding; the remaining elements form typical ionic compounds in which ionization, M^+X^- is regarded as complete. The slightly anomalous position of lithium is illustrated by the similarity of its chemistry to that of magnesium, in accordance with their diagonal relationship in the periodic table. For example, lithium hydroxide is much less soluble than the hydroxides of the other group 1 elements; lithium perchlorate is soluble in several organic solvents. Because of the higher lattice energies associated with smaller ions lithium hydride and nitride are fairly stable compared to NaH, which decomposes at 345°C. Na_2N, K_3N etc., are not obtained pure and decompose below room temperature.

The oxides also display the trend in properties as lithium forms M_2O with only traces of M_2O_2, sodium forms M_2O_2 and at high temperatures and pressures MO_2, potassium, rubidium, and cesium form M_2O_2 if oxygen is restricted but MO_2 if burnt in air. Hydrolysis of the oxides or direct reaction of the metal with water leads to the formation of the hydroxide ion.

Salts of the bases MOH are known for all acids and these are generally white crystalline solids. The ions M^+ are hydrated in water and remain unchanged in most reactions of alkali metal salts.

Because of the ease of formation of the ions M^+ there are very few complexes of the type ML_n^+ apart from solvated species of very low correlation times.

Francium is formed only by radioactive decay and in nuclear reactions; all the isotopes of francium have short half-lives, the longest of which (^{223}Fr) is 21 minutes. The few chemical studies that have been carried out on francium indicate that it has similar properties to those of the other alkali metals.

alkalimetry *See* acidimetry.

alkaline earth *See* alkaline-earth metals.

alkaline-earth metals (group 2 elements) A group of moderately reactive metals, harder and less volatile than the alkali metals. They were formerly classified in subgroup IIA of the periodic table. The term *alkaline earth* strictly refers to the oxides, but is often used loosely for the elements themselves. The electronic configurations are all those of a rare-gas structure with an additional two electrons in the outer s orbital. The elements are beryllium (Be), magnesium (Mg), calcium (Ca), strontium (Sr), barium (Ba), and radium (Ra). The group shows an increasing tendency to ionize to the divalent state M^{2+}. The first member, beryllium, has a much higher ionization potential than the others and the smallest atomic radius. Thus it has a high charge/size ratio and consequently the bonding in beryllium compounds is largely covalent. The chemistry of the heavier members of the group is largely that of divalent ions.

The group displays a typical trend toward metallic character as the group is descended. For example, beryllium hydroxide is amphoteric; magnesium hydroxide is almost insoluble in water and is slightly basic; calcium hydroxide is sparingly soluble and distinctly basic; and strontium and barium hydroxides are increasingly soluble in water and strongly basic. The group also displays a smooth trend in the solubilities of the sulfates ($MgSO_4$ is soluble, $CaSO_4$ sparingly soluble, and $BaSO_4$ very insoluble). The trend to increasing metallic character is also shown by the increase in thermal stabilities of the carbonates and nitrates with increasing relative atomic mass.

The elements all burn in air (beryllium must be finely powdered) to give the oxide MO (covalent in the case of beryllium) and for barium the peroxide BaO_2 in addition to BaO. The heavier oxides, CaO, SrO, and BaO, react with water to form hydroxides, $M(OH)_2$; magnesium oxide reacts only at high temperatures and beryllium oxide not at all. The metals Ca, Sr, and Ba all react readily with water to give the hydroxide:

$$M + 2H_2O \rightarrow M^{2+} + 2OH^- + H_2$$

In contrast, magnesium requires dilute acids in order to react (to the salt plus hydrogen), and beryllium is resistant to acid attack. A similar trend is seen in the direct reaction of hydrogen: under mild conditions calcium, strontium, and barium give ionic hydrides, high pressures are required to form magnesium hydride, and beryllium hydride can not be prepared by direct combination.

Because of its higher polarizing power, beryllium forms a range of complexes in which the beryllium atom should be treated as an electron acceptor (i.e. the vacant p orbitals are being used). Complexes such as etherates, acetylethanoates, and the tetrafluoride (BeF_4^{2-}) are formed, all of which are tetrahedral. In contrast Mg^{2+}, Ca^{2+}, Sr^{2+}, and Ba^{2+} have poor acceptor properties and form only weak complexes, even with donors such as ammonia or edta.

All isotopes of radium are radioactive and radium was once widely used for radiotherapy. The half-life of ^{226}Ra (formed by decay of ^{238}U) is 1600 years.

allotropy The ability of certain elements to exist in more than one physical form (*allotrope*). Carbon, sulfur, and phosphorus are the most common examples. Allotropy is more common in groups 14, 15, and 16 of the periodic table than in other groups. *See also* enantiotropy; monotropy. *Compare* polymorphism.

alloy A mixture of two or more metals (e.g. BRONZE or BRASS) or a metal with small amounts of nonmetals (e.g. STEEL). Alloys may be completely homogeneous mixtures or may contain small particles of one PHASE in the other phase. Alloys are stronger, harder, and often more corrosion resistant than their components, but exhibit reduced ductility and lower electrical conductivity. Most metals encountered in everyday life are actually alloys.

Alnico (*Trademark*) Any of a group of very hard brittle alloys used to make powerful permanent magnets. They contain nickel, aluminum, cobalt, and copper in various proportions. Iron, titanium, and niobium can also be present. They magnetize strongly when exposed to an exciting magnetic field and resist demagnitization even when exposed to a reverse magnetizing force.

alpha particle A He^{2+} ion emitted with high kinetic energy by a radioactive substance undergoing *alpha decay*. Alpha particles are emitted at high velocity, and are used to cause nuclear disintegration reactions.

alum A type of double salt. Alums are double sulfates obtained by crystallizing mixtures in the correct proportions. They have the general formula:

$$M_2SO_4.M'_2(SO_4)_3\ .24H_2O$$

where M is a univalent metal or ion, and M′ is a trivalent metal. Thus, ALUMINUM POTASSIUM SULFATE (called *potash alum*, or simply *alum*) is

$$K_2SO_4.Al_2(SO_4)_3.24H_2O$$

and aluminum ammonium sulfate (called *ammonium alum*) is

$$(NH_4)_2SO_4.Al_2(SO_4)_3.24H_2O$$

The name alum originally came from the presence of Al^{3+} as the trivalent ion, but is now also applied to other double salts containing trivalent ions, thus, chromium(III) potassium sulfate (*chrome alum*) is

$$K_2SO_4.Cr_2(SO_4)_3.24H_2O$$

alumina *See* aluminum oxide.

aluminate *See* aluminum hydroxide.

aluminosilicate *See* silicates.

aluminum A soft moderately reactive metal; the second element in group 13 (formerly IIIA) of the periodic table. Aluminum has the electronic structure of neon plus three additional outer electrons. There are numerous minerals of aluminum; it is the most common metallic element in the Earth's crust (8.1% by mass) and the third in order of abundance. Commercially important minerals are bauxite (hydrated Al_2O_3), corundum (anhydrous Al_2O_3), cryolite (sodium hexafluroaluminate Na_3AlF_6), and clays and mica (aluminosilicates).

The metal is produced on a massive scale by the Hall-Heroult method in which aluminum oxide, a nonelectrolyte, is dissolved in molten cryolite and electrolyzed in a large cell. The bauxite contains iron oxide and other impurities, which would contaminate the product, so the bauxite is dissolved in hot alkali, the impurities are removed by filtration, and the pure aluminum oxide then precipitated by acidification. In the cell, molten aluminum is tapped off from the base and carbon dioxide evolved at the graphite anodes, which are consumed in the process. The aluminum atom is much bigger than boron (the first member of group 13) and its ionization potential is not particularly high. Consequently aluminum forms positive Al^{3+} ions. However, aluminum also has nonmetallic chemical properties. Thus, it is amphoteric and also forms a number of covalently bonded compounds.

Unlike boron, aluminum does not form a vast range of hydrides – AlH_3 and Al_2H_6 may exist at low pressures, and the only stable hydride, $(AlH_3)_n$, must be prepared by reduction of aluminum trichloride. The ion AlH_4^- is widely used in the form of $LiAlH_4$ as a powerful reducing agent.

The reaction of aluminum metal with oxygen is very exothermic but at ordinary temperatures an impervious film of the oxide protects the bulk metal from further attack. This oxide film also protects aluminum from oxidizing acids. There is only one oxide, Al_2O_3, but a variety of polymorphs and hydrates are known. It is relatively inert and has a high melting point, and for this reason is widely used as a furnace lining and for general refractory brick. Aluminum metal will react with alkalis, releasing hydrogen to initially produce $Al(OH)_3$, then $Al(OH)_4^-$.

Aluminum reacts readily with the halogens; in the case of chlorine thin sheets of the metal will burst into flame. Aluminum fluoride has a high melting point (1290°C) and is ionic. The other halides are dimers in the vapor phase (two halogen bridges). Aluminum also forms a sulfide (Al_2S_3), nitride (AlN), and carbide (Al_4C), the latter two at extremely high temperatures.

Because of aluminum's ability to expand its coordination number and tendency towards covalence it forms a variety of complexes such as AlF_6^{2-} and $AlCl_4^-$.

Symbol: Al; m.p. 660.37°C; b.p. 2470°C; r.d. 2.698 (20°C); p.n. 13; r.a.m. 26.981539.

aluminum acetate *See* aluminum ethanoate.

aluminum bromide ($AlBr_3$) A white solid soluble in water and many organic solvents.

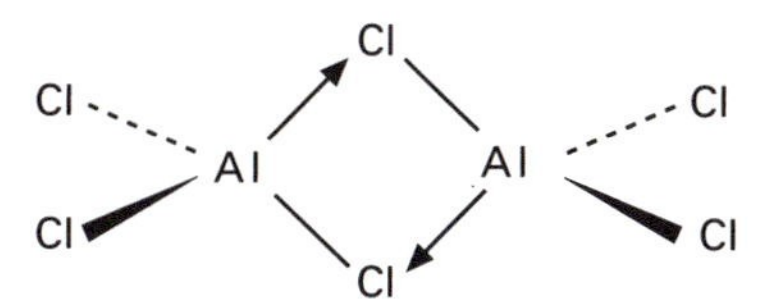

Aluminum chloride: the dimer Al_2Cl_6

aluminum chloride ($AlCl_3$) A white covalent solid that fumes in moist air and reacts violently with water according to the equation:

$$AlCl_3 + 3H_2O \rightarrow Al(OH)_3 + 3HCl$$

It is prepared by heating aluminum in dry chlorine or dry hydrogen chloride or industrially by heating aluminum oxide and carbon in the presence of chlorine. Vapor-density measurements show that its structure is a dimer; it consists of Al_2Cl_6 molecules in the vapor. Aluminum chloride is used as a catalyst in various organic reactions, and in the cracking of petroleum.

aluminum ethanoate (aluminum acetate; $Al(OOCCH_3)_3$) A white solid soluble in water. It is usually obtained as the dibasic salt, *basic aluminum ethanoate*, $Al(OH)(CH_3COO)_2$. It is prepared by dissolving aluminum hydroxide in ethanoic acid and is used extensively as a mordant in dyeing, as a size for paper and cardboard products, and in tanning. The solution is hydrolyzed and contains various complex aluminum-hydroxyl species and colloidal aluminum hydroxide.

aluminum fluoride (AlF_3) A white crystalline solid that is slightly soluble in water but insoluble in most organic solvents. Its primary use is as an additive to the cryolite (Na_3AlF_6) electrolyte in the production of aluminum.

aluminum hydroxide (aluminate; $Al(OH)_3$) A white powder prepared as a colorless gelatinous precipitate by adding ammonia solution or a small amount of sodium hydroxide solution to a solution of an aluminum salt. It is an amphoteric hydroxide and is used as a foaming agent in fire extinguishers and as a mordant in dyeing.

Its amphoteric nature causes it to dissolve in excess sodium hydroxide solution to form the *aluminate* (tetrahydroxoaluminate(III) ion):

$$Al(OH)_3 + OH^- \rightarrow Al(OH)_4^- + H_2O$$

When precipitating from solution, aluminum hydroxide readily absorbs colored matter from dyes to form lakes.

aluminum nitrate ($Al(NO_3)_3.9H_2O$) A hydrated white crystalline solid prepared by dissolving freshly prepared aluminum hydroxide in nitric acid. It is used as a mordant. It cannot be prepared by the action of dilute nitric acid on aluminum because the metal is rendered passive by a thin surface layer of oxide.

aluminum oxide (alumina; Al_2O_3) A white crystalline powder that is almost insoluble in water, occurring in two main forms, one of which is weakly acidic, and the other amphoteric. It occurs naturally as bauxite, corundum, and emery, and with minute amounts of chromium and cobalt as ruby and sapphire, respectively. It is manufactured by heating aluminum hydroxide. It is used in the extraction by electrolysis of aluminum, as an abrasive (corundum), in furnace linings (because of its refractory properties), and as a catalyst (e.g. in the dehydration of alcohols).

aluminum potassium sulfate (potash alum; $Al_2(SO_4)_3.K_2SO_4.24H_2O$) A white crystalline solid, soluble in water but insoluble in alcohol, prepared by mixing solutions of ammonium and aluminum sulfates followed by crystallization. It is used as a mordant for dyes, as a waterproofing agent, and as a tanning additive.

aluminum sulfate ($Al_2(SO_4)_3.18H_2O$, $A1_25O_4$) A white crystalline solid. Both the hydrated and anhydrous forms are soluble in water, but only the anhydrous form is soluble in ethanol, and to only a slight degree. It is used as a size for paper, a precipitating agent in sewage and water treatment, a foaming agent in fire control, and as a fireproofing agent. Its solutions are acidic by hydrolysis, containing such species as $Al(H_2O)_5(OH)^{2+}$. It is prepared by dissolving $Al(OH)_3$ in sulfuric acid.

aluminum trimethyl *See* trimethylaluminum.

amalgam An alloy of mercury with at least one other metal. Amalgams may be liquid or solid. An amalgam of sodium (Na/Hg) with water is used as a source of NASCENT HYDROGEN:

$$2Na/Hg + 2H_2O\ (l) \rightarrow 2NaOH(aq) + H_2(g) + Hg$$

An amalgam of Ha/Ag/Sn was used in dentistry.

amatol A high explosive that consists of a mixture of ammonium nitrate and TNT (trinitrotoluene).

ambidentate ligand *See* isomerism.

americium A highly toxic radioactive silvery element of the actinoid series of metals. A transuranic element, it is found naturally on Earth in trace amounts in uranium ore. It can also be synthesized by bombarding ^{239}Pu with neutrons. The metal can be obtained by reducing the trifluoride with barium. It reacts with oxygen, steam, and acids. ^{241}Am has been used in gamma-ray radiography and in smoke alarms.

Symbol: Am; m.p. 1172°C; b.p. 2607°C; r.d. 13.67 (20°C); p.n. 95; most stable isotope ^{243}Am (half-life 7.37×10^3 years).

amethyst A purple form of the mineral quartz (silicon(IV) oxide, SiO_2) used as a semiprecious gemstone. The color comes from impurities such as oxides of iron.

ammine A complex in which ammonia molecules are coordinated to a metal ion; e.g. $[Cu(NH_3)_4]^{2+}$.

ammonia (NH_3) A colorless gas with a characteristic pungent odor. On cooling and compression it forms a colorless liquid, which becomes a white solid on further cooling. Ammonia is very soluble in water (a saturated solution at 0°C contains 36.9% of ammonia): the aqueous solution is alkaline and contains a proportion of free ammonia. Ammonia is also soluble in

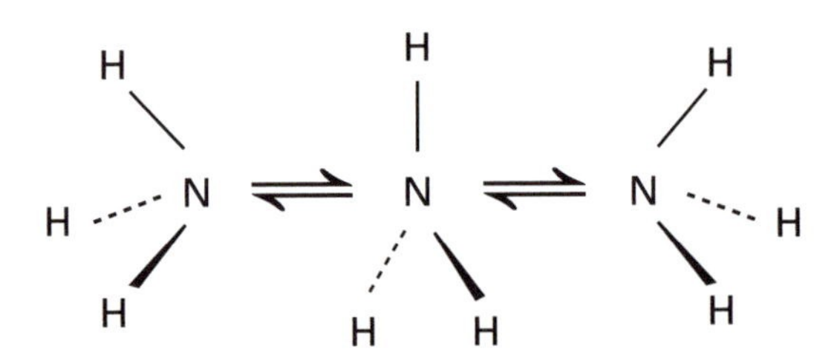

Ammonia: umbrella inversion of the molecule

ethanol. It occurs naturally to a small extent in the atmosphere, and is usually produced in the laboratory by heating an ammonium salt with a strong alkali. Ammonia is synthesized industrially from hydrogen and atmospheric nitrogen by the HABER PROCESS.

The compound does not burn readily in air, but ignites – giving a yellowish-brown flame – in oxygen. It will react with atmospheric oxygen in the presence of platinum or a heavy metal catalyst – a reaction used as the basis of the commercial manufacture of nitric acid, which involves the oxidation of ammonia to nitrogen monoxide and then to nitrogen dioxide. Ammonia coordinates readily to form ammines and reacts with sodium or potassium to form inorganic amides and with acids to form ammonium salts; for example, it reacts with hydrogen chloride to form ammonium chloride:

$$NH_3(g) + HCl(g) \rightarrow NH_4Cl(g)$$

Ammonia is used commercially in the manufacture of fertilizers, mainly ammonium nitrate, urea, and ammonium sulfate. It is also used to make explosives, resins, and dyes. As a liquefied gas it is used in the refrigeration industry. Liquid ammonia is an excellent solvent for certain substances, which ionize in the solutions to give ionic reactions similar to those occurring in aqueous solutions. Ammonia is marketed as the liquid, compressed in cylinders ('anhydrous ammonia'), or as aqueous solutions of various strengths. *See also* ammonium hydroxide.

ammoniacal Describing a solution in aqueous ammonia.

ammonia-soda process *See* Solvay process.

ammonia solution *See* ammonium hydroxide.

ammonium alum *See* alum.

ammonium carbonate (sal volatile; $(NH_4)_2CO_3$) A white solid that crystallizes as plates or prisms. It is very soluble in water and readily decomposes on heating to ammonia, carbon dioxide, and water. The white solid sold commercially as ammonium carbonate is actually a double salt of both ammonium hydrogencarbonate (NH_4HCO_3) and ammonium aminomethanoate ($NH_2CO_2NH_4$). This salt is manufactured from ammonium chloride and calcium carbonate. It decomposes on exposure to damp air into ammonium hydrogencarbonate and ammonia, and it reacts with ammonia to give the true ammonium carbonate. Commercial ammonium carbonate is used in baking powders, smelling salts, in the dyeing and wool-scouring industries, and in cough medicines.

ammonium chloride (sal ammoniac; NH_4Cl) A white crystalline solid with a characteristic saline taste. It is very soluble in water. Ammonium chloride can be manufactured by the action of ammonia on hydrochloric acid. It sublimes on heating according to the equilibrium reaction:

$$NH_4Cl(s) \rightleftharpoons NH_3(g) + HCl(g)$$

Ammonium chloride is used in galvanizing, as a flux for soldering, as a mordant in dyeing and calico printing, and in the manufacture of Leclanché and 'dry' cells.

ammonium hydroxide (ammonia solution; NH_4OH) An alkali that is formed when ammonia dissolves in water. It probably contains hydrated ammonia molecules as well as some NH_4^+ and OH^- ions. It is a useful reagent and cleansing agent.

ammonium ion The ion NH_4^+, formed by coordination of NH_3 to H^+. It has tetrahedral symmetry.

ammonium nitrate (NH_4NO_3) A colorless crystalline solid that is very soluble in water and also soluble in ethanol. It is usually manufactured by the action of ammonia gas on nitric acid. It is used in fertilizers because of its high nitrogen content, and in the manufacture of explosives and rocket propellants.

ammonium phosphate (triammonium phosphate(V); $(NH_4)_3PO_4$) A colorless crystalline salt made from ammonia and phosphoric(V) acid, used as a fertilizer to add both nitrogen and phosphorus to the soil.

ammonium sulfate ($(NH_4)_2SO_4$) A colorless crystalline solid that is soluble in water but not in ethanol. When heated carefully it gives ammonium hydrogensulfate, which on stronger heating yields nitrogen, ammonia, sulfur(IV) oxide (sulfur dioxide), and water. Ammonium sulfate is manufactured by the action of ammonia on sulfuric acid. It is an important ammonium salt because of its widespread use as a fertilizer. Its only drawback as a fertilizer is that it tends to leave an acidic residue in the soil.

amorphous Describing a solid substance that has no 'long-range' regular arrangement of atoms; i.e. is not crystalline. Amorphous materials can consist of minute particles that possess order over very short distances. Glasses are amorphous, because the atoms in the solid have a random arrangement. X-ray diffraction analysis has shown that many substances that were once described as amorphous are in fact composed of very small crystals. For example, charcoal, coke, and soot (all forms of carbon) are made up of small graphitelike crystals.

amount of substance Symbol: n A measure of the number of elementary entities present in a substance. It is measured in MOLES. *See also* Avogadro constant.

ampere Symbol: A The SI base unit of electric current, defined as the constant current that, maintained in two straight parallel infinite conductors of negligible circular cross section placed one meter apart in vacuum, would produce a force between the conductors of 2×10^{-7} newton per meter.

amphiprotic *See* acid; solvent.

ampholyte ion *See* zwitterion.

amphoteric Describing material that can display both acidic and basic properties. The term is most commonly applied to the oxides and hydroxides of metals that can form both cations and complex anions. For example, zinc oxide dissolves in acids to form zinc salts and also dissolves in alkalis to form zincates, $[Zn(OH)_4]^{2-}$. The amino acids are also considered to be amphoteric because they contain both acidic and basic groups.

amu *See* atomic mass unit.

analysis The process of determining the constituents or components of a sample. There are two broad major classes of analysis, QUALITATIVE ANALYSIS – essentially answering the question 'what is it?' – and QUANTITATIVE ANALYSIS – answering the question 'how much of such and such a component is present?' There is a vast number of analytical methods that can be applied, depending on the nature of the sample and the purpose of the analysis. These include GRAVIMETRIC ANALYSIS, VOLUMETRIC, and systematic qualitative analysis (classical wet methods); and instrumental methods, such as CHROMATOGRAPHY, SPECTROSCOPY, NUCLEAR MAGNETIC RESONANCE, POLAROGRAPHY, and fluorescence techniques.

ångstrom Symbol: Å A unit of length defined as 10^{-10} meter, formerly used to measure wavelengths of radiation, including those of visible light, and inter-molecular distances. The preferred SI unit for such measurements is the nanometer. One ångstrom equals 0.1 nanometer. The ångstrom was named for Anders Jonas Ångstrom (1814–74), a Swedish physicist and astronomer.

anhydride A compound formed by removing water from an acid or, less commonly, a base. Many nonmetal oxides are anhydrides of acids: for example CO_2 is the anhydride of H_2CO_3 and SO_3 is the anhydride of H_2SO_4.

anhydrite *See* calcium sulfate.

anhydrous Describing a substance that lacks moisture, or a salt lacking water of crystallization. For example, on strong heating, blue crystals of copper(II) sulfate pentahydrate, $CuSO_4.5H_2O$, form white anhydrous copper(II) sulfate, $CuSO_4$.

anion A negatively charged ion, formed by the addition of electrons to atoms or molecules. In electrolysis anions are attracted to the positive electrode or anode. *Compare* cation.

anionic detergent *See* detergents.

anionic resin An ION-EXCHANGE material that can exchange anions, such as Cl^- and OH^-, for anions in the surrounding medium. Such resins are often produced by the addition of a quaternary ammonium group ($-N(CH_3)_3^+$) or a phenolic group ($-OH^-$) to a stable polyphenylethene resin. A typical exchange reaction is:

$$\text{resin}-N(CH_3)_3^+Cl^- + KOH \rightleftharpoons \text{resin}-N(CH_3)_3^+OH^- + KCl$$

Anionic resins can be used to separate mixtures of halide ions. Such mixtures can be attached to the resin and recovered separately by ELUTION.

anisotropic Describing certain substances that have one or more physical properties, such as refractive index, that differ according to direction. Most crystals are anisotropic.

annealing A type of heat treatment applied to metals to change their physical properties. The metal is heated to, and held at, an appropriate temperature before being cooled at a suitable rate to produce the desired grain structure. Annealing is most commonly used to remove the stresses that have arisen during rolling, to increase the softness of the metal, and to make it easier to machine. Objects made of glass can also be annealed to remove strains.

anode In electrolysis, the electrode that is at a positive potential with respect to the cathode, and to which anions are therefore attracted. In any electrical system, such as a discharge tube or electronic device, the anode is the terminal from which electrons flow out of the system.

anode sludge *See* electrolytic refining.

anodizing An industrial process for protecting aluminum with an oxide layer. The aluminum object is made the anode in an electrolytic cell containing an oxidizing acid (e.g. sulfuric(VI) acid). The layer of Al_2O_3 formed is porous and can be colored with certain dyes.

anthracite The highest grade of COAL, with a carbon content of between 92% and 98%. It burns with a hot blue flame, gives off little smoke and leaves hardly any ash.

antibonding orbital *See* orbital.

***anti*-isomer** *See* isomerism.

antimonic Designating an antimony(V) compound.

antimonous Designating an antimony(III) compound.

antimony A toxic metalloid element of group 15 (formerly VA) of the periodic table. It exists in three allotropic forms; the most stable is a hard, brittle, silvery-blue metal. Yellow and black antimony, the other two allotropes, are unstable and nonmetallic. Antimony is found in many minerals, principally stibnite (Sb_2S_3), from which it is recovered by reduction with iron or by first roasting it to yield the oxide. It is also a poor conductor of heat and electricity, making it useful in the manufacture of semiconductors. Antimony compounds are used in pigments, flame retardants, medical treatments, ceramics, and glass.

Symbol: Sb; m.p. 630.74°C; b.p. 1750°C; r.d. 6.691; p.n. 51; r.a.m. 112.74.

antimony(III) chloride (antimony trichloride; $SbCl_3$) A white deliquescent solid, formerly known as butter of antimony. It is prepared by direct combination of antimony and chlorine. It is readily hydrolyzed by cold water to form a white precipitate of *antimony(III) chloride oxide* (antimonyl chloride, SbOCl):

$$SbCl_3 + H_2O = SbOCl + 2HCl$$

antimony(III) chloride oxide *See* antimony(III) chloride.

antimonyl chloride *See* antimony(III) chloride.

antimony(III) oxide (antimony trioxide; Sb_2O_3) A white insoluble solid. It is an amphoteric oxide with a strong tendency to act as a base. It can be prepared by direct oxidation by air, oxygen, or steam and is formed when antimony(III) chloride is hydrolyzed by excess boiling water. It is used as a flame retardant in plastics and as an additive in paints to make them more opaque.

antimony(V) oxide (antimony pentoxide; Sb_2O_5) A yellow solid. It is usually formed by the action of concentrated nitric acid on antimony or by the hydrolysis of antimony(V) chloride. Although an acidic oxide, it is only slightly soluble in water. It is used as a flame retardant.

antimony pentoxide *See* antimony(V) oxide.

antimony trichloride *See* antimony(III) chloride.

antimony trioxide *See* antimony(III) oxide.

antioxidant A substance that inhibits oxidation. Antioxidants are added to such products as foods, paints, plastics, and rubber to delay their oxidation by atmospheric oxygen. Some work by forming CHELATES with metal ions, thus neutralizing the catalytic effect of the ions in the oxidation process. Other types remove intermediate oxygen FREE RADICALS. Naturally occurring antioxidants can limit tissue or cell damage in the body. They include vitamins C and E, and β-carotene.

antiparallel spins Spins of two neighboring electrons in which the magnetic moments associated with electron spin are aligned in opposite directions.

apatite A common, naturally occurring phosphate of calcium, $Ca_5(PO_4)_3(OH,F,Cl)$, that occurs in several color varieties. Crystals are hexagonal and have a greasy luster. Apatite occurs in rocks of igneous and metamorphic origin and is mined as a source of phosphorus for use in fertilizers. Gemstone quality crystals are known from several locations.

aprotic *See* solvent.

aqua fortis An old name for nitric acid, HNO_3.

aqua regia A mixture of concentrated nitric acid and three to four parts of concentrated hydrochloric acid. With the exception of silver, with which it forms an insoluble chloride, it dissolves all metals, including gold. The mixture contains chlorine and NOCl (nitrosyl chloride). The name means 'royal water'.

aqueous Describing a solution in water.

aragonite An anhydrous mineral form of calcium carbonate, $CaCO_3$, which occurs associated with limestone and in some metamorphic rocks. It is also the main ingredient of pearls. It is not as stable as calcite, into which it may change over time. Pure aragonite is colorless or white, but impurities such as strontium, zinc, or lead may tint it various colors.

argentic oxide *See* silver(II) oxide.

argentous oxide *See* silver(I) oxide.

argon An inert colorless odorless monatomic element of the rare-gas group. It forms 0.93% by volume of air, from which it is obtained by fractional distillation. Argon is used to provide an inert atmosphere in electric and fluorescent lights, in aluminum welding, and in titanium and silicon extraction. The element forms no known compounds.

Symbol: Ar; m.p. −189.37°C; b.p. −185.86°C; r.d. 0.0001 784 (0°C); p.n. 18; r.a.m. 39.95.

Arrhenius, Svante August (1859–1927) Swedish physical chemist who first postulated that the electrical conductivity of electrolytes is due to the dissolved substance being dissociated into electrically charged particles (ions). He put forward this idea in 1883 and developed it in 1887. Arrhenius also made a major contribution to the theory of chemical reactions in 1889 when he suggested that a molecule can take part in a chemical reaction only if its energy is higher than a certain value. This gave rise to the *Arrhenius equation* relating the rate of a chemical reaction to the absolute temperature. He also performed calculations that led him to the idea of the greenhouse effect. Arrhenius won the 1903 Nobel Prize for chemistry for his theory of electrolytes.

Arrhenius equation An equation, proposed by Svante Arrhenius in 1889, that relates the rate constant of a chemical reaction to the temperature at which the reaction is taking place:

$$k = A\exp(-E_a/RT)$$

where A is a constant for the given reaction, k the rate constant, T the thermodynamic temperature in kelvins, R the gas constant, and E_a the activation energy of the reaction.

Reactions proceed at different rates at different temperatures, i.e. the magnitude of the rate constant is temperature dependent. The Arrhenius equation is often written in a logarithmic form, i.e.

$$\log_e k = \log_e A - E_a/RT$$

This equation enables the ACTIVATION ENERGY for a reaction to be determined. The equation can also be applied to problems dealing with diffusion, viscosity, electrolytic conduction, etc.

Arrhenius theory *See* acid; base.

arsenate(III) (arsenite) A salt of the hypothetical arsenic(III) acid, formed by reacting arsenic(III) oxide, A_2O_3, with alkalis. Arsenate(III) salts contain the ion AsO_3^{3-}. Copper arsenate(III) is used as an insecticide.

arsenate(V) A salt of arsenic(V) acid, made by reacting arsenic(III) oxide, As_2O_3, with nitric acid. Arsenate(V) salts contain the ion AsO_4^{3-}. Disodiumhydrogenarsenate(V) is used in printing calico.

arsenic A toxic metalloid element of group 15 (formerly VB) of the periodic table. It exists in three allotropic forms; the most stable is a brittle gray metal. Yellow and black arsenic, the other allotropes, are not metallic. Arsenic is found native and in several ores including mispickel or arsenopyrite (FeSAs), realgar (As_4S_4), and orpiment (As_2S_3). The ores are roasted to produce arsenic(III) oxide, which is then reduced with carbon or hydrogen to recover the element. Arsenic reacts with hot acids and molten sodium hydroxide but is unaffected by water, acids, or alkalis at normal temperatures. It is used in semiconductors, alloys, lasers, and fireworks. Arsenic and its compounds are poisonous and therefore also find use in insecticides, rodentricides, and herbicides.

Symbol: As; m.p. 817°C (gray) at 3 MPa pressure; sublimes at 616°C (gray); r.d. 5.78 (gray at 20°C); p.n. 33; r.a.m. 74.92159.

arsenic(III) chloride (arsenious chloride; $AsCl_3$) A poisonous oily liquid. It fumes in moist air due to hydrolysis with water vapor:

$$AsCl_3 + 3H_2O = As_2O_3 + 6HCl$$

Arsenic(III) chloride is covalent and exhibits nonmetallic properties.

arsenic hydride *See* arsine.

arsenic(III) oxide (white arsenic; arsenious oxide; As_2O_3) A colorless crystalline solid that is very poisonous (0.1 g would be a lethal dose). Analysis of the solid and vapor states suggests a dimerized structure of As_4O_6. An amphoteric oxide, arsenic(III) oxide is sparingly soluble in water, producing an acidic solution. It is formed when arsenic is burned in air or oxygen. It is used as an insecticide, herbicide, and defoliant, and in the manufacture of glass and ceramics having a milky iridescence.

arsenic(V) oxide (arsenic oxide; As_2O_5) A white amorphous deliquescent solid. It is an acidic oxide prepared by dissolving arsenic(III) oxide in hot concentrated nitric acid, followed by crystallization then heating to 210°C.

arsenide A compound of arsenic and another metal. For example, with iron arsenic forms iron(III) arsenide, $FeAs_2$, while with gallium arsenic forms gallium arsenide, GaAs. Gallium arsenide is an important semiconductor.

arsenious chloride *See* arsenic(III) chloride.

arsenious oxide *See* arsenic(III) oxide.

arsenite *See* arsenate(III).

arsine (arsenic hydride; AsH_3) A highly poisonous colorless gas with an unpleasant smell. It is produced by reacting mineral acids with arsenides or by reducing arsenic compounds with nascent hydrogen. Arsine decomposes to arsenic and hydrogen at 230°C. This phenomenon is put to use in *Marsh's test* for arsenic, in which arsine generated from a sample is fed through a glass tube. Here the arsine, if present, decomposes to leave a brown deposit of arsenic, which can be distinguished from antimony by the fact that antimony will not dissolve in NaOC1. Arsine is used to make *n*-type semiconductors doped with trace amounts of arsenic.

artificial radioactivity Radioactivity induced by bombarding stable nuclei with high-energy particles. For example:

$$^{27}_{13}Al + ^{1}_{0}n \rightarrow ^{24}_{11}Na + ^{4}_{2}He$$

represents the bombardment of aluminum with neutrons to produce an isotope of sodium and helium nuclei (alpha particles). All transuranic elements above curium, atomic number 96, are artificially radioactive because they do not occur in nature. Even neptunium, plutonium, americium, and curium, elements 93 through 96, are only found in very minute amounts in nature and thus are also usually produced artificially.

asbestos Any of several fibrous varieties of various rock-forming silicate minerals, such as the amphiboles and chrysotile. Asbestos has many uses that employ its properties of exceptional heat-resistance, chemical inertness, and electrical resistance. It is, for example, spun and woven into fabric that is used to make fireproof clothing and brake linings, uses for which there are currently no adequate substitutes. However, prolonged exposure to asbestos dust may cause asbestosis – a serious, progressive disease that eventually leads to respiratory failure. Mesothelioma, a malignant cancer of the membrane enclosing the lung, may also result.

aspirator An apparatus for sucking a gas or liquid from a vessel or body cavity.

associated liquids *See* association.

association The combination of molecules of a substance with those of another to form more complex species. An example is a mixture of water and ethanol (which are termed *associated liquids*), the molecules of which combine via hydrogen bonding.

astatine A radioactive halogen element of group 17 (formerly VIIA) of the periodic table. It occurs in minute quantities in uranium ores as the result of radioactive decay, and can be artifically created by bombarding ^{200}Bi with alpha particles. Many short-lived radioisotopes are known, all alpha-particle emitters. Due to its rapid decay it is used as a radioactive tracer in medicine.

Symbol: At; m.p. 302°C (est.); b.p. 337°C (est.); p.n. 85; most stable isotope ^{210}At (half-life 8.1 hours).

Aston, Francis William (1877–1945) British chemist and physicist. Aston's main contribution to science was the development of the mass spectrograph. This enabled atomic masses to be determined and demonstrated the existence of isotopes. From experiments soon after World War I Aston was able to explain Prout's hypothesis and the exceptions to it. He discovered that neon consists of two isotopes neon-20 and neon-22, with the former being ten times more common. This gives an average of 20.2 for the weight of a large collection of neon atoms. The atomic weights (relative atomic masses) of other elements were explained in a similar way. Between the mid-1920s and mid-1930s he was able to determine atomic weights more accurately using a new mass spectrograph. He found small discrepancies from the whole numbers postulated by Prout due to the binding energy of nuclei. He was awarded the 1922 Nobel Prize for chemistry.

asymmetric atom *See* isomerism; optical activity.

atmolysis The separation of gases by using their different rates of diffusion through a porous membrane or partition.

atmosphere A unit of atmospheric pressure, equal to 101.325 kilopascals in SI units. It is used in chemistry only for rough indications of high pressure; in particular, those of high-pressure industrial processes.

atom The smallest part of an element that can exist as a stable entity. Atoms consist of a small, dense, positively charged nucleus, made up of neutrons and protons, with electrons in a cloud around this nucleus. The chemical reactions of an element are determined by the number of its electrons, which is equal to the number of protons in its nucleus. All atoms of a given element have the same number of protons, an identifying quantity known as the element's PROTON NUMBER. A given element may also have two or more isotopes, which differ in the number of neutrons in their nuclei.

The electrons surrounding the nucleus are grouped into energy levels traditionally called SHELLS – i.e. main orbits around the nucleus. Within these main orbits there may be subshells corresponding to atomic ORBITALS.

An electron in an atom is specified by four quantum numbers:

1. The *principal quantum number* (n), which specifies the main energy levels. The principal quantum number is a positive integer which can have values 1, 2,

Quantum property	Integer values	Quantum number
Principal, n	1,2,3, ...	1 2 3
Orbital, l	0 to $n-1$	0 0 1 0 1 2
Magnetic, m	$-l$, ..., 0, ..., $+l$	0 0 −1 0 +1 0 −1 0 +1 −2 −1 0 +1 +2

Atom: quantum numbers

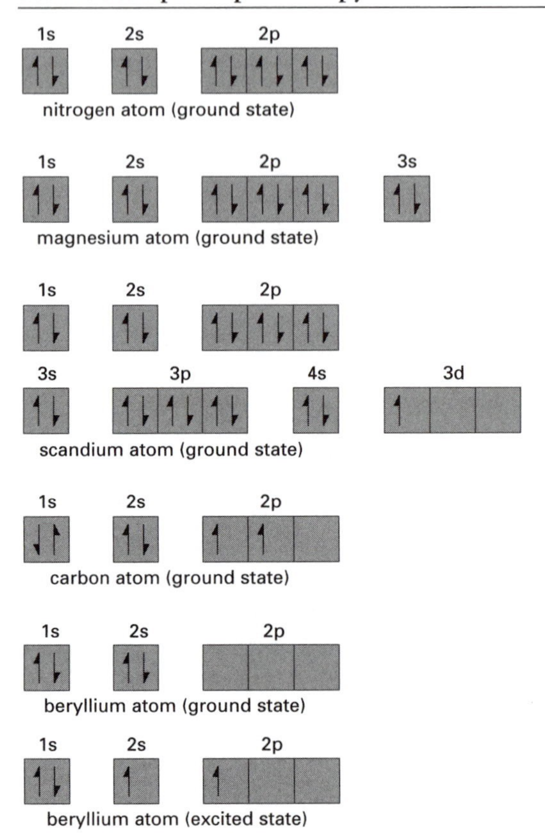

Atom: examples of box-and-arrow diagrams

3, etc. The corresponding shells are denoted by letters K, L, M, etc., the K shell (n = 1) being the one nearest to the nucleus. The maximum number of electrons in a given shell is $2n^2$.

2. The *orbital quantum number* (l), which specifies the angular momentum of the electron and thus the shape of the orbital. For a given value of n, *1* can have possible integer values of n–1, n–2, ... 2, 1, 0. For instance, the M shell (n = 3) has three subshells with different values of l (0, 1, and 2). Subshells with angular momentum 0, 1, 2, and 3 are designated by letters *s*, *p*, *d*, and *f*.
3. The *magnetic quantum number* (m) which determines the orientation of the electron orbital in a magnetic field. This number can have integer values $-l, -(l-1) \ldots 0 \ldots + (l-1), +l$.
4. The *spin quantum number* (m_s), which specifies the intrinsic angular momentum of the electron. It can have values +½ and –½.

According to the *exclusion principle* enunciated by Pauli in 1925 (the Pauli exclusion principle), no two electrons in an atom can have the same set of four quantum numbers. Each electron's unique set of four quantum numbers therefore specifies its QUANTUM STATE and helps to explain the electronic structure of atoms. *See also* Bohr theory.

atomic absorption spectroscopy (AAS) A method of chemical analysis in which a sample is vaporized and an absorption spectrum is taken of the vapor. The elements present are identified by their characteristic absorption lines.

atomic emission spectroscopy (AES) A technique of chemical analysis that involves vaporizing a sample of material, with atoms in their excited states emitting electromagnetic radiation at particular frequencies characteristic of that type of atom.

atomic force microscope (AFM) An instrument used to investigate surfaces. A small probe consisting of a very small chip of diamond is held just above a surface of a sample by a spring-loaded cantilever. As the probe is slowly moved over the surface the force between the surface and the tip is measured and recorded and the probe is automatically raised and lowered to keep this force constant. Scanning the surface in this way enables a contor map of the surface to be generated with the help of a computer. An atomic force microscope closely resembles a SCANNING TUNNELLING MICROSCOPE (STM) in some ways, although it uses forces rather than electrical signals to investigate the surface. Like an STM, it can resolve individual molecules. Unlike an STM, it can be used to investigate nonconducting materials, a feature that is useful in investigating biological samples.

atomic heat *See* Dulong and Petit's law.

atomicity The number of atoms per molecule of an element. Helium (He), for example, has an atomicity of one, nitrogen (N_2) two, and ozone (O_3) three.

atomic mass unit (amu) Symbol: u A unit of mass used to indicate the relative atomic mass of atoms and molecules, equal to 1/12 of the mass of an atom of carbon-12. It is equal to 1.660540×10^{-27} kilogram. In biochemistry it is sometimes known as the *dalton*.

atomic number *See* proton number.

atomic orbital *See* orbital.

atomic weight *See* relative atomic mass (r.a.m.).

atto- Symbol: a A prefix used with SI units denoting 10^{-18}. For example, 1 attometer (am) = 10^{-18} meter (m).

Aufbau principle A principle that governs the order in which the atomic orbitals are filled in elements of successive proton number; i.e. in order of increasing energy. The name is from the German *aufbauen*, meaning 'to build up'. The order is as follows:

$1s^2$, $2s^2$, $2p^6$, $3s^2$, $3p^6$, $4s^2$, $3d^{10}$, $4p^6$, $5s^2$, $4d^{10}$, $5p^6$, $6s^2$, $4f^{14}$, $5d^{10}$, $6p^6$, $7s^2$, $5f^{14}$, $6d^{10}$

with the superscript indicating the maximum number of electrons for each level.

Note that degenerate orbitals are occupied singly before spin pairing occurs. Note also the unexpected position of the d levels, the filling of which give rise to the first, second, and third transition series, and the unusual position of the f levels, the filling of which give rise to the lanthanoids and actinoids.

Auger effect An effect in which an excited ion decays by emission of an electron rather than by a photon. For example, if a substance is bombarded by high-energy electrons or gamma rays, an electron from an inner shell may be ejected. The result is a positive ion in an excited state. This ion will decay to its ground state as an outer electron falls to an inner shell. The energy released may result in the emission of a photon in the x-ray region of the electromagnetic spectrum (x-ray fluorescence). Alternatively, the energy may be released in the form of a second electron ejected from the atom resulting in a doubly charged ion. The emitted electron, known as an *Auger electron*, has a characteristic energy corresponding to the difference in energy levels in the ion. The Auger effect is a form of autoionization. It is named for French physicist Pierre Auger (1899–1994), who discovered it in 1925.

Auger electron *See* Auger effect.

auric Designating a compound of gold(III).

auric chloride *See* gold(III) chloride.

aurous Designating a compound of gold(I).

autocatalysis *See* catalyst.

autoclave An apparatus consisting of an airtight container whose contents are heated by high-pressure steam; the apparatus may also have a mechanism by which to agitate the contents. Autoclaves are used to react substances under pressure at high temperature, to sterilize objects and substances, and to carry out industrial processes.

autoionization The spontaneous ionization of excited atoms, ions, or molecules, as in the AUGER EFFECT.

average atomic mass *See* relative atomic mass. Count of Quaregna and Cerreto

Avogadro, Lorenzo Romano Amedeo Carlo, Count of Quaregna and Cerreto (1776–1856) Italian scientist. Avogadro is mainly remembered for a paper he wrote in 1811 in which he put forward what is now known as AVOGADRO'S LAW. Avogadro was able to use this idea to show that molecules of hydrogen and oxygen are diatomic and that the formula of water is H_2O. He was also able to give an explanation of Gay-Lussac's law. Avogadro's work made little impact in his lifetime and was revived later by Stanislao CANNIZZARO.

Avogadro constant (Avogadro's number) Symbol: N_A The number of particles in one MOLE of a substance. Its value is 6.02242×10^{23} mol^{-1}.

Avogadro's law (Avogadro's hypothesis) The principle that equal volumes of all gases at the same temperature and pressure contain equal numbers of molecules. It is often called Avogadro's hypothesis because it was first proposed by the Italian chemist and physicist Amedeo Avogadro (1776–1856) in 1811. It is strictly true only for ideal gases.

Avogadro's number *See* Avogadro constant.

azide An inorganic compound containing the ion N_3^-, or an organic compound having the general formula RN_3. Heavy metal azides are highly explosive.

B

back e.m.f. An e.m.f. that opposes the normal flow of electric charge in a circuit or circuit element. In some electrolytic cells a back e.m.f. is caused by the layer of hydrogen bubbles that builds up on the cathode as hydrogen ions pick up electrons and form gas molecules (i.e. as a result of POLARIZATION of the electrode).

baking powder A mixture of sodium hydrogencarbonate (sodium bicarbonate, baking soda) and a weakly acidic substance, such as tartaric acid or potassium hydrogentartrate (cream of tartar). The addition of moisture or heat causes a reaction that produces bubbles of carbon dioxide gas, which make dough or cake mixture rise. It is used as a yeast substitute in baking certain types of bread.

baking soda *See* sodium hydrogencarbonate.

ball mill A device commonly used in the chemical industry for grinding solid material. Ball mills usually have slowly rotating steel-lined drums containing steel balls. The material is crushed by the tumbling action of the balls in the drum. *Compare* hammer mill.

Balmer series A series of lines in the spectrum of radiation emitted by excited hydrogen atoms. The lines correspond to the atomic electrons falling into the second lowest energy level, emitting energy as radiation. The wavelengths (λ) of the radiation in the Balmer series are given by:

$$1/\lambda = R(1/2^2 - 1/n^2)$$

where n is an integer and R is the Rydberg constant. The series is named for J. J. Balmer (1825–98), who discovered a general equation for spectral lines in 1885. *See also* Bohr theory; spectral series.

banana bond (bent bond) A multicenter bond of the type present in BORON HYDRIDES. The term is used in a quite separate sense for the bonds in certain strained-ring organic compounds.

band spectrum A SPECTRUM that appears as a number of bands of emitted or absorbed radiation. Band spectra are characteristic of molecules. Often each band can be resolved into a number of closely spaced lines. The different bands correspond to changes of electron orbit in the molecules. The closely spaced lines in each band, seen under higher resolution, are the result of different vibrational states of the molecule.

Barff process A process formerly used for protecting iron from corrosion by heating it in steam to form a layer of tri-iron tetroxide (Fe_3O_4). It is also known as the Bower–Barff process and is now only of historical interest.

barites *See* barium sulfate.

barium A dense, low-melting reactive metal; the fifth member of group 2 (formerly IIA) of the periodic table and a typical alkaline-earth element. The electronic configuration is that of xenon with two additional outer 6s electrons. Barium is of low abundance; it is found in mineral form as *witherite* ($BaCO_3$) and *barytes* ($BaSO_4$). The metal is obtained by the electrolysis of the fused chloride using a cooled cathode which is slowly withdrawn from the melt. Because of its low melting point barium is readily purified by vacuum distillation.

Barium metal is used as a 'getter', i.e., a compound added to a vacuum system to remove the last traces of oxygen. It is also a constituent of certain bearing alloys.

Barium has a low ionization potential and a large radius. It is therefore strongly electropositive and its properties, and those of its compounds, are very similar to those of the other alkaline-earth elements calcium and strontium. Notable differences in the chemistry of barium from the rest of the group are:

1. The much higher stability of the carbonate.
2. The formation of a peroxide below 800°C. Barium peroxide decomposes on strong heating to give oxygen and barium oxide:

$$BaO_2 \rightleftharpoons BaO + O$$

Barium is also notable for the very low solubility of the sulfate. Barium compounds give a characteristic green color to flames, a phenomenon that is used in qualitative analysis. Barium salts are all highly toxic with the exception of the most insoluble materials. Barium sulfate is used as a radiopaque material in making x-ray photographs of the gastric and intestinal tract.

Symbol: Ba; m.p. 729°C; b.p. 1640°C; r.d. 3.594 (20°C); p.n. 56; r.a.m. 137.327.

barium bicarbonate *See* barium hydrogencarbonate.

barium carbonate ($BaCO_3$) A white insoluble salt that occurs naturally as the mineral *witherite.* It can be precipitated by adding an alkali carbonate to a solution of a barium salt. On heating it decomposes reversibly with the formation of the oxide and carbon dioxide:

$$BaCO_3 \rightleftharpoons BaO + CO_2$$

The compound is highly toxic and is used as a rat poison.

barium chloride ($BaCl_2$) A white solid that can be prepared by dissolving barium carbonate in hydrochloric acid and crystallizing out the dihydrate ($BaCl_2.2H_2O$). Barium chloride is used as the electrolyte in the extraction of barium, as a rat poison, and in the leather industry. In the laboratory barium chloride solution is used as a test for sulfates, with which it gives a white precipitate.

barium hydrogencarbonate (barium bicarbonate; $Ba(HCO_3)_2$) A compound that occurs only in aqueous solution. It is formed by the action of cold water containing carbon dioxide on barium carbonate.

$$BaCO_3 + CO_2 + H_2O \rightleftharpoons Ba(HCO_3)_2$$

barium hydroxide (baryta; $Ba(OH)_2$) A white solid usually obtained as the octahydrate, $Ba(OH)_2.8H_2O$. It can be made by adding water to barium oxide. Barium hydroxide is the most soluble of the group 2 hydroxides. It is used in volumetric analysis for the estimation of weak acids using phenolphthalein as an indicator.

barium oxide (BaO) A white powder prepared by heating barium in oxygen or by heating barium carbonate. It has been used in the manufacture of additives for lubricating oils, and in pigments.

barium peroxide (BaO_2) A dense off-white powder that can be prepared by carefully heating barium oxide in oxygen. The reaction, which is reversible, was the basis of the now obsolete BRIN PROCESS for obtaining oxygen. Barium peroxide is used for bleaching straw and silk and is also used in the laboratory preparation of hydrogen peroxide.

barium sulfate ($BaSO_4$) A white solid that occurs naturally as the mineral *barites*, also known as *heavy spar*. Barium sulfate is very insoluble in water and can be prepared easily as a precipitate by adding sulfuric acid to barium chloride. It is an important industrial chemical. Under the name of *blanc fixe* it is used as a pigment extender in surface coating compositions. It is also used in the glass and rubber industries. Because barium compounds are opaque to x-rays barium sulfate is used in medicine for taking radiographs of the digestive system.

Bartlett, Neil (1932–) British–

American chemist. Bartlett was studying metal fluorides and found that the compound platinum hexafluoride (PtF_6) has an extremely high electron affinity, and reacts with molecular oxygen to form the novel compound $O_2^+PtF_6^-$. This was the first example of a compound containing the oxygen cation. At the time it was an unquestioned assumption of chemistry that the rare gases – helium, neon, argon, krypton, and xenon – were completely inert. Bartlett knew that the ionization potential of xenon was not too much greater than the ionization potential of the oxygen molecule and was able to produce xenon hexafluoroplatinate ($XePtF_6$) by direct reaction – the first compound of a rare gas. Once the first compound had been detected xenon was soon shown to form other compounds, such as xenon fluoride (XeF_4) and oxyfluoride ($XeOF_4$). Krypton and radon were also found to form compounds although the lighter rare gases have so far remained inactive.

baryta *See* barium hydroxide.

basalt A dark-colored basic igneous rock derived from solidified volcanic lava. It consists mainly of fine crystals of pyroxine and plagioclase feldspar.

base In the *Arrhenius theory*, a compound that releases hydroxide ions, OH^-, in aqueous solution. Basic solutions have a pH greater than 7. In the *Lowry-Brønsted theory* of ACIDS and bases a base is a substance that tends to accept a proton. Thus OH^- is basic because it accepts H^+ to form water, but H_2O is also a base (although somewhat weaker) because it can accept a further proton to form H_3O^+. In this treatment the ions of classical mineral acids such as SO_4^{2-} and NO_3^- are weak *conjugate bases* of their respective conjugate acids; i.e. H_2SO_4 and HNO_3, respectively. Ammonia is also a base. In solution it donates an electron pair to a proton to form an ammonium ion:

$$NH_3 + H_2O \rightleftharpoons NH_4^+ + OH^-$$

Here the ammonium ion is the conjugate acid. Other nitrogenous compounds, such as trimethylamine ($(CH_3)_3N$) and pyridine, are examples of organic bases. *See also* acid; Lewis acid.

base-catalyzed reaction A reaction catalyzed by bases; i.e. by hydroxide ions.

base metal A common metal such as iron, lead, or copper, as distinguished from a precious metal such as gold, silver, or platinum.

base unit A unit within a system of measurement from which other units may be derived by combining it with one or more other base units. For example the joule, a derived unit, can be defined in terms of base units as one metre (m) squared, kilogram (kg) per second (s) squared, or m^2kgs^{-2}. With the current exception of the kilogram, base SI UNITS are defined in terms of reproducible physical phenomena.

basic Behaving as a base. Any solution in which the concentration of OH^- ions is greater than that in pure water at the same temperature is described as basic; i.e. the pH of a basic solution is greater than 7.

basic aluminum ethanoate *See* aluminum ethanoate.

basic oxide An oxide of a metal that reacts with water to form a base, or with an acid to form a salt and water. Calcium oxide is a typical example. It reacts with water to form calcium hydroxide:

$$CaO + H_2O \rightarrow Ca(OH)_2$$

and with hydrochloric acid to produce calcium chloride and water:

$$CaO + 2HCl \rightarrow CaCl_2 + H_2O$$

Compare acidic oxide.

basic oxygen process *See* Bessemer process.

basic salt A compound intermediate between a normal salt and a hydroxide or oxide. The term is often restricted to hydroxyhalides (such as $Pb(OH)Cl$, $Mg_2(OH)_3Cl_2.4H_2O$, and $Zn(OH)F$) and hydroxy-oxy salts (for example, $2PbCO_3.$-$Pb(OH)_2$, $Cu(OH)_2.CuCl_2$, $Cu(OH)_2.$-

$CuCO_3$). The hydroxyhalides are essentially closely packed assemblies of OH^- with metal or halide ions in the octahedral holes. The hydroxy-oxy salts usually have more complex structures than those implied by the formulae.

basic slag *See* slag.

battery A number of electric cells working together. Many dry 'batteries' used in radios, flashlights, etc., are in fact single cells. If a number of identical cells are connected in series, the total e.m.f. of the battery is the sum of the e.m.f.s of the individual cells. If the cells are in parallel, the e.m.f. of the battery is the same as that of one cell, but the current drawn from each is less (the total current is split among the cells).

bauxite A mineral hydrated form of aluminum hydroxide; the principal ore of aluminum.

b.c.c. *See* body-centered cubic crystal.

Beckmann thermometer A type of mercury thermometer designed to measure small differences in temperature rather than scale degrees. Beckmann thermometers have a larger bulb than common thermometers and a stem with a small internal diameter, so that a range of 5°C covers about 30 centimeters in the stem. The mercury bulb is connected to the stem in such a way that the bulk of the mercury can be separated from the stem once a particular 5° range has been attained. The thermometer can thus be set for any particular range. It is named for the German chemist Ernst Beckmann (1853–1923).

becquerel Symbol: Bq The SI unit of activity equal to the activity of a radioactive substance that has one spontaneous nuclear change per second; $1 \text{ Bq} = 1 \text{ s}^{-1}$. It is named for the discoverer of radioactivity, the French physicist Antoine Henri Bequerel (1852–1908).

Beer–Lambert law A law that relates the intensity of electromagnetic radiation passing through a material to the length of the path of the radiation through the material. It has the form $\log I_0/I = E c l$, where I_0 is the incident intensity of light, I is the intensity of light after it has passed through a sample of the material of length l, c is the concentration of absorbing species in the material and E is a constant known as the *absorption coefficient*. The law is named for the German mathematician Johann Heinrich Lambert (1728–77) and the German astronomer Wilhelm Beer (1797–1850). The physical significance of the Beer–Lambert law is that the intensity of electromagnetic radiation passing through a sample of material decreases exponentially with the thickness and concentration of the sample (for a given wavelength of electromagnetic radiation). If deviations from the Beer–Lambert law occur this is due to such phenomena as dissociation or the formation of complexes.

Belousov–Zhabotinskii reaction (B–Z reaction) An oscillating chemical reaction in which there are periodic oscillations in the color of a mixture of sulfuric acid, potassium bromate, cerium (or iron) sulfate, and propanedioic acid. The period of oscillation is about one minute. The color changes are caused by repeated oxidations and reductions of cerium (or iron) ions. The reaction was first observed by the Russian chemist B. P. Belousov in the case of cerium and modified to iron by A. M. Zhabotinskii in 1963. The mechanism of the B–Z reaction is highly complicated and involves a large number of individual steps.

bench dilute acid *See* dilute.

beneficiation The process of separating an ore into a useful component and waste material (known as *gangue*). The process is sometimes described as *ore dressing*.

bent bond *See* banana bond.

bentonite A type of clay that is used as an absorbent in making paper and as a precipitating agent to remove proteins from beer and wine. The gelatinous suspension it forms with water is also used to bind together the sand used for making castings. Chemically bentonite is an aluminosilicate of variable composition.

bent sandwich compound *See* sandwich compound.

Bergius, Friedrich Karl Rudolph (1884–1949) German industrial chemist. Bergius worked with Hermann Nernst at Berlin and Fritz Haber at Karlsruhe, where he became interested in high-pressure chemical reactions. He is noted for his development of the BERGIUS PROCESS. He shared the Nobel Prize for chemistry with Carl Bosch in 1931.

Bergius process A process formerly used for making hydrocarbon fuels from coal. A powdered mixture of coal, heavy oil, and a catalyst was heated with hydrogen at high pressure. The process was used by Germany as a source of motor fuel in World War II. It was developed by Friedrich Bergius in 1912.

berkelium A silvery radioactive transuranic element of the actinoid series of metals, not found naturally on Earth. The first samples were prepared by bombarding ^{291}Am with alpha particles. Several radioisotopes have since been synthesized. The metal reacts with oxygen, steam, and acids. It tends to collect in bone and thus may one day find use in medical research.

Symbol: Bk; m.p. 1050°C; b.p. unknown; r.d. 14.79 (20°C); p.n. 97; most stable isotope ^{247}Bk (half-life 1400 years).

beryl A mineral, $3BeO.Al_2O_3.6SiO_2$, used as a source of beryllium. Crystals are hexagonal. There are several color varieties including the gemstones emerald, which is colored green by traces of chromium oxide, and aquamarine, which has a blue-green color.

beryllate *See* beryllium; beryllium hydroxide.

beryllium A light metallic element, similar to aluminum but somewhat harder; the first element in group 2 (formerly IIA) of the periodic table. It has the electronic configuration of helium with two additional outer 2s electrons.

Beryllium occurs in a number of minerals such as beryllonite ($NaBePO_4$), chrysoberyl ($Be(AlO_2)_2$), bertrandite ($4BeO.2SiO_2$), and beryl ($3BeO.Al_2O_3.6SiO_2$). The element accounts for only 0.0006% by mass of the Earth's crust. The metal is obtained by conversion of the ore to the sulfate at high temperature and pressure with concentrated sulfuric acid, then to the chloride, followed by electrolysis of the fused chloride. Alternatively, it is possible to treat the ore with hydrogen fluoride followed by electrolysis of the fused fluoride. The metal has a much lower general reactivity than other elements in group 2. It is used as an antioxidant and hardener in copper, steel, bronze, and nickel alloys.

Beryllium has the highest ionization potential of group 2 and the smallest atomic radius. Consequently it is less electropositive and more polarizing than other members of the group. Thus, Be^{2+} ions do not exist as such in either solids or solutions, and there is partial covalent character in the bonds, even with the most electronegative elements. The metal reacts directly with oxygen, nitrogen, sulfur, and the halogens at various elevated temperatures, to form the oxide BeO, nitride Be_3N_2, sulfide BeS, and halides BeX_2, all of which are covalent. Beryllium does not react directly with hydrogen but a polymeric hydride $(BeH_2)_n$ can be prepared by reduction of $(CH_3)_2Be$ using lithium tetrahydroaluminate(III) ($LiAlH_4$).

Beryllium is amphoteric forming *beryllate* species, such as $[Be(OH)_4]^{2-}$ and $[Be(OH)]_3^{3+}$. The hydroxide is only weakly basic. The element does not form a true carbonate; the basic beryllium carbonate, $BeCO_3.Be(OH)_2$ is formed when sodium carbonate is added to solutions of beryllium compounds.

Beryllium hydride, chloride, and dimethylberyllium form polymeric bridged species but, whereas the bridging in the chloride is via an electron pair on chlorine atoms and can be regarded as an electron-pair donor bond, the bonding in the hydride and in the methyl compound involves two-electron three-center bonds similar to those found in the BORON HYDRIDES. COMPLEXES are quite common with beryllium; some examples include $[BeCl_4]^{2-}$, $(R_2O)_2$-$BeCl_2$, and $[Be(NH_3)_4]Cl_2$. Beryllium and its compounds are very toxic and may cause serious respiratory diseases if inhaled.

Symbol: Be; m.p. 1278±5°C; b.p. 2970°C (660 kPa); r.d. 1.85 (20°C); p.n. 4; r.a.m. 9.012182.

beryllium bicarbonate *See* beryllium hydrogencarbonate.

beryllium bronze *See* bronze.

beryllium carbonate ($BeCO_3$) An unstable solid prepared by prolonged treatment of a suspension of beryllium hydroxide with carbon dioxide. The resulting solution is evaporated and filtered in an atmosphere of carbon dioxide.

beryllium chloride ($BeCl_2$) A white solid obtained by passing chlorine over a heated mixture of beryllium oxide and carbon. The compound is not ionic: it is a poor conductor in the fused state and in the solid form consists of a covalent polymeric structure. It is readily soluble in organic solvents but is hydrolyzed in the presence of water to give the hydroxide. The anhydrous salt is used as a catalyst.

beryllium hydrogencarbonate (beryllium bicarbonate; $Be(HCO_3)_2$) A compound formed in solution by the action of carbon dioxide on a suspension of the carbonate, to which it reverts on heating:

$$BeCO_3 + CO_2 + H_2O \rightleftharpoons Be(HCO_3)_2$$

beryllium hydroxide ($Be(OH)_2$) A white solid that can be precipitated from beryllium-containing solutions by an alkali. Beryllium hydroxide, like aluminum hydroxide, is amphoteric. In an excess of alkali, beryllium hydroxide dissolves to give a *beryllate*, $Be(OH)_4^{2-}$.

beryllium oxide (BeO) A white solid formed by heating beryllium in oxygen or by the thermal decomposition of beryllium hydroxide or carbonate. Beryllium oxide is insoluble in water but it shows basic properties by dissolving in acids to form beryllium salts:

$$BeO + 2H^+ \rightarrow Be^{2+} + H_2O$$

However, beryllium oxide also resembles acidic oxides by reacting with alkalis to form beryllates:

$$BeO + 2OH^- + H_2O \rightarrow Be(OH)_4^{2-}$$

It is thus an amphoteric oxide. Beryllium oxide is used in the production of refractory materials, high-output transistors, and printed circuits. Its chemical properties of beryllium oxide are similar to those of aluminum oxide.

beryllium sulfate ($BeSO_4$) An insoluble salt prepared by the reaction of beryllium oxide with concentrated sulfuric acid. On heating, it breaks down to give the oxide.

Berzelius, Jöns Jacob (1779–1848) Swedish chemist who had a major influence on the development of modern chemistry. He introduced the symbols now used to denote chemical elements and coined a number of words still used in chemistry, such as 'catalysis', 'halogen', 'isomerism', and 'protein'. He also determined relative atomic masses (atomic weights) accurately, thereby helping establish Dalton's atomic theory. Berzelius was also involved in the discovery of several elements: cerium, silicon, selenium, and thorium, and the introduction of some experimental apparatus, such as rubber tubes and filter paper, into chemistry.

Bessemer process A process for making steel from PIG IRON. A vertical cylindrical steel vessel (*converter*) is used, lined with a refractory material. Air is blown over and through the molten iron to oxidize and thereby remove impurities such as carbon, silicon, sulfur, and phosphorus by converting them to slag. For instance, irons containing large amounts of phosphorus are treated in a converter lined with a basic material, so that a phosphate slag is formed. The required amount of carbon is then added to the iron to produce the desired type of steel. The process was invented in 1856 by the British engineer, Sir Henry Bessemer (1813–98). In the *basic oxygen process*, a modern version of the Bessemer process, the air is replaced by a mixture of oxygen and steam in order to minimize the amount of nitrogen that is absorbed by the steel.

beta particle An electron or positron emitted by a radioactive substance as it decays.

Bethe, Hans Albrecht (1906–) German physicist who initiated the subject of crystal field theory in 1929 when he analysed the splitting of atomic energy levels by surrounding ligands in terms of group theory. This work led to an understanding of the optical, magnetic, and spectroscopic properties of transition-metal and rare-earth complexes. Bethe's main interest was nuclear physics. In particular, he explained the energy of stars in terms of nuclear reactions that fuse hydrogen into helium. Bethe won the 1967 Nobel Prize for physics for this work.

bi- Prefix used formerly in naming acid salts. The prefix indicates the presence of hydrogen; for instance, sodium bisulfate ($NaHSO_4$) is sodium hydrogensulfate, etc. In organic chemistry it is sometimes used with the meaning 'two'; e.g. biphenyl.

bicarbonate *See* hydrogencarbonate.

bimolecular Describing a reaction or step that involves two molecules, ions, etc. A common example of this type of reaction is the decomposition of hydrogen iodide:

$$2HI \rightarrow H_2 + I_2$$

This takes place between two molecules and is therefore a bimolecular reaction. Other common examples are:

$$H_2O_2 + I_2 \rightarrow OH^- + HIO$$
$$OH^- + H^+ \rightarrow H_2O.$$

binary compound A chemical compound formed from two elements; e.g. Fe_2O_3 or NaCl.

biochemistry The study of chemical compounds and reactions occurring in living organisms.

bioinorganic chemistry The study of biological molecules that contain metal atoms or ions. Many enzymes have active metal atoms and bioinorganic compounds are important in other roles such as protein folding, oxygen transport, and electron transfer. Two important examples of bioinorganic compounds are hemoglobin (containing iron) and chlorophyll (containing magnesium).

Bipyridine: 2,2′-bipyridine

bipy Abbreviation for the bidentate ligand 2,2′-bipyridine.

bipyridine *See* bipy.

Birkeland–Eyde process An industrial process for fixing nitrogen (as nitrogen monoxide) by passing air through an electric arc:

$$N_2 + O_2 \rightarrow 2NO$$

The process was invented by two Norwegian chemists, Kristian Birkeland 1867–1913) and Samuel Eyde (1866–1940), who introduced it in 1903 to exploit the cheap sources of hydroelectricity then available in Scandinavia. *See also* nitrogen fixation.

bismuth A brittle pinkish metallic element belonging to group 15 (formerly VA) of the periodic table. It occurs native and in the ores bismuthinite (Bi_2S_3) and bismite (Bi_2O_3). The element does not react with oxygen or water under normal temperatures. It can be dissolved by concentrated nitric acid. It is the most diamagnetic of the metals and has less thermal conductivity than any metal except mercury. Bismuth is widely used in alloys, especially low-melting alloys, which find use in heat-activated sprinkler systems. The element has the unusual property of expanding when it solidifies, making it a desirable component of alloys used to make detailed castings. Compounds of bismuth are used in cosmetics and medicines. It is also used as a catalyst in the textile industry.

Symbol: Bi; m.p. 271.35°C; b.p. 1560±5°C; r.d. 9.747 (20°C); p.n. 83; r.a.m. 208.980 37.

bismuth(III) carbonate dioxide (bismuthyl carbonate; $Bi_2O_2CO_3$) A white solid prepared by mixing solutions of bismuth nitrate and ammonium carbonate. It contains the $(BiO)^+$ ion (sometimes known as the *bismuthyl ion*).

bismuth(III) chloride (bismuth trichloride; bismuth(III) chloride oxide; $BiCl_3$) A white deliquescent solid. It can be prepared by direct combination of bismuth and chlorine. Bismuth(III) chloride dissolves in excess dilute hydrochloric acid to form a clear liquid, but if diluted it produces a white precipitate of *bismuth(III) chloride oxide* (*bismuthyl chloride*, BiOCl):

$$BiCl_3 + H_2O \rightleftharpoons BiOCl + 2HCl$$

This reaction is often discussed in chemistry as a typical example of a reversible reaction. It is also a confirmatory test for bismuth in analysis. Bismuth(V) chloride does not exist.

bismuth(III) chloride oxide *See* bismuth(III) chloride.

bismuth(III) nitrate oxide (bismuthyl nitrate; bismuth subnitrate; $BiONO_3$) A white insoluble solid precipitated when bismuth(III) nitrate is diluted and contains the $(BiO)^+$ ion. Bismuth(III) nitrate oxide is used in pharmaceutical preparations.

bismuth subnitrate *See* bismuth(III) nitrate oxide.

bismuth trichloride *See* bismuth(III) chloride.

bismuthyl carbonate *See* bismuth(III) carbonate dioxide.

bismuthyl chloride *See* bismuth(III) chloride.

bismuthyl compound A compound containing the $(BiO)^+$ ion or BiO grouping.

bismuthyl ion *See* bismuth(III) carbonate dioxide.

bismuthyl nitrate *See* bismuth(III) nitrate oxide.

bistability The ability of a chemically reacting system to occur in two steady states. For an OSCILLATING REACTION to occur it is necessary for there to be bistability. The two steady states are not states of thermodynamic equilibrium There is a sudden jump from one of the bistable states to the other when a certain critical concentration of one of the reactants occurs. *See also* oscillating reaction.

bisulfate *See* hydrogensulfate.

bisulfite *See* hydrogensulfite.

bittern The liquid that is left after sodium chloride has been crystallized from sea water. It is a source of iodine, bromine, and some magnesium salts.

bitumen A mixture of solid or semisolid hydrocarbons obtained naturally or from coal, oil, etc.

bituminous coal A type of second-grade COAL, containing more than 65% carbon but also quantities of gas, coal tar, and water. It is the most abundant type of coal. Domestic use of coal is practically nonexistent in the U.S.A. and Canada today.

bivalent *See* divalent.

Black, Joseph (1728–99) Scottish chemist and physicist, one of the first to use quantitative methods in developing modern chemistry. He discovered carbon dioxide (which he called 'fixed air') when he investigated the chemical reactions of limestone in a quantitative way, reporting the work in 1756. The idea of latent heat occurred to him in 1757 and in subsequent years he determined the latent heat in the formation of ice and steam experimentally. He was careful to distinguish between the concepts of heat and temperature, a distinction that led him to the concept of specific heat.

blackdamp *See* firedamp.

blanc fixe *See* barium sulfate.

blast furnace A tall furnace for SMELTING iron from various iron ores, especially hematite ($Fe_2)_3$) and magnetite (Fe_3O_4). A mixture of the ore with coke and a flux is fed into the top of the furnace and heated by pre-heated air blown into the bottom of the furnace, where temperatures are much higher. The flux is often calcium oxide (from limestone). As it descends the furnace the ore is first reduced to iron(II) oxide (FeO) and then to molten iron, both by the action of carbon monoxide (CO) obtained by burning coke in air. Molten PIG IRON is then run off from the bottom of the furnace.

bleach Any substance used to remove color from materials such as cloth and paper. All bleaches are oxidizing agents, and include chlorine, sodium chlorate(I) solution (NaClO, sodium hypochlorite), hydrogen peroxide, and sulfur(IV) oxide (SO_2, sulfur dioxide). Sunlight also has a bleaching effect.

bleaching powder A white solid that can be regarded as a mixture of calcium chlorate(I) (calcium hypochlorite ($Ca(ClO)_2$)), calcium chloride, and calcium hydroxide. It is prepared commercially by passing chlorine through a tilted cylinder down which is passed calcium hydroxide. Bleaching powder has been used for bleaching paper pulp and fabrics and for sterilizing water. Its activity arises from the formation, in the presence of air containing carbon dioxide, of the oxidizing agent chloric(I) acid (hypochlorous acid, HClO):

$$Ca(ClO)_2.Ca(OH)_2.CaCl_2 + 2CO_2 \rightarrow 2CaCO_3 + CaCl_2 + 2HClO$$

blende A sulfide ore of a metal; e.g. zinc blende (ZnS_2).

Bloch, Felix (1905–83) Swiss-born American physicist who invented the technique of nuclear magnetic resonance (NMR) in 1946 (as did Edward Purcell, independently). This led to the extensive use of NMR in investigating the structue of complex molecules. Bloch and Purcell shared the 1952 Nobel Prize for physics for their work in this field. Bloch also worked on the quantum theory of solids.

blue vitriol Copper(II) sulfate pentahydrate ($CuSO_4.5H_2O$).

body-centered cubic crystal (b.c.c.) A crystal structure in which the unit cell has an atom, ion, or molecule at each corner of a cube and also at the center of the cube. In this type of structure the coordination number is 8. It is less close-packed than the face-centered cubic structure. The alkali metals form crystals with body-centered cubic structures. *See also* cubic crystal.

Bohr, Niels Hendrik David (1885–1962) Danish physicist. Niels Bohr was responsible for a key development in our understanding of atomic structure when he showed (1913) how the structure of the atom could be explained by imposing 'quantum conditions' on the orbits of electrons, thus allowing only certain orbits. This theory accounted for details of the hydrogen spectrum. Bohr also contributed to nuclear physics, particularly the theory of nuclear fission. He was awarded the 1922 Nobel Prize for physics for his work on atomic theory.

bohrium A synthetic radioactive element first detected by bombarding a bismuth target with chromium nuclei. Only a small number of atoms have ever been produced.

Symbol: Bh; p.n. 107; most stable isotope ^{262}Bh (half life 0.1s).

Bohr magneton A unit of magnetic moment that is convenient at the atomic level. It is denoted by μ_B and given by μ_B =

$eh/(4\pi m_e)$, where e is the charge of an electron, h is the PLANCK CONSTANT, and m_e is the rest mass of an electron. The Bohr magneton has a value of 9.274×10^{-24} A m^2.

Bohr theory A theory introduced by Niels Bohr (1911) to explain the spectrum of atomic hydrogen. The model he used was that of a nucleus with charge +e, orbited by an electron with charge –e, moving in a circle of radius r. If v is the velocity of the electron, the centripetal force, mv^2/r is equal to the force of electrostatic attraction, $e^2/4\pi\varepsilon_0 r^2$. Using this, it can be shown that the total energy of the electron (kinetic and potential) is $-e^2/8\pi\varepsilon_0 r$.

If the electron is considered to have wave properties, there must be a whole number of wavelengths around the orbit, otherwise the wave would be a progressive wave. For this to occur

$$n\lambda = 2\pi r$$

where n is an integer, 1, 2, 3, 4, ... The wavelength, λ, is h/mv, where h is the Planck constant and mv the momentum. Thus for a given orbit:

$$nh/2\pi = mvr$$

This means that orbits are possible only when the angular momentum (mvr) is an integral number of units of $h/2\pi$. Angular momentum is thus quantized. In fact, Bohr in his theory did not use the wave behavior of the electron to derive this relationship. He assumed from the beginning that angular momentum was quantized in this way. Using the above expressions it can be shown that the electron energy is given by:

$$E = -me^4/8\varepsilon_0^2 n^2 h^2$$

Different values of n (1, 2, 3, etc.) correspond to different orbits with different energies; n is the principal quantum number. In making a transition from an orbit n_1 to another orbit n_2 the energy difference ΔW is given by:

$$\Delta W = W_1 - W_2$$
$$= me^4(1/n_2^2 - 1/n_1^2)/8\varepsilon_0^2 h^2$$

This is equal to hv where v is the frequency of radiation emitted or absorbed. Since $v\lambda = c$, then

$$1/\lambda = me^4(1/n_1^2 - 1/n_2^2)/8\varepsilon_0^2 ch^3$$

The theory is in good agreement with experiment in predicting the wavelengths of lines in the hydrogen spectrum, although it is less successful for larger atoms. Different values of n_1 and n_2 correspond to different spectral series, with lines given by the expression:

$$1/\lambda = R(1/n_1^2 - 1/n_2^2)$$

R is the *Rydberg constant*. Its experimental value is 1.09678×10^7 m^{-1}. The value from Bohr theory ($me^4/8\pi\varepsilon_1^2 ch^2$) is $1.097\,00 \times 10^7$ m^{-1}. *See also* atom.

boiling The process by which a liquid is converted into a gas or vapor by heating at its *boiling point*; i.e. the temperature at which the vapor pressure of a liquid is equal to atmospheric pressure. This temperature is always the same for a particular pure liquid at a given pressure (for reference purposes usually taken as standard pressure).

Boltzmann constant Symbol: k The constant 1.38054 J K^{-1}, equal to the gas constant (R) divided by the Avogadro constant (N_A). It is named for the Austrian physicist Ludwig Edward Boltzmann (1844–1906).

Boltzmann formula A fundamental result in statistical mechanics stating that the entropy S of a system is related to the number W of distinguishable ways in which the system can be realised by the equation: $S = k \ln W$, where k is the Boltzmann constant. This formula is a quantitative expression of the idea that the entropy of a system is a measure of its disorder. It was discovered by the Austrian physicist Ludwig Boltzmann in the late 19th century in the course of his investigations into the foundations of statistical mechanics.

bomb calorimeter A sealed insulated container, used for measuring energy released during combustion of substances (e.g. foods and fuels). A known amount of the substance is ignited inside the calorimeter in an atmosphere of pure oxygen, and undergoes complete combustion at constant volume. The resultant rise in temperature is related to the energy released by the reaction. Such energy values (*calorific values*) are often quoted in joules per kilogram (J kg^{-1}).

BORAX-BEAD COLORS (H = hot; C = cold)

Metal	*Oxidizing flame*	*Reducing flame*
chromium	green H+C	green H+C
cobalt	blue H+C	blue H+C
copper	green H, blue C	often opaque
iron	brown-red H, yellow C	green H+C
manganese	violet H+C	colourless H+C
nickel	red–brown C	gray–black C

bond *See* chemical bond.

bond dissociation energy *See* bond energy.

bond energy The energy involved in forming a chemical bond. For ammonia, for instance, the energy of the N–H bond is one third of the energy involved for the process

$$NH_3 \rightarrow N + 3H$$

It is thus one third of the heat of atomization. This is also known as the *mean bond energy.*

The *bond dissociation energy* is measured in the opposite direction. It is the energy required to break a particular bond in a compound, e.g.:

$$NH_3 \rightarrow NH_2 + H$$

More formally, it is common to specify the *bond enthalpy.*

bond enthalpy *See* bond energy.

bonding orbital *See* orbital.

bond length The length of a chemical bond, i.e. the distance between the centers of the nuclei of two atoms joined by a chemical bond. Bond lengths may be measured by electron or x-ray diffraction.

boracic acid *See* boric acid.

borane *See* boron hydride.

borate *See* boron.

borax *See* disodium tetraborate decahydrate.

borax-bead test A preliminary test in qualitative inorganic analysis that can be a guide to the presence of certain metals. A bead is formed by heating a little disodium tetraborate decahydrate (borax) on a loop in a platinum wire. A minute sample of the compound to be tested is introduced into the bead and the color observed in both the oxidizing and reducing areas of a Bunsen-burner flame. The color is also noted when the bead is cold.

boric acid (boracic acid; orthoboric acid; trioxoboric(III) acid; H_3BO_3) A white crystalline solid soluble in water; in solution it is a very weak acid. Boric acid is used as a mild antiseptic eye lotion and was formerly used as a food preservative. It is used in glazes for enameled objects and is a constituent of borosilicate glass.

Trioxoboric(III) acid is the full systematic name for the solid acid and its dilute solutions; in more concentrated solutions polymerization occurs to give *polydioxoboric(III) acid.*

boric anhydride *See* boron oxide.

boric(III) oxide *See* boron oxide.

boride A compound of boron, especially one with a more electropositive element. Borides have a wide range of stoichiometries, from M_4B through to MB_6, and can exist in close-packed arrays, chains, and two-dimensional nets. Due to their high melting points and unreactivity with nonoxidizing acids, metal borides are used in refractory materials. Borides are also good abrasives.

Born, Max (1882–1970) German physicist who was one of the founders of quantum mechanics in the 1920s. In particular, he put forward the 'Born interpretation' for the wave-function of an electron in terms of probability in 1926. Born also made major contributions to the theory of crystals and to the quantum theory of molecules. He was awarded a share of the 1954 Nobel Prize in physics (together with Walther Bothe) for his work on quantum mechanics.

Born–Haber cycle A cycle used in calculating the lattice energies of solids. The steps involved are:

Atomization of sodium:

$$Na(s) \rightarrow Na(g)\ \Delta H_1$$

Ionization of sodium:

$$Na(g) \rightarrow Na^+(g)\ \Delta H_2$$

Atomization of chlorine:

$$Cl_2(g) \rightarrow 2Cl(g)\ \Delta H_3$$

Ionization of chlorine:

$$Cl(g) + e^- \rightarrow Cl^-(g)\ \Delta H_4$$

Formation of solid:

$$Na^+(g) + Cl^-(g) \rightarrow NaCl(s)\ \Delta H_5$$

This last step involves the lattice energy ΔH_5. The sum of all these enthalpies is equal to the heat of the reaction:

$$Na(s) + \tfrac{1}{2}Cl_2(g) \rightarrow NaCl(s)$$

See also Hess's law.

Born–Oppenheimer approximation An approximation used in the quantum theory of molecules in which it is assumed that the motion of atomic nuclei is very much slower than the motion of electrons and that it is a good approximation to take the nuclei to be in fixed positions when discussing electron transitions. The German physicist Max Born and the American physicist Julius Robert Oppenheimer first used this approximation in 1927.

borohydride ions *See* boron hydride.

boron A hard, rather brittle metalloid element of group 3 (formerly IIIA) of the periodic table. It exists in two forms: as an amorphous yellow-brown powder and as a black metallic crystal. It has the electronic structure $1s^22s^22p^1$. Boron is of low abundance (0.0003%) but occurs in very concentrated forms in its natural minerals, which include borax ($Na_2B_4O_7.10H_2O$), ulexite ($NaCaB_5O_9.H_2O$), howlite ($Ca_2B_5SiO_9(OH)_5$), and colemanite ($Ca_2B_6O_{11}$). The element is obtained by conversion to boric acid followed by dehydration to B_2O_3 then reduction with magnesium. High-purity boron for semiconductor applications is obtained by conversion to BORON TRICHLORIDE, which can be purified by distillation, then reduction using hydrogen. Only small quantities of elemental boron are needed commercially; the vast majority of boron is used in the form of disodium tetraborate decahydrate (borax) or boric acid.

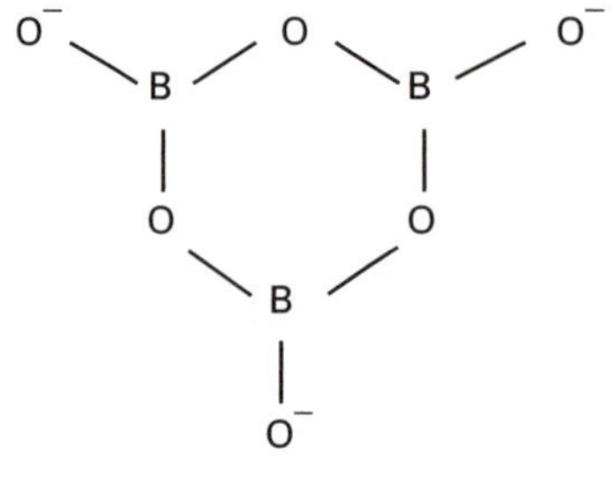

$B_3O_6^{3-}$ ion

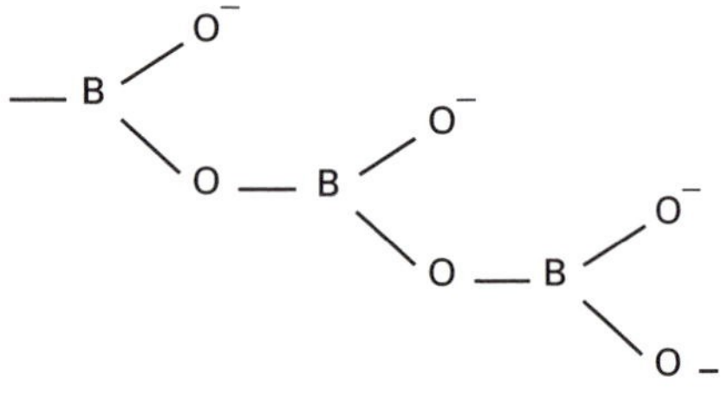

$(BO_2)_n^{n-}$ ion

Borate: examples of cyclic and linear borate ions

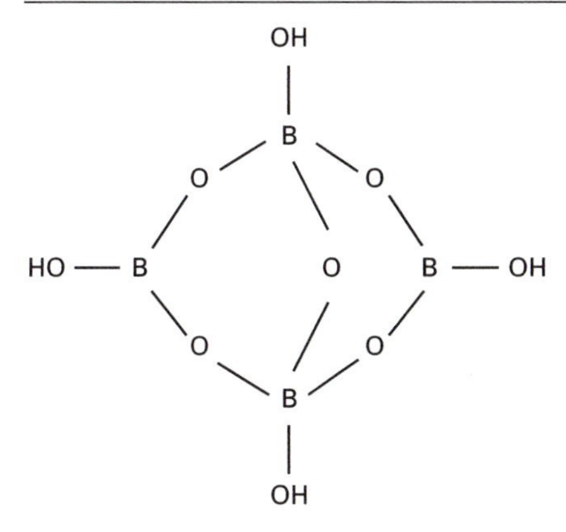

The ion $[B_4O_5(OH)_4]^{2-}$ present in borax

Because boron has a small atom and has a relatively high ionization potential its compounds are predominantly covalent; the ion B^{3+} does not exist. Boron does not react directly with hydrogen but the hydrolysis of magnesium boride does produce a range of BORON HYDRIDES. The species BH_3 is only a short-lived reaction intermediate.

Finely divided boron burns in oxygen above 600°C to give BORON OXIDE, B_2O_3, an acidic oxide which will dissolve slowly in water to give boric acid (H_3BO_3) and rapidly in alkalis to give *borates* such as $Na_2B_4O_7$. A number of polymeric species with B–B and B–O links are known, e.g. a lower oxide $(BO)_x$, and a polymeric acid $(HBO_2)_n$. Although the parent acid is weak, many salts containing borate anions are known but their stoichiometry gives little indication of their structures, many of which are cyclic or linear polymers. These polymers contain both BO_3 planar groups and BO_4 tetrahedra. Boric acid and the borates give a range of glassy substances on heating; these contain cross-linked B–O–B chains and nets. In a procedure known as the BORAX-BEAD TEST, these materials in the molten state react with metal ions to form borates, which on cooling give characteristic colors to the glass.

Boron reacts with nitrogen on strong heating (1000°C) to give BORON NITRIDE. Elemental boron reacts directly with fluorine and chlorine but for practical purposes the halides are obtained via the BF_3 route from boron oxide:

$$B_2O_3 + 3CaF_2 + 3H_2SO_4 \rightarrow 3CaSO_4 + 3H_2O + 2BF_3$$

$$BF_3 + AlCl_3 \rightarrow AlF_3 + BCl_3$$

The halides are all covalent molecules, which are all planar and trigonal in shape. Boron halides are industrially important as catalysts or promoters in a variety of organic reactions. The decomposition of boron halides in atmospheres of hydrogen at elevated temperatures is also used to deposit traces of pure boron in semiconductors.

Boron forms a range of compounds with elements that are less electronegative than itself, called BORIDES. Natural boron consists of two isotopes, ^{10}B (18.83%), used in steel alloys for control rods in nuclear reactors, and ^{11}B (81.17%). These percentages are sufficiently high for their detection by splitting of infrared absorption or by nuclear magnetic resonance spectroscopy.

Both borax and boric acid are used as mild antiseptics and are not generally regarded as toxic; boron hydrides are, however, highly toxic.

Symbol: B; m.p. 2300°C; b.p. 2658°C; r.d. 2.34 (20°C); p.n. 5; r.a.m. 10.811.

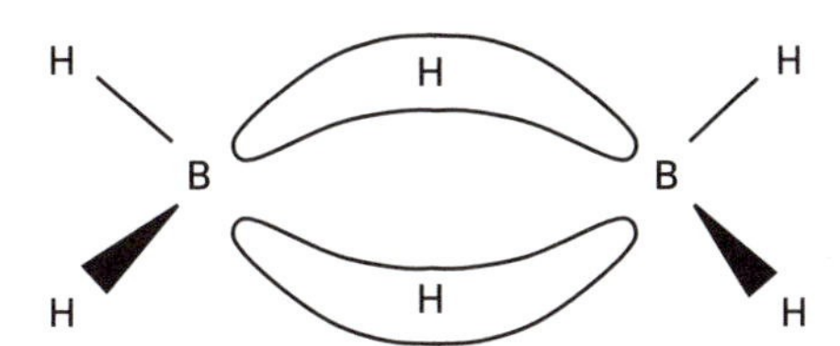

Boron hydride: the bonding in diborane

boron hydride (borane) Any of a number of binary compounds of boron and hydrogen. A mixture of boron hydrides can be obtained by the action of acid on magnesium boride (MgB_2); others can be made by controlled PYROLYSIS of *diborane* (B_2H_6), which is the simplest member of this class of compounds. Many boranes have formulae of the type B_nH_{n+4} and B_nH_{n+6}, where n is an integer. Examples are B_5H_9, B_6H_{10}, B_4H_{10}, and B_5H_{11}. Typically, the boron hydrides are volatile, highly reactive compounds that oxidize in air, some explosively.

The boron hydrides are examples of ELECTRON-DEFICIENT COMPOUNDS, in which there are not enough electrons to form classical electron-pair bonds. For example, diborane (B_2H_6) has a formular resembling that of ethane (C_2H_6), but its molecular structure and bonding are quite different. The two boron atoms are not linked directly but rather are joined through two hydrogen bridges. These B-H-B links cannot be explained in terms of two localized bonds. Instead, the bonding involves the idea of a MULTICENTER BOND in which three atoms are bound by two electrons. These are known as 3c,2e bonds, i.e. having three centers and two electrons, in contrast to conventional electron-pair bonds, which are 2c,2e bonds. In more complex boron hydrides it is possible for three boron atoms to be at the vertices of a triangle with three orbitals overlapping at the center, the resulting molecular orbital containing two electrons.

In addition to the neutral boron hydrides there are numbers of negative *borohydride ions*. Typically these have formulae of the type $B_nH_n^{2-}$, e.g. $B_6H_6^{2-}$. There are various types of boron hydride and borohydride ion. Those with a closed polyhedron of boron atoms are designated closo compounds. Ones in which there is an incomplete polyhedron (by removal of one vertex) are called nido compounds (from the Greek word for 'nest'). Those with a more open structure (removal of two or more vertices) are arachno compounds (from the Greek 'spider').

boron nitride (BN) A compound formed by heating boron in nitrogen in the vicinity of 1000°C. The material has an extremely high melting point and is thermally very stable but there is sufficient bond polarity in the B–N links to permit slow hydrolysis by water to give ammonia. It has two crystalline forms. One is a slippery white solid with a layer structure similar to that of graphite, i.e. hexagonal rings of alternating B and N atoms. The other is a 'diamond-like' form of B–N that some claim to be as hard or even harder than diamond.

boron oxide (boric anhydride; boric(III) oxide; diboron trioxide; B_2O_3) A glassy hygroscopic solid that eventually forms boric acid. It has amphoteric properties.

boron tribromide (BBr_3) A colorless liquid. *See* boron; boron trichloride.

boron trichloride (BCl_3) A fuming liquid made by passing dry chlorine over heated boron. It is rapidly hydrolysed by water:

$$BCl_3 + 3H_2O \rightarrow 3HCl + H_3BO_3$$

As there are only three pairs of shared electrons in the outer shell of the boron atom, boron halides form very stable addition compounds with ammonia by the acceptance of a lone electron pair in a coordinate bond to complete a shared octet. *See also* boron.

boron trifluoride (BF_3) A colorless fuming gas made by heating a mixture of boron oxide, calcium fluoride, and concentrated sulfuric acid. *See* boron; boron trichloride.

borosilicates Complex compounds similar to silicates, but containing BO_3 and BO_4 units in addition to the SiO_4 units. Certain crystalline borosilicate minerals such as danburite, $CaB_2Si_2O_8$, are known. In addition, borosilicate glasses can be made by using boron oxide in addition to silicon(IV) oxide. These glasses tend to be tougher and more heat resistant than standard silicate glass. Pyrex, for example, is commonly used in laboratory glassware and in cookware.

Bosch, Carl (1874–1940) German industrial chemist. Bosch joined the large German dyestuffs company, Badische Anilin und Soda Fabrik (BASF), in 1899. Following Fritz Haber's successful small-scale ammonia synthesis in 1909, Bosch began to develop a high-pressure ammonia plant at Oppau for BASF. The plant was opened in 1912 – a successful application of the HABER PROCESS on a large scale. Bosch also introduced the use of the water-gas shift reaction as a source of hydrogen for the process:

$$CO + H_2O = CO_2 + H_2$$

After World War I the large-scale ammonia fertilizer industry was established and the high-pressure technique was extended by BASF to the synthesis of methanol from carbon monoxide and hydrogen in 1923. Bosch shared the Nobel Prize for chemistry with Friedrich Bergius in 1931.

Bosch process The reaction between carbon monoxide and steam over a hot catalyst:

$$CO + H_2O \rightarrow CO_2 + H_2$$

It has been used as a source of hydrogen for the Haber process. The process was developed by the German chemist Carl Bosch (1874–1940).

bowl classifier A device that separates solid particles in a mixture of solids and liquid into fractions according to particle size. Feed enters the center of a shallow bowl, which contains revolving blades. The coarse solids collect on the bottom, fine solids at the edge.

Boyle, Robert (1627–91) English scientist who is generally regarded as one of the founders of modern chemistry. He established the law known as BOYLE'S LAW relating the pressure and volume of gases experimentally and gave an explicit statement of it in 1662. In 1661 he published a book entitled *The Sceptical Chymist* in which he put forward the idea that all matter is corpuscular in nature, with the corpuscules having different sizes and shapes. This idea enabled him to distinguish between compounds and mixtures. Boyle was also able to distinguish between acids and bases by using vegetable dyes as indicators.

Boyle's law At a constant temperature, the pressure of a fixed mass of a gas is inversely proportional to its volume: i.e.

$$pV = K$$

Here K is a constant whose value depends on the temperature and on the nature of the gas. The law holds strictly only for ideal gases. Real gases follow Boyle's law at low pressures and high temperatures. *See* gas laws.

Brackett series *See* hydrogen atom spectrum.

Bragg, Sir William Henry (1862–1942) and **Bragg, Sir William Lawrence** (1890–1971) British physicists. The Braggs (father and son) realized very soon after the discovery of x-ray diffraction in 1912 that this phenomenon could be used to determine the structure of crystals. Lawrence Bragg formulated the *Bragg equation* in 1912. This initiated the subject of x-ray crystallography. One of their early results was to show that sodium chloride consists of sodium ions and chloride ions rather than NaCl molecules. Lawrence Bragg subsequently studied silicates and encouraged research on complex molecules of biological interest. The Braggs shared the 1915 Nobel Prize for physics.

Bragg equation An equation used to deduce the crystal structure of a material using data obtained from x-rays directed at its surface. The conditions under which a crystal will reflect a beam of x-rays with maximum intensity is:

$$n\lambda = 2d\sin\theta$$

where θ is the angle of incidence and reflection (*Bragg angle*) that the x-rays make with the crystal planes, n is a small integer, λ is the wavelength of the x-rays, and d is the distance between the crystal planes. The equation was discovered by Lawrence Bragg in 1912.

branched chain *See* chain.

brass Any of a group of copper–zinc alloys containing up to 50% of zinc. The color of brass changes from red-gold to golden to silvery-white with increasing zinc content. Brasses are easy to work and resist corrosion well. Brasses with up to 35% zinc can be worked cold and are specially suited for rolling into sheets, drawing into wire, and making into tubes. Brasses with 35–46% zinc are harder and stronger but less ductile; they require hot working (e.g. forging). The properties of brass can be improved by the addition of other elements; lead improves its ability to be machined, while aluminum and tin increase its corrosion resistance. *See also* bronze.

bremsstrahlung *See* x-radiation.

brine A concentrated solution of sodium chloride or calcium chloride. The term is also applied to naturally occurring solutions of other salts.

Brin process An obsolete process for producing oxygen by first heating barium oxide in air to produce the peroxide (BaO_2); then heating the peroxide at higher temperatures to release oxygen.

bromic(I) acid (hypobromous acid; HBrO) A pale yellow liquid made by reacting mercury(II) oxide with BROMINE WATER. Bromic(I) acid is a weak acid in aqueous solution but is a good oxidizing agent with strong bleaching power.

bromic(V) acid ($HBrO_3$) A colorless liquid made by the addition of dilute sulfuric acid to barium bromate. It is a strong acid in aqueous solution.

bromide *See* halide.

bromination *See* halogenation.

bromine A deep red, moderately reactive element belonging to the halogens; i.e. group 17 (formerly VIIA) of the periodic table. Bromine is a volatile liquid at room temperature (mercury being the only other element with this property). It occurs in small amounts in seawater, salt lakes, and salt deposits but is much less abundant than chlorine. Bromine reacts with most metals but generally with less vigor than chlorine. It has a lower oxidizing power than chlorine and consequently can be released from solutions of bromides by reaction with chlorine gas. The laboratory method is the more convenient oxidation of bromides using manganese dioxide. Industrially most bromine is produced by displacement with chlorine from naturally occurring brines. Bromine and its compounds are used in pharmaceuticals, photography, chemical synthesis, and fumigants. Bromine compounds are also used as flame retardants and refrigerants.

The electropositive elements form electrovalent bromides and the nonmetals form fully covalent bromides. Like chlorine, bromine forms oxides, Br_2O and BrO_2, both of which are unstable. The related oxo-acid anions hypobromite (BrO^-) and bromate (BrO_3^-) are formed by the reaction of bromine with cold aqueous alkali and hot aqueous alkali respectively, but the bromine analogs of chlorite and perchlorate are not known.

Bromine and the *interhalogens*, chemical compounds formed between two haogens, are highly toxic. Liquid bromine and bromine solutions are also very corrosive and goggles and gloves should always be worn when handling such compounds.

Symbol: Br; m.p. –7.25°C; b.p. 58.78°C; r.d. 3.12 (20°C); p.n. 35; r.a.m. 79.904.

bromine trifluoride (BrF_3) A colorless fuming liquid made by direct combination of fluorine and bromine. It is a very reactive compound, its reactions being similar to those of its component halogens.

bromine water A yellow solution of bromine in water, which contains bromic(I) acid (hypobromous acid, HBrO). It is a weak acid and a strong oxidizing agent, which decomposes to give a mixture of bromide (Br^-) and bromate(V) (BrO_3^-) ions.

Brønsted acid *See* acid.

bronze Any of a group of copper-tin alloys usually containing 0.5–15% of tin. They are generally harder, stronger in compression, and more corrosion resistant than brass. Zinc is often added, as in *gunmetal* (2–4% zinc), to increase strength and corrosion-resistance; bronze coins often contain more zinc (2.5%) than tin (0.5%). The presence of lead improves its machining qualities.

Some copper-rich alloys containing no tin are also called bronzes. *Aluminum bronzes*, for example, with up to 10% aluminum, are strong, resistant to corrosion and wear, and can be worked cold or hot; *silicon bronzes*, with 1–5% silicon, have high corrosion-resistance; *beryllium bronzes*, with about 2% beryllium, are very hard and strong. *See also* brass.

brown coal *See* lignite.

brown-ring test A qualitative test used for the detection of an ionic nitrate. A freshly prepared solution of iron(II) sulfate is mixed with the sample and concentrated sulfuric acid is introduced slowly to the bottom of the tube using a dropping pipette, so that two layers are formed. A brown ring formed at the interface where the liquids meet indicates the presence of nitrate. The brown color is caused by the presence of $[Fe(NO)]SO_4$, which breaks down on shaking.

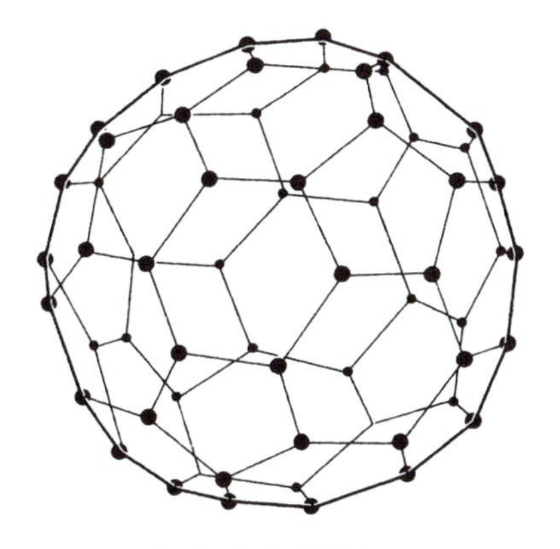

Buckminsterfullerene

buckminsterfullerene (fullerene) An allotrope of carbon discovered in soot in 1985, that contains clusters of 60 carbon atoms bound in a highly symmetric polyhedral structure. The C_{60} polyhedron has a combination of pentagonal and hexagonal faces similar to the panels on a soccer ball. The molecule was named after the American architect Richard Buckminster Fuller (1895–1983) because its structure resembles the geodesic dome, which was invented by Fuller. The C_{60} polyhedra are informally called *bucky balls*. The original method of making the allotrope was to fire a high-power laser at a graphite target. This procedure also produces less stable carbon clusters, such as C_{70}. Buckminsterfullerene can be produced more conveniently using an electric arc between graphite electrodes in an inert gas. The allotrope is soluble in benzene, from which it can be crystallized to give yellow crystals. This solid form is known as *fullerite*.

The discovery of buckminsterfullerene led to a considerable amount of research into its properties and compounds. Particular interest has been shown in trapping metal ions inside the carbon cage to form enclosure compounds. The term fullerene also applies to derivatives of buckminsterfullerene and to similar clusters (e.g. C_{70}). Carbon structures similar to that in C_{60} can also form small tubes, known as *bucky tubes*.

bucky ball *See* buckminsterfullerene.

bucky tube *See* buckminsterfullerene.

buffer A solution in which the pH remains reasonably constant when acids or alkalis are added to it; i.e. it acts as a buffer against (small) changes in pH. Buffer solutions generally contain a weak acid and one of its salts derived from a strong base; e.g. a solution of ethanoic acid and sodium ethanoate. If an acid is added, the H^+ reacts with the ethanoate ion (from dissociated sodium ethanoate) to form undissociated ethanoic acid; if a base is added the OH^- reacts with the ethanoic acid to form water and the ethanoate ion. The effectiveness of the buffering action is determined by the concentrations of the acid–anion pair:

$$K = [H^+][CH_3COO^-]/[CH_3COOH]$$

where K is the dissociation constant.

Phosphate, oxalate, tartrate, borate, and carbonate systems can also be used for buffer solutions.

bumping Violent boiling of a liquid caused when bubbles form at a pressure above atmospheric pressure.

Bunsen, Robert Wilhelm (1811–99) German chemist. Along with Gustav Kirchoff, Bunsen pioneered the use of spectroscopy to identify chemical elements.

This work was started in 1859 and soon led them to the discovery of rubidium and cesium. The Bunsen burner is named for Robert Bunsen, although it was probably developed by Bunsen's technician Peter Desdega.

Bunsen burner A gas burner consisting of a vertical metal tube with an adjustable air-inlet hole at the bottom. Gas is admitted into the bottom of the tube and the gas–air mixture is burnt at the top. With too little air the flame is yellow and sooty. Correctly adjusted, the burner gives a flame with a pale blue inner cone of incompletely burnt gas, and an almost invisible outer flame where the gas is fully oxidized and reaches a temperature of about 1500°C. The inner region is the reducing part of the flame and the outer region the oxidizing part. The Bunsen burner is named for Robert Bunsen, who helped make it popular, although it was probably invented by Michael Faraday and improved by Bunsen's technician Peter Desdega. *See also* borax-bead test.

Bunsen cell A type of primary cell in which the positive electrode is formed by carbon plates in a nitric acid solution and the negative electrode consists of zinc plates in sulfuric acid solution.

burette A piece of apparatus used for the addition of variable volumes of liquid in a controlled and measurable way. The burette is a long cylindrical graduated tube of uniform bore fitted with a stopcock and a small-bore exit jet, enabling a drop of liquid at a time to be added to a reaction vessel. Burettes are widely used for titrations in volumetric analysis. Standard burettes permit volume measurement to 0.005 cm^3 and have a total capacity of 50 cm^3; a variety of smaller *microburette* is available. Similar devices are used to introduce measured volumes of gas at regulated pressure in the investigation of gas reactions.

by-product A substance obtained during the manufacture of a main chemical product. For example, calcium chloride is a by-product of the Solvay process for making sodium carbonate. Some metals are obtained as by-products in processes to extract other metals. Cadmium, for instance, is a by-product in the extraction of zinc.

B–Z reaction *See* Belousov–Zhabotinskii reaction.

C

cadmium A transition metal, an element in group 12 (formerly IIB) of the periodic table, obtained as a by-product during the extraction of zinc. It is used to protect other metals from corrosion, as a neutron absorber in nuclear reactors, in alkali batteries, and in certain pigments. It is highly toxic.

Symbol: Cd; m.p. 320.95°C; b.p. 765°C; r.d. 8.65 (20°C); p.n. 48; r.a.m. 112.411.

cadmium cell *See* Weston cadmium cell.

caesium *See* cesium.

cage compounds *See* clathrate.

calamine **1.** A mineral consisting of hydrated zinc silicate ($2ZnO.SiO_2.H_2O$). It is also known as *hemimorphite*.
2. Zinc carbonate ($ZnCO_3$), which occurs naturally as a mineral also known as *smithsonite*. The carbonate is used medicinally in suspension as a soothing lotion for sunburn and skin complaints. Formerly, the natural mineral was used but now medicinal calamine is made by precipitating a basic zinc carbonate from a solution of a zinc salt. It is usually colored pink by the addition of small quantities of iron(III) oxide.

calcination The formation of a calcium carbonate deposit from hard water.

calcite A mineral form of calcium carbonate occurring in limestone, chalk, and marble.

calcium A moderately soft, low-melting reactive metal; the third element in group 2 (formerly IIA) of the periodic table. The electronic configuration is that of argon with an additional pair of 4s electrons.

Calcium is widely distributed in the Earth's crust and is the third most abundant element. Large deposits occur as chalk or marble, $CaCO_3$; gypsum, $CaSO_4$.-$2H_2O$; anhydrite, $CaSO_4$; fluorspar, CaF; and apatite, $CaF_2.Ca_3(PO_4)_3$. However sufficiently large quantities of calcium chloride are available as waste from the Solvay process to satisfy industrial requirements for the metal, which is produced by electrolysis of the fused salt. Large quantities of lime, $Ca(OH)_2$, and quicklime, CaO, are produced by decomposition of the carbonate for use in both building and agriculture. Several calcium minerals are mined as a source of other substances. Thus, limestone is a cheap source of carbon dioxide, gypsum and anhydrite are used in the manufacture of sulfuric acid, rock phosphate is a source of phosphoric acid, and fluorspar for a range of fluorochemicals.

Calcium has a low ionization potential and a relatively large atomic radius. It is therefore a very electropositive element. The metal is very reactive and the compounds contain the divalent ion Ca^{2+}. Calcium forms the oxide (CaO), a white ionic solid, on burning in air, but for practical purposes the oxide is best prepared by heating the carbonate, which decomposes at about 800°C. Both the oxide and the metal itself react with water to give the basic hydroxide ($Ca(OH)_2$). On heating with nitrogen, sulfur, or the halogens, calcium reacts to form the nitride (Ca_3N_2), sulfide (CaS), or the halides (CaX_2). Calcium also reacts directly with hydrogen to give the hydride CaH_2 and borides, arsenides, carbides, and silicides can be prepared in a similar way. Both the carbonate and sulfate are insoluble. Calcium salts impart a characteristic brick-red color to flames which is an aid to qualitative analysis. At ordinary temperatures calcium has the face-centered cubic structure with a transition at 450°C to the close-packed hexagonal structure.

Symbol: Ca; m.p. 839°C; b.p. 1484°C; r.d. 1.55 (20°C); p.n. 20; r.a.m. 40.0878.

calcium acetylide *See* calcium dicarbide.

calcium bicarbonate *See* calcium hydrogencarbonate.

calcium carbide *See* calcium dicarbide.

calcium carbonate ($CaCO_3$) A white solid that occurs naturally in two crystalline forms: calcite and aragonite. These minerals make up the bulk of such rocks as marble, limestone, and chalk. Calcium carbonate also occurs in the mineral dolomite ($CaCO_3.MgCO_3$). It is sparingly soluble in water but dissolves in rainwater containing carbon dioxide to form calcium hydrogencarbonate, which causes temporary hardness of water. Calcium carbonate is a basic raw material in the Solvay process and is used for making glass, mortar, and cement.

calcium chloride ($CaCl_2$) A white solid that occurs in a number of hydrated forms and is readily available as a by-product of the Solvay process. It is readily soluble in water and the solution, known as *brine*, is used in refrigerating plants. Other applications that depend on its water-absorbing property and the low freezing point of the aqueous solution include the suppression of dust on roads and in mines and the melting of snow. Calcium chloride is also used as the electrolyte in the production of calcium.

calcium cyanamide ($CaCN_2$) A solid prepared by heating calcium dicarbide to a temperature in excess of 800°C in an atmosphere of nitrogen. It is used as a fertilizer because ammonia and calcium carbonate are slowly formed when water is added:

$$CaCN_2 + 3H_2O \rightarrow CaCO_3 + 2NH_3$$

Other uses include the defoliation of cotton plants and the production of melamine.

calcium dicarbide (calcium acetylide; calcium carbide; CaC_2) A colorless solid when pure. In countries where electricity is cheap, calcium dicarbide is produced on a large scale by heating calcium oxide with coke at a temperature in excess of 2000°C in an electric-arc furnace. Water is then added to give ethyne (acetylene, C_2H_2), an important industrial organic chemical:

$$CaC_2 + 2H_2O \rightarrow C_2H_2 + Ca(OH)_2$$

The structure of calcium dicarbide is interesting because the carbon is present as carbide ions (C_2^{2-}). *See* ethyne.

calcium fluoride (CaF_2) A white crystalline compound found naturally as *fluorite* (*flourspar*).

calcium-fluoride structure (fluorite structure) A form of crystal structure in which each calcium ion is surrounded by eight fluorine ions arranged at the corners of a cube and each fluorine ion is surrounded tetrahedrally by four calcium ions.

calcium hydrogencarbonate (calcium bicarbonate, $Ca(HCO_3)_2$) A solid formed when water containing carbon dioxide dissolves calcium carbonate:

$$CaCO_3 + H_2O + CO_2 \rightarrow Ca(HCO_3)_2$$

Calcium hydrogencarbonate is a cause of temporary hard water. The solid is unknown at room temperature. *See* hardness.

calcium hydroxide (slaked lime; caustic lime; $Ca(OH)_2$) A white solid that dissolves sparingly in water to give the alkali known as *limewater*. Calcium hydroxide is manufactured by adding water to the oxide (lime or quicklime), a process known as *slaking*, which evolves much heat. If just sufficient water is added so that the oxide turns to a fine powder, the product is *slaked lime*. If more water is added, a thick suspension called *milk of lime* is formed.

Calcium hydroxide has many uses. As a base, it is used to neutralize acid soil and in industrial processes such as the Solvay process. It is also used in the manufacture of mortar, whitewash, and bleaching powder and for the softening of temporary hard water.

calcium nitrate ($Ca(NO_3)_2$) A deliquescent salt that is very soluble in water. It is usually crystallized as the tetrahydrate $Ca(NO_3)_2.4H_2O$. When the hydrate is heated, the anhydrous salt is first produced and this subsequently decomposes to give calcium oxide, nitrogen dioxide, and oxygen. Calcium nitrate is used as a nitrogenous fertilizer.

calcium octadecanoate (calcium stearate; $Ca(CH_3(CH_2)_{16}COO)_2$) An insoluble salt of octadecanoic acid. It is formed as 'scum' when soap, containing the soluble salt sodium octadecanoate, is mixed with hard water containing calcium ions. *See* detergents.

calcium oxide (quicklime; CaO) A white solid formed by heating calcium in oxygen or, more widely, by the thermal decomposition of calcium carbonate. On a large scale, limestone (calcium carbonate) is heated in a tall tower called a lime kiln to a temperature of 550°C. The reversible reaction:

$$CaCO_3 \rightleftharpoons CaO + CO_2$$

proceeds in a forward direction as the carbon dioxide is carried away by the upward current through the kiln. Calcium oxide is used in extractive metallurgy to produce a slag with the impurities in metal ores; it is also used as a drying agent and it is an intermediate for the production of calcium hydroxide.

calcium phosphate ($Ca_3(PO_4)_2$) A solid that occurs naturally in the mineral apatite ($CaF_2.Ca_3(PO_4)_3$) and in rock phosphate. It is the chief constituent of animal bones and is used extensively in fertilizers.

calcium silicate (Ca_2SiO_4) A white insoluble crystalline compound found in various cements and minerals. It is also a component of the slag produced in smelting iron and other metals in a blast furnace.

calcium stearate *See* calcium octadecanoate.

calcium sulfate (anhydrite; $CaSO_4$) A white solid that occurs abundantly as the mineral anhydrite and as the dihydrate ($CaSO_4.2H_2O$), known as *gypsum* or *alabaster*. When heated, gypsum loses water to form the hemihydrate ($2CaSO_4.H_2O$), which is *plaster of Paris*. If the water is replaced, gypsum reforms and sets as a solid. Plaster of Paris is therefore used for taking casts and for setting broken limbs. Calcium sulfate is sparingly soluble in water and is a cause of permanent hardness in water. It is used in ceramics, paint, and in paper making.

Calgon (*Trademark*) A substance often added to detergents to remove unwanted chemicals that have dissolved in water and would otherwise react with soap to form a scum. Calgon consists of complicated sodium polyphosphate molecules, which absorb dissolved calcium and magnesium ions. The metal ions become trapped within the Calgon molecules.

caliche An impure commercial form of sodium nitrate.

californium A silvery radioactive transuranic element of the actinoid series of metals, not found naturally on Earth. Several radioisotopes have been synthesized, including californium-252, which is used as an intense source of neutrons in certain types of portable detector and in the treatment of cancer.

Symbol: Cf; m.p. 900°C; p.n. 98; most stable isotope ^{251}Cf (half-life 900 years).

calixarene *See* host–guest chemistry.

calomel *See* mercury(I) chloride.

calomel electrode A half cell having a mercury electrode coated with mercury(I) chloride (called *calomel*), in an electrolyte consisting of potassium chloride and (saturated) mercury(I) chloride solution. Its standard electrode potential against the hydrogen electrode is accurately known (–0.2415 V at 25°C) and it is a convenient secondary standard.

calorie Symbol: cal A unit of energy approximately equal to 4.2 joules. It was formerly defined as the energy needed to raise the temperature of one gram of water by one degree Celsius. Because the specific thermal capacity of water changes with temperature, this definition is not precise. The mean or thermochemical calorie (cal_{TH}) is defined as 4.184 joules. The international table calorie (cal_{IT}) is defined as 4.1868 joules. Formerly the mean calorie was defined as one hundredth of the heat needed to raise one gram of water from 0°C to 100°C, and the 15°C calorie as the heat needed to raise it from 14.5°C to 15.5°C.

calorific value *See* bomb calorimeter.

calorimeter A device or apparatus for measuring thermal properties such as specific heat capacity, calorific value, etc. *See* bomb calorimeter.

calx A metal oxide obtained by heating an ore to high temperatures in air.

candela Symbol: cd The SI base unit of luminous intensity, equal to that, in a given direction, of a source emitting monochromatic radiation having a frequency of 540 terahertz and a radiant intensity of 1/683 watt.

Cannizzaro, Stanislao (1826–1910) Italian chemist. Cannizzaro was responsible for reviving interest in the ideas of Amedeo AVOGADRO in a pamphlet published in 1858 and in an address to the Chemical Congress at Karlsruhe, Germany. His pamphlet clarified the concepts of atomic and molecular weights and also showed how the molecular weight (now termed relative molecular mass) of a compound could be determined by measuring its vapor density. He also discovered the organic reaction known as *Cannizzaro's reaction* in 1853.

canonical form *See* resonance.

carbide A compound of carbon with a more electropositive element. The carbides of the elements are classified into:

1. *Ionic carbides*, which contain the carbide ion C^{4-}. An example is aluminum carbide, Al_4C_3. Compounds of this type react with water to give methane (they were formerly also called *methanides*). The *dicarbides* are also ionic carbon compounds but contain the dicarbide ion $^-C:C^-$. The best-known example is calcium dicarbide, CaC_2, also known as calcium carbide, or simply *carbide.* Compounds of this type give ethyne (acetylene) with water. They were formerly called *acetylides* or *ethynides.* Ionic carbides are formed with very electropositive metals. They are crystalline.
2. *Covalent carbides*, which have giant molecular structures, as in SILICON CARBIDE (SiC) and boron carbide (B_4C_3). These are hard solids with high melting points.
3. *Interstitial carbides*, which are INTERSTITIAL COMPOUNDS of carbon with transition metals. Titanium carbide (TiC) is an example. These compounds are all hard solids with high melting points and metallic properties. Some carbides (e.g. nickel carbide Ni_3C) have properties intermediate between those of interstitial and ionic carbides.

carbon The first element of group 14 (formerly IVA) of the periodic table. Carbon is a universal constituent of living matter and the principal deposits of carbon compounds, i.e. chalk and limestone, from which we obtain carbonates, and coal, oil, and gas fields, from which we obtain fossil fuels, are derived from once living organisms. Carbon also occurs in the mineral dolomite. The element forms only 0.032% by mass of the Earth's crust. Minute quantities of elemental carbon also occur as the allotropes graphite and diamond. A third allotrope, BUCKMINSTERFULLERENE (C_{60}), was discovered in the mid-1980s.

The industrial demand for graphite is such that it is manufactured in large quantities using the *Acheson process* in which coke and small amounts of asphalt or clay are raised to high temperatures. Large quantities of impure carbon are also consumed in the reductive extraction of metals. In addition to the demand for diamond as a gemstone there is a large industrial demand for small low-grade diamonds for use in drilling and grinding machinery. Studies show that graphite and diamond can be interconverted at 3000°C under extremely high pressures, but that commercial exploitation of this process would not be viable.

Carbon burns in oxygen to form carbon dioxide and carbon monoxide. Carbon dioxide is soluble in water, forming the weakly acidic carbonic acid, which is the parent acid of the metal carbonates. In contrast CO is barely soluble in water but will react with alkali to give the methanoate (formate) ion:

$$CO + OH^- \rightarrow HCO_2^-$$

Carbon will react readily with sulfur at red heat to form carbon disulphide, CS_2, but it does not react directly with nitrogen. Cyanogen, $(CN)_2$, must be prepared by heating covalent metal cyanides such as CuCN. Carbon will also react directly with many metals at elevated temperatures to give CARBIDES. Carbides can also be obtained by heating the metal oxide with carbon or heating the metal with a hydrocarbon. There is a bewilderingly wide range of metal carbides, both saltlike with electropositive elements (for example, CaC_2) and covalent with the metalloids (for example, SiC), and there are also many interstitial carbides formed with metals such as Cr, Mn, Fe, Co, and Ni.

Compounds with C–N bonds form a significant branch of the inorganic chemistry of carbon: these comprise hydrogen cyanide (HCN) and the cyanides, cyanic acid (HNCO) and the cyanates, and thiocyanic acid (HNCS) and the thiocyanates.

Naturally occurring carbon has the isotopic composition ^{12}C (98.89%), ^{13}C (1.11%), and radioactive ^{14}C (minute traces of which are produced in the upper atmosphere as the result of the capture of high-energy neutrons by ^{14}N atoms, which each then release a proton to decay into ^{14}C). Carbon-14 (^{14}C) is used for radiocarbon dating because it is taken up in minute quantities by organisms when they are alive and decays over the comparatively long half-life of 5780 years after their death.

Symbol: C; m.p. 3550°C; b.p. 4830°C (sublimes); C_{60} sublimes at 530°C; r.d. 3.51 (diamond), 2.26 (graphite), 1.65 (C_{60}) (all at 20°C); p.n. 6; r.a.m. 12.011.

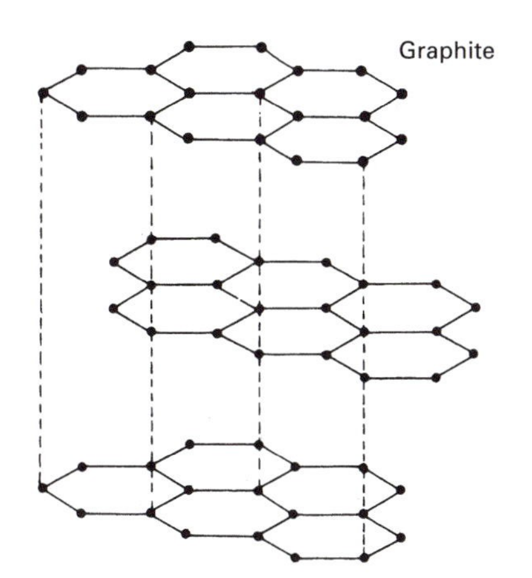

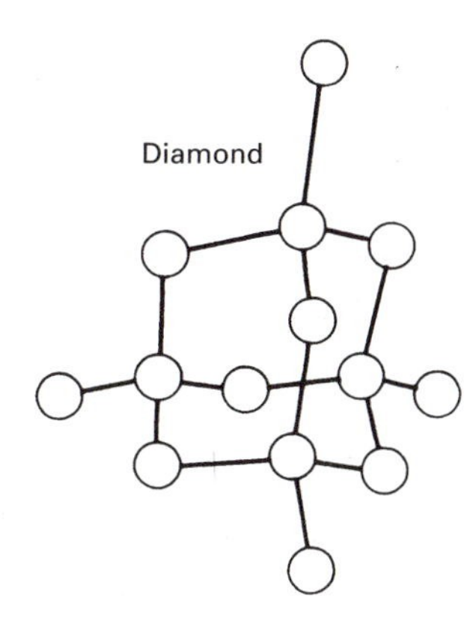

Carbon

carbonate Any salt of carbonic acid (i.e. containing the ion CO_3^{2-}). Many alkali and alkaline-earth carbonates are important industrially.

carbonation **1.** The solution of carbon dioxide in a liquid under pressure, as in carbonated soft drinks.
2. The addition of carbon dioxide to compounds.

carbon black A finely divided form of carbon produced by the incomplete combustion of such hydrocarbon fuels as natural gas or oil. It is used as a black pigment in inks, paints, and plastics, and as a filler for rubber in tire manufacture.

carbon dioxide (CO_2) A colorless odorless nonflammable gas formed when carbon burns in excess oxygen. It is also produced by respiration. Carbon dioxide is present in the atmosphere (0.03% by volume) and is converted in plants to carbohydrates by photosynthesis. In the laboratory it is made by the action of dilute acid on metal carbonates. Industrially, it is obtained as a by-product in certain processes, such as fermentation and the manufacture of calcium oxide. The main uses of carbon dioxide are as a refrigerant in its solid state, called *dry ice*, and in fire extinguishers and carbonated drinks. Increased levels of carbon dioxide in the atmosphere from the combustion of fossil fuels are thought to contribute to the greenhouse effect.

Carbon dioxide is the anhydride of the weak acid carbonic acid, which is formed in water:

$$CO_2 + H_2O \rightarrow H_2CO_3$$

carbon disulfide (CS_2) A colorless poisonous flammable liquid made from methane (derived from natural gas) and sulfur. It is a solvent for waxes, oils, and rubber, and is used in the manufacture of viscose rayon. The pure compound is virtually odorless, but CS_2 usually has a revolting smell because of the presence of other sulfur compounds.

carbon fibers Fibers of graphite, which are used, for instance, to strengthen POLYMERS. They are made by heating stretched textile fibers and have an orientated crystal structure.

carbonic acid (H_2CO_3) A DIBASIC ACID formed in small amounts in solution when carbon dioxide dissolves in water:

$$CO_2 + H_2O \rightleftharpoons H_2CO_2$$

It forms two series of salts: hydrogencarbonates (HCO_3^-) and carbonates (CO_3^{2-}). The pure acid cannot be isolated.

carbonize (carburize) To convert an organic compound into carbon by incomplete oxidation at high temperature.

carbon monoxide (CO) A colorless flammable toxic gas formed by the incomplete combustion of carbon or carbon-containing compounds. Industrially, it is produced by the oxidation of carbon or methane, or by the reaction used to produce WATER GAS. It is a powerful reducing agent and is used as such in metallurgy.

Carbon monoxide is neutral and only sparingly soluble in water. It forms metal carbonyls with transition metals. Its toxicity derives from the fact that hemoglobin (in the blood) has a far stronger affinity for carbon monoxide than for oxygen, which is essential for cell metabolism.

carbonyl A complex in which carbon monoxide ligands are coordinated to a metal atom. A common example is tetracarbonyl nickel(0), $Ni(CO)_4$.

carbonyl chloride (phosgene; $COCl_2$) A colorless, toxic gas with a choking smell, made by reacting carbon monoxide and chlorine in the presence of light or a catalyst. It is used as a chlorinating agent and to make certain plastics and insecticides; it was formerly employed as a war gas.

carbonyl group The group =C=O. It occurs in organic compounds, and in carbonyl complexes of transition metals.

carborundum *See* silicon carbide.

carboxylate ion The ion $-COO^-$, produced by ionization of a carboxyl group. In a carboxylate ion the negative charge is generally delocalized over the O–C–O grouping and the two C–O bonds have the same length, intermediate between that of a double C=O and a single C–O.

carboxyl group The organic group –CO.OH, present in carboxylic acids.

carboxylic acid A type of organic compound containing the CARBOXYL GROUP. Simple carboxylic acids have the general formula RCOOH. Many carboxylic acids occur naturally in plants and (in the form of esters) in fats and oils, hence the alternative name *fatty acids*.

carburizing *See* case hardening.

carcinogen An agent that causes or promotes the development of cancer in animals, such as tobacco smoke or ionizing radiation.

carnallite A mineral chloride of potassium and magnesium, $KCl.MgCl_2.6H_2O$. It is used as a source of potassium salts for fertilizers.

Carnot cycle The idealized reversible cycle of four operations occurring in a perfect heat engine. These are the successive adiabatic compression, isothermal expansion, adiabatic expansion, and isothermal compression of the working substance. The cycle returns to its initial pressure, volume, and temperature, and transfers energy to or from mechanical work. The efficiency of the Carnot cycle is the maximum attainable in a heat engine. It was published in 1824 by the French physicist Nicolas L. S. Carnot (1796–1832). *See* Carnot's principle.

Carnot's principle (Carnot theorem) The efficiency of any heat engine cannot be greater than that of a reversible heat engine operating over the same temperature range. Carnot's principle follows directly from the second law of thermodynamics, and means that all reversible heat engines have the same efficiency, independent of the working substance. If heat is absorbed at temperature T_1 and given out at T_2, then the efficiency according to Carnot's principle is $(T_1 - T_2)/T_1$. *See* Carnot cycle.

Carnot's theorem *See* Carnot's principle.

Caro's acid *See* peroxosulfuric(VI) acid.

carrier gas The gas used to carry the sample in gas chromatography.

case hardening Processes for increasing the hardness of the surface or 'case' of the steel used to make such components as gears and crankshafts. The oldest method is *carburizing*, in which the carbon content of the surface layer is increased by heating the component in a carbon-rich environment. *Nitriding* involves the diffusion of nitrogen into the surface layer of the steel, thus forming intensely hard NITRIDE particles in the structure. A combination of both carburizing and nitriding is sometimes employed.

cast iron Any of various alloys of iron and carbon made by remelting the crude iron produced in a blast furnace. The carbon content is usually between 2.4 and 4.0% and may be present as iron carbide (FeC) or as graphite. In the former case the product is known as *white cast iron*, and in the latter as *gray cast iron*. Other metals may be added to improve the properties of the alloy. Additional elements such as phosphorus, sulfur, and manganese are also present as impurities. Cast iron is cheap and has an extensive range of possible properties. It is used on a very large scale.

catalyst A substance that alters the rate of a chemical reaction without itself being changed chemically in the reaction. The catalyst can, however, undergo physical change; for example, large lumps of catalyst can, without loss in mass, be converted into a powder. Small amounts of catalyst are often sufficient to alter the rate of reaction considerably. A *positive catalyst* increases the rate of a reaction and a *negative* *catalyst* reduces it. *Homogeneous catalysts* are those that act in the same phase as the reactants (i.e. in gaseous and liquid systems). For example, nitrogen(II) oxide gas will catalyze the reaction between sulfur(IV) oxide and oxygen in the gaseous phase. *Heterogeneous catalysts* act in a different phase from the reactants. For example, finely divided nickel (a solid) will catalyze the hydrogenation of oil (liquid).

The function of a catalyst is to provide a new pathway for the reaction, along which the rate-determining step has a lower activation energy than in the uncatalyzed reaction. A catalyst does not change the products in an equilibrium reaction and their concentration remains identical to the concentration of the products in an uncatalyzed reaction; i.e. the position of the equilibrium remains unchanged. The catalyst simply changes the rate at which equilibrium is attained.

In *autocatalysis*, one of the products of the reaction itself acts as a catalyst. In this type of reaction the reaction rate increases with time up to a maximum value and finally slows down. *See also* promoter.

catalytic converter A device fitted to the exhaust system of gasoline-fueled vehicles to remove pollutant gases from the exhaust. It typically consists of a structure honeycombed to provide maximum area and coated with platinum, palladium, and rhodium catalysts. Such devices can convert carbon monoxide to carbon dioxide, oxides of nitrogen to nitrogen, and unburned fuel to carbon dioxide and water.

catenation The formation of chains of atoms in molecules.

cathode In electrolysis, the electrode that is at a negative potential with respect to the anode, and to which CATIONS are attracted. In any electrical system, such as a discharge tube or electronic device, the cathode is the terminal at which electrons enter the system. *Compare* anode.

cation A positively charged ion, formed by removal of electrons from atoms or molecules. In electrolysis, cations are attracted to the negatively charged electrode or cathode. *Compare* anion.

cationic detergent *See* detergents.

cationic resin An ION-EXCHANGE material that can exchange cations, such as H^+ and Na^+, for cations in the surrounding medium. Such materials are often produced by adding a sulfonic acid group ($-SO_3^-H^+$) to a stable resin. A typical exchange reaction is:

$$\text{resin–}SO_3^-H^+ + NaCl \rightleftharpoons \text{resin–}SO_3^-Na^+ + HCl$$

Cationic resins have been used to great effect to separate mixtures of cations of similar size having the same charge. During this procedure, the mixtures are attached to the resins, after which progressive ELUTION is used to recover the cations in order of decreasing ionic radius. Promethium was first isolated using this technique.

caustic lime *See* calcium hydroxide.

caustic potash *See* potassium hydroxide.

caustic soda *See* sodium hydroxide.

Cavendish, Henry (1731–1810) English chemist and physicist. Cavendish performed pioneering work on gases. He showed that ordinary air is a mixture of oxygen and nitrogen, with nitrogen being four times more common. In 1781 he also demonstrated that when hydrogen and oxygen are mixed in the proportions of 2:1 and the resulting mixture exploded (using an electric spark) then water is produced, thus showing that water is a compound and not an element. He also used a spark to combine nitrogen with oxygen and dissolved the resulting gas in water, thus generating nitric acid. Cavendish also did important work in physics, particularly in electricity and gravitation.

celestine (celestite) A naturally occurring mineral sulfate of strontium, $SrSO_4$, found in sedimentary rocks. It is a source of strontium and is also used to make special glasses.

celestite *See* celestine.

cell A system having two ELECTRODES in a conducting liquid or paste ELECTROLYTE. An *electrolytic cell* is used to produce a chemical reaction by means of a current passing through the electrolyte (i.e. by ELECTROLYSIS). Conversely, a *voltaic* (or *galvanic*) *cell* produces an e.m.f. by means of chemical reactions that occur at each electrode. Electrons are transferred to or from the electrodes, giving each a net charge.

There is a convention for writing cell reactions in voltaic cells. The DANIELL CELL, for instance, consists of a zinc electrode in a solution of Zn^{2+} ions connected, via a porous partition, to a solution of Cu^{2+} ions in which a copper electrode is placed. The reactions at the electrodes are

$$Zn \rightarrow Zn^{2+} + 2e$$

i.e. oxidation of the zinc to zinc(II), and

$$Cu^{2+} + 2e \rightarrow Cu$$

i.e. reduction of copper(II) to copper. A cell reaction of this type is written:

$$Zn|Zn^{2+}(aq)|Cu^{2+}(aq)|Cu$$

The e.m.f. is the potential of the electrode on the right minus the potential on the left. Copper is positive in this case and the e.m.f. of the cell is stated as +1.10 volts. *See also* accumulator.

Celsius scale (centigrade scale) A temperature scale in which the temperature of pure melting ice at standard pressure is fixed at 0° and the temperature of pure boiling water at standard pressure is fixed at 100°. The temperature interval unit on the Celsius scale is the *degree Celsius* (°C), which is equal in magnitude to the kelvin. Before 1948 the Celsius scale was known as the *centigrade scale*. Celsius's original scale had 0° as the water steam fixed point and 100° as the ice/water fixed point. *See also* temperature scale; Fahrenheit scale.

cement A generic term for chemical preparations typically used as binding agents to bond disparate particles together into a strong, solid mass, e.g. concrete. Such cements commonly consist of a powdered mixture of calcium silicates and aluminates, which is made by heating limestone ($CaCO_3$) with clay and grinding the result. When mixed with water, reactions occur with the water (hence the name *hydraulic cement*) and a hard solid aluminosilicate is formed.

cementite (Fe_3C) A constituent of certain cast irons and steels. The presence of cementite increases the hardness of the alloy.

centi- Symbol: c A prefix used with SI units denoting 10^{-2}. For example, 1 centimeter (cm) = 10^{-2} meter (m).

centigrade scale *See* Celsius scale.

centrifugal pump A device commonly used for transporting fluids around a chemical plant. Centrifugal pumps usually have 6–12 curved vanes that rotate inside a fixed circular casing. As the blades rotate, the fluid is impelled out of the pump along a discharge pipe. Centrifugal pumps do not produce high pressures but they have the advantages of being relatively cheap due to their simple design, have no valves, and work at high speeds. In addition they are not damaged if a blockage develops. *Compare* displacement pump.

centrifuge An apparatus for rotating a container at high speeds, used to increase the rate of sedimentation of suspensions or the separation of two immiscible liquids. *See also* ultracentrifuge.

ceramics Useful high-melting inorganic materials. Ceramics include silicates and aluminosilicates, refractory metal oxides, and metal nitrides, borides, etc. Pottery and porcelain are examples of ceramics. Various types of glass are also sometimes included.

cerium A ductile, malleable, gray element of the lanthanoid series of metals. It occurs in association with other lanthanoids in many minerals, including monazite and bastnaesite. The metal reacts rapidly with hot water, tarnishes in air, and will ignite under friction. It is used as a catalyst in several alloys (especially for lighter flints), in tracer bullets and in gas mantles, and in compound form in carbon-arc searchlights, etc., and in the glass industry.

Symbol: Ce; m.p. 799°C; b.p. 3426°C; r.d. 6.7 (hexagonal structure, 25°C); p.n. 58; r.a.m. 140.15.

cermet A synthetic composite material made by combining a ceramic and a sintered metal. Cermets have better temperature and corrosion resistance than straight ceramics. For example, a chromium–alumina cermet is used to make blades for gas-turbine engines.

cerussite A naturally occurring form of lead(II) carbonate ($PbCO_3$) that is an important lead ore. It forms orthorhombic crystals and is often found together with galena (PbS).

cesium (caesium) A soft, silvery, highly reactive, low-melting element of the alkali-metal group, the sixth element of group 1 (formerly IA) of the periodic table. It is found in several silicate minerals, including pollucite ($CsAlSi_2O_6$). The metal oxidizes in air and reacts violently with water. Cesium is used in photocells, as a catalyst, and in the cesium atomic clock. The radioactive isotopes ^{134}Cs (half life 2.065 years) and ^{137}Cs (half life 30.3 years) are produced in nuclear reactors and are potentially dangerous atmospheric pollutants. Cesium-137 is also used as a radiation source in cancer therapy, and as a medical tracer.

Symbol: Cs; m.p. 28.4°C; b.p. 678.4°C; r.d. 1.873 (20°C); p.n. 55; r.a.m. 132.91.

cesium-chloride structure A form of crystal structure that consists of alternate layers of cesium ions and chloride ions with the center of the lattice occupied by a cesium ion in contact with eight chloride ions (i.e. four chloride ions in the plane above and four in the plane below).

c.g.s. system An early metric system of units based on the centimeter, the gram, and the second as the fundamental mechanical units. Much early scientific work was performed using this system, but it has now almost been abandoned in favor of SI units.

chain When two or more atoms form bonds with each other in a molecule, a chain of atoms results. This chain may be a *straight chain*, in which each atom is added to the end of the chain, or it may be a *branched chain*, in which the main chain of atoms has one or more smaller *side chains* branching off it.

chain reaction A self-sustaining chemical reaction consisting of a series of steps, each of which is initiated by the one before it. An example is the reaction between hydrogen and chlorine:

$$Cl_2 \rightarrow 2Cl\bullet$$
$$H_2 + Cl\bullet \rightarrow HCl + H\bullet$$
$$H\bullet + Cl_2 \rightarrow HCl + Cl\bullet$$
$$2H\bullet \rightarrow H_2$$
$$2Cl\bullet \rightarrow Cl_2$$

The first stage, chain initiation, is the dissociation of chlorine molecules into atoms; this is followed by two chain propagation reactions. Two molecules of hydrogen chloride are produced and the ejected chlorine atom is ready to react with more hydrogen. The final steps, chain termination, stop the reaction.

Induced nuclear fission reactions also depend on chain reactions; the fission reaction is maintained by the two or three neutrons set free in each fission.

chalcogens *See* group 16 elements.

chalk A soft, natural, fine-grained form of calcium carbonate ($CaCO_3$) formed from the skeletal remains of ancient marine organisms. So-called blackboard chalk is made from a different compound, calcium sulfate ($CaSO_4$).

chamber process *See* lead-chamber process.

chaotic reaction A chemical reaction in which the concentrations of the reactants show chaotic behavior, i.e. the evolution of the reaction may become unpredictable. An example is the BELOUSOV–ZHABOTINSKII REACTION. Reactions of this type ususally involve a large number of complex interlinked steps. *See also* bistability; oscillating reaction.

charcoal An amorphous form of carbon made by heating wood or other organic material in the absence of air. *Activated charcoal* is charcoal heated to drive off absorbed gas. It is used for absorbing gases and for removing impurities from liquids.

Charles' law (Gay-Lussac's law) For a given mass of gas at constant pressure, the volume increases by a constant fraction of the volume at 0°C for each Celsius degree rise in temperature. The constant fraction (α) has almost the same value, about 1/273, for all gases and Charles' law can thus be written in the form

$$V = V_0(1 + \alpha_v\theta)$$

where V is the volume at temperature θ°C and V_0 the volume at 0°C. The constant α_v is the thermal expansivity of the gas. For an ideal gas its value is 1/273.15.

A similar relationship exists for the pressure of a gas heated at constant volume:

$$p = p_0(1 + \alpha_p\theta)$$

Here, α_p is the pressure coefficient. For an ideal gas

$$\alpha_p = \alpha_v$$

although they differ slightly for real gases. It follows from Charles' law that for a gas heated at constant pressure,

$$V/T = K$$

where T is the thermodynamic temperature and K is a constant. Similarly, at constant volume, p/T is a constant.

Charles' law is named for the French chemist and physicist J. A. C. Charles, who discovered it by means of experiments begun in 1787. The law was published in 1802, however, by J. Gay-Lussac, another French chemist and physicist, is sometimes known as Gay-Lussac's law.

chelate A metal coordination COMPLEX in which one molecule or ion forms a coordinate bond at two or more points to the same metal ion, thus becoming a *ligand*. The resulting complex contains rings of atoms that include the metal atom. Chelates may be used to deliver essential metal ions in fertilizer preparations, while *chelating agents*, or compounds capable of forming chelates, may be used to trap heavy metals and thus render them harmless in the case of metal poisonings. 'Chelate' comes from the Greek word meaning 'claw'. Ligands forming chelates are classified according to the number of sites at which they can coordinate: bidentate, tridentate, etc. *See also* sequestration.

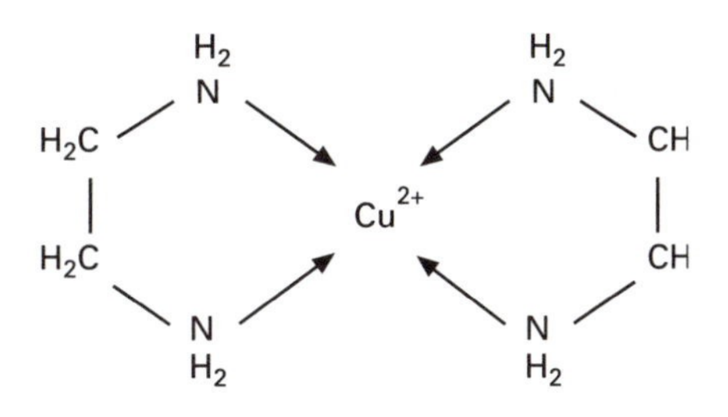

Chelate

chemical bond A force holding atoms together in a molecule or in a crystal. Chemical bonds are formed by transfer or sharing of electrons (*see* electrovalent bond; covalent bond) and typically have energies of about 1000 kJ mol^{-1}. Weaker interactions, such as HYDROGEN BONDS and VAN DER WAALS FORCES, are not regarded as chemical bonds. *See also* coordinate bond; metallic bond.

chemical combination, laws of A group of chemical laws, developed during the late 18th and early 19th centuries, which arose from the recognition of the importance of quantitative (as opposed to qualitative) study of chemical reactions. The laws are:

1. The law of conservation of mass (matter);
2. The law of constant (definite) proportions;
3. The law of multiple proportions;
4. The law of equivalent (reciprocal) proportions.

These laws played a significant part in Dalton's development of his atomic theory in 1808.

chemical dating A method of using chemical analysis to find the age of an archaeological specimen in which compositional changes have taken place over time. For example, the determination of the amount of fluorine in bone that has been buried gives an indication of the time the bone has been underground because phosphate in the bone is gradually replaced by fluoride ions from groundwater. Another dating technique depends on the fact that, in living organisms, amino acids are optically active. After death a slow RACEMIZATION reaction occurs and a mixture of L- and D-isomers forms. The age of bones can be accurately determined by measuring the relative amounts of optically active versus optically inactive acids present.

chemical engineering The branch of engineering concerned with the design and maintenance of a chemical plant and its ability to withstand extremes of temperature, pressure, corrosion, and wear. It enables laboratory processes producing grams of material to be converted into a large-scale plant processes yielding tonnes of material. Chemical engineers plan large-scale chemical processes by linking together the appropriate unit processes and by studying such parameters as heat and mass transfer, separations, and distillations.

chemical equation A method of representing a chemical reaction using chemical FORMULAS. The formulas of the reactants are given on the left-hand side of the equation, with the formulas of the products given on the right. The two halves are separated by a directional arrow or arrows, or an equals sign. A number preceding a formula, called a *stoichiometric coefficient*, indicates the number of molecules of that substance involved in the reaction. The equation must balance – that is, the number of atoms of any one element must be the same on both sides of the equation. A simple example is the equation for the reaction between hydrogen and oxygen to form water:

$$2H_2 + O_2 \rightarrow 2H_2O$$

A more complex equation represents the reaction between disodium tetraborate decahydrate (borax) and aqueous hydrochloric acid to give boric acid and sodium chloride:

$$Na_2B_4O_7 + 2HCl + 5H_2O \rightarrow 4H_3BO_3 + 2NaCl$$

chemical equilbrium *See* equilibrium.

chemical formula *See* formula.

chemical potential Symbol: μ. For the *i*th component of a mixture the chemical potential μi is defined by the partial derivative of the Gibbs free energy G of the system with respect to the amount n_i of the component, when the temperature, pressure, and amounts of other components are constant, i.e. $\mu i = \partial G/\partial n_i$. If the chemical potentials of components are equal then the components are in equilibrium. Also, in a one-component system with two phases it is necessary for the chemical potentials to be equal in the two phases for there to be equilibrium.

chemical reaction A process in which one or more elements or chemical compounds (the reactants) react to produce a different substance or substances (the products).

chemical shift *See* nuclear magnetic resonance.

chemical symbol A letter or pair of letters or, in the case of unnamed elements, a triplet of letters that stand for a chemical element, as used for example in equations and chemical formulas. Unlike abbreviations, symbols do not take periods, are written exactly the same in all languages and applications, and are designed to be universally understood to represent specific elements, units, quantities, etc.

chemiluminescence The emission of light during a chemical reaction.

chemisorption *See* adsorption.

Chile saltpeter *See* sodium nitrate.

China clay (kaolin) A white powdery clay obtained from the natural decomposition of granites. It is used as a filler in paints and paper-making, in the pottery industries, and in pharmaceuticals. *See* kaolinite.

chiral Having the property of CHIRALITY.

chiral center *See* isomerism.

chirality The property of existing in left- and right-handed forms; i.e. mirror images that are not superimposable. In chemistry the term is applied to the existence of optical isomers. 'Chiral' comes from the Greek word 'cheir', meaning hand. *See* isomerism; optical activity.

chlorate A salt of chloric(V) acid.

chloric(I) acid (hypochlorous acid; HClO) A colorless liquid produced when chlorine dissolves in water. It is a bleach and gives chlorine water its disinfectant properties. To increase the yield of acid, the chlorine water can be shaken with a small amount of mercury(II) chloride. The Cl–O bond is broken more easily than the O–H bond in aqueous solution; the acid is consequently a poor proton donor and hence a weak acid.

chloric(III) acid (chlorous acid; $HClO_2$) A pale yellow liquid produced by reacting chlorine dioxide with water. It is a weak acid and oxidizing agent.

chloric(V) acid (chloric acid; $HClO_3$) A colorless liquid with a pungent odor, formed by the action of dilute sulfuric acid on barium chlorate. It is a strong acid and has bleaching properties. Chloric(V) acid is also a strong oxidizing agent; in concentrated solution it will ignite organic substances, such as paper and sugar.

chloric(VII) acid (perchloric acid; $HClO_4$) A colorless liquid that fumes strongly in moist air. It is made by vacuum distillation of a mixture of potassium perchlorate and concentrated sulfuric acid. It is a strong acid and oxidizing agent, and in contact with organic material it is dangerously explosive.

The hydrate ($HClO_4.H_2O$) of chloric(VII) acid is a white crystalline solid at room temperature and has an ionic lattice structure of the form $(H_3O)^+(ClO_4)^-$.

chloric acid *See* chloric(V) acid.

chloride *See* halide.

chlorination **1.** Treatment with chlorine; for instance, the use of chlorine to disinfect water.
2. *See* halogenation.

chlorine A green reactive gaseous element belonging to the halogens; i.e. group 17 (formerly VIIA) of the periodic table, of which it is the second element, occurring in seawater, salt lakes, and underground deposits of the mineral halite as NaCl. It accounts for about 0.055% of the Earth's crust.

Chlorine is strongly oxidizing and can be liberated from its salts only by strong oxidizing agents, such as manganese(IV) oxide, potassium permanganate(VII), or potassium dichromate; sulfuric acid is not sufficiently oxidizing to release chlorine from chlorides. Industrially, chlorine is prepared by the electrolysis of brine or magnesium chloride, although in some processes chlorine is recovered by the high-temperature oxidation of waste hydrochloric acid. Chlorine is used in large quantities, both as the element to produce chlorinated organic solvents, for the production of polyvinyl chloride (PVC), the major thermoplastic in use today, and in the form of hypochlorites for bleaching.

Chlorine reacts directly and often vigorously with many elements; it reacts explosively with hydrogen in sunlight to form hydrogen chloride, HCl, and combines with the electropositive elements to form metal chlorides. The metals of main groups 1 and 2 form electrovalent chlorides but an increase in the metal charge/size ratio leads to the chlorides becoming increasingly covalent. For example, CsCl is totally electrovalent, $AlCl_3$ has a layer lattice, and $TiCl_4$ is essentially covalent. The electronegative elements form volatile molecular chlorides characterized by the single covalent bond to chlorine. With the exception of Pb^{2+}, Ag^{+}, and Hg_2^{2+}, the electrovalent chlorides are soluble in water, dissolving to give the hydrated metal ion and the chloride ion Cl^-. Chlorides of metals other than the most electropositive are hydrolyzed if aqueous solutions are evaporated; for example,

$$ZnCl_2 + H_2O \rightarrow Zn(OH)Cl + HCl$$

$$FeCl_3 + 3H_2O \rightarrow Fe(OH)_3 + 3HCl$$

Chlorine forms four main oxides, dichlorine oxide, Cl_2O; chlorine dioxide, ClO_2; dichlorine hexoxide, Cl_2O_6; and dichlorine heptoxide, Cl_2O_7, all of which are highly reactive and explosive. Chlorine dioxide finds commercial application as an active oxidizing agent, but because of its explosive nature it is usually diluted by air or other gases. The chloride ion is able to function as a ligand with a large variety of metal ions forming such species as $[FeCl_4]^-$, $[CuCl_4]^{3-}$, and $[Co(NH_3)_4Cl_2]^+$. The formation of anionic chloro-complexes is applied to the separation of metals by ion-exchange methods.

Because of the hydrolysis of many metal chlorides when solutions are evaporated, special techniques must be used to prepare anhydrous chlorides. These are:

1. Reaction of dry chlorine with the hot metal;
2. For metals having lower valences, reaction of dry hydrogen chloride with the hot metal;
3. Reaction of dry hydrogen chloride on the hydrated chloride.

The solubility of inorganic metal chlorides is such that they are not an environmental problem unless the metal ion itself is toxic. However, chlorine and hydrogen chloride are both highly toxic. Thus chlorides that hydrolyze to release HCl should also be regarded as toxic. They should not be handled without gloves and precautions should be taken against inhalation.

Symbol: Cl; m.p. –100.38°C; b.p. –33.97°C; d. 3.214 kg m^{-3} (0°C); p.n. 17; r.a.m. 35.4527.

chlorine dioxide (ClO_2) An orange gas formed in the laboratory by the action of concentrated sulfuric acid on potassium chlorate. It is a powerful oxidizing agent, and its explosive properties in the presence of a reducing agent were used to make one of the first combustible matches. It is widely used in the purification of water and as a bleach in the flour and wood-pulp industry. On an industrial scale an aqueous solution of chlorine dioxide is made by passing nitrogen dioxide up a tower packed with a fused mixture of aluminum oxide and clay, down which a solution of sodium chlorate flows.

chlorine monoxide *See* dichlorine oxide.

chlorine(I) oxide *See* dichlorine oxide.

chlorite A a salt of chloric(III) acid (chlorous acid).

chloroplatinic(IV) acid (platinic chloride; H_2PtCl_6) A reddish compound prepared by dissolving platinum in aqua regia. When crystallized from the resulting solution, crystals of the hexahydrate ($H_2PtCl_6.6H_2O$) are obtained. The crystals are needle-shaped and deliquesce on exposure to moist air. Chloroplatinic acid is a relatively strong acid, giving rise to the family of chloroplatinates.

chlorous acid *See* chloric(III) acid.

chromate Any oxygen-containing derivative of chromium, most particularly the chromate(VI) species.

chromatography A technique used to separate or analyze complex mixtures. A number of related techniques exist; all depend on two phases: a *mobile phase*, which may be a liquid or a gas, and a *stationary phase*, which is either a solid or a liquid held by a solid. The sample to be separated or analyzed is carried by the mobile phase through the stationary phase. Different components of the mixture are absorbed or dissolved to different extents by the stationary phase, and consequently move along at different rates. In this way the components are separated. There are many different forms of chromatography depending on the techniques used and the nature of the partition process between mobile and stationary phases. The main classification is into *column chromatography* and *planar chromatography*.

A simple example of column chromatography is in the separation of liquid mixtures. A vertical column is packed with an absorbent material, such as alumina (aluminum oxide) or silica gel, representing the stationary phase. The sample is introduced into the top of the column and washed down it (the mobile phase) using a solvent. This process is known as ELUTION. If the components are colored, visible bands appear down the column as the sample separates out. The components are separated as they emerge from the bottom of the column. In this particular example of chromatography the partition process is adsorption on the particles of alumina or silica gel. Column chromatography can also be applied to mixtures of gases, in which case it is known as GAS CHROMATOGRAPHY.

Planar chromatography works in similar fashion, except that the stationary phase is a flat sheet of absorbent material such as paper. Components of the mixture are held back by the stationary phase either by adsorption (e.g. on the surface of alumina) or because they dissolve in it (e.g. in the moisture within paper). *See* paper chromatography; thin-layer chromatography.

chrome alum *See* alum.

chrome iron ore *See* chromite.

chromic anhydride *See* chromium(VI) oxide.

chromic oxide *See* chromium(III) oxide.

chromite (chrome iron ore; iron chromium oxide; $FeCr_2O_4$) A mineral that consists of mixed oxides of chromium and iron, the principal ore of chromium. It occurs as brownish black masses with a metallic luster in rocks of igneous origin. Large deposits are found in Zimbabwe and the western United States.

chromium A hard, silver-gray transition metal of group 6 (formerly VIB) of the periodic table. It occurs naturally as CHROMITE ($FeCr_2O_4$). The ore is first converted into sodium dichromate(VI) and then reduced with carbon to chromium(III) oxide, after which it is finally reduced to metallic chromium with aluminum. Chromium is used to make strong alloy steels and stainless steel, and for decorative electroplated coatings. It resists corrosion at normal temperatures. It reacts slowly with dilute hydrochloric and sulfuric acids to give hydrogen and blue chromium(II) compounds, which quickly oxidize in air to green chromium(III) ions. The bright colors of many chromium compounds make them useful as pigments. The oxidation states are +6 in CHROMATES (CrO_4^{2-}) and *dichromates* ($Cr_2O_7^{2-}$), +3 (the most stable), and +2. In acidic solutions the yellow chromate(VI) ion changes to the orange dichromate(VI) ion. Dichromates are strong oxidizing agents and are used as such in the laboratory. For example they are used as a test for sulfur(IV) oxide (sulfur dioxide) and to oxidize alcohols. *See* potassium dichromate.

Symbol: Cr; m.p. 1860±20°C; b.p. 2672°C; r.d. 7.19 (20°C); p.n. 24; r.a.m. 51.9961.

chromium dioxide *See* chromium(IV) oxide.

chromium(II) oxide (chromous oxide; CrO) A black powder prepared by the oxidation of chromium amalgam with dilute nitric acid. At high temperatures (around 1000°C) chromium(II) oxide is reduced by hydrogen.

chromium(III) oxide (chromic oxide; chromium sesquioxide; Cr_2O_3) A green powder that is almost insoluble in water. It has the same crystal structure as iron(III) oxide and aluminum(III) oxide. Chromium(III) oxide is prepared by gently heating chromium(III) hydroxide or by heating ammonium dichromate. Alternative preparations include the heating of a mixture of ammonium chloride and potassium dichromate or the decomposition of chromyl chloride by passing it through a red-hot tube. Chromium(III) oxide is used as a pigment in the paint and glass industries.

chromium(IV) oxide (chromium dioxide; CrO_2) A black solid prepared by heating chromium(III) hydroxide in oxygen at a temperature of 300–350°C. Chromium(IV) oxide is very unstable.

chromium(VI) oxide (chromium trioxide; chromic anhydride; CrO_3) A red crystalline solid formed when concentrated sulfuric acid is added to a cold saturated solution of potassium dichromate. The long prismatic needle-shaped crystals that are produced are extremely deliquescent. Chromium(VI) oxide is readily soluble in water, forming a solution that contains several of the *polychromic acids* i.e. acids containing more than one atom of chromium. On heating it decomposes to give chromium(III) oxide. Chromium(VI) oxide is used as an oxidizing reagent.

chromium sesquioxide *See* chromium(III) oxide.

chromium trioxide *See* chromium(VI) oxide.

chromophore A group of atoms in a molecule that is responsible for the color of the compound.

chromous oxide *See* chromium(II) oxide.

chromyl chloride (CrO_2Cl_2) A dark red covalent liquid prepared either by distilling the vapors evolved when a dry mixture of potassium dichromate and sodium chloride is treated with concentrated sulfuric acid or by the action of concentrated sulfuric acid on chromium(VI) oxide dissolved in concentrated hydrochloric acid. Chromyl chloride is violently hydrolyzed by water, and with solutions of alkalis it undergoes immediate hydrolysis to produce chromate ions. It is used as a potent oxidizing agent.

cinnabar A red mineral form of mercury(II) sulfide (HgS), associated with areas of volcanic activity. It is the principal ore of mercury.

cis-isomer *See* isomerism.

cis-trans isomerism *See* isomerism.

Clark cell A type of cell formerly used as a standard source of e.m.f. It consists of a mercury cathode coated with mercury sulfate, and a zinc anode. The electrolyte is zinc sulfate solution. The e.m.f. produced is 1.4345 volts at 15°C. The Clark cell has been superseded as a standard by the WESTON CADMIUM CELL.

clathrate (cage compound; enclosure compound) A substance in which small 'guest' molecules are trapped within the lattice of a crystalline 'host' compound. Clathrates are formed when suitable host compounds are crystallized in the presence of molecules of the appropriate size. Although the term 'clathrate compound' is often used, clathrates are not true compounds; no chemical bonds are formed, and the guest molecules interact by weak van der Waals forces. The clathrate is maintained by the cagelike lattice of the host. The host lattice must be broken down, for example by heating or dissolution, in order to release the guest. In ZEOLITES, in comparison, the holes in the host lattice are large enough to permit entrance or emergence of the guest without breaking bonds in the host lattice. Water (ice), for instance, forms a clathrate with xenon.

Clausius, Rudolf Julius Emmanuel (1822–88) German physicist. Clausius is best remembered as one of the founders of thermodynamics. In papers published in the 1850s he stated the second law of thermodynamics and he coined the word *entropy* in 1865. Clausius made other contributions to thermodynamics and also to the development of the kinetic theory of gases. In electrochemistry he pioneered the idea of dissociation of substances into ions in solution. He also studied the dielectric properties of materials.

clays Naturally occurring aluminosilicates that form pastes and gels with water.

cleavage The splitting of a crystal along planes of atoms, to form smooth surfaces.

close packing The arrangement of particles (usually atoms) in crystalline solids in which the UNIT CELL has an atom, ion, or molecule at each corner and also at the center of each face of a cube. Each particle thus has 12 nearest neighbors: six in the same layer (or plane) as itself and three each in the layer above and below for a COORDINATION NUMBER of 12. This arrangement provides the most economical use of space (74% efficiency). The two principal types of close packing are FACE-CENTERED CUBIC CRYSTAL and HEXAGONAL CLOSE PACKING.

cluster A three-dimensional structure consisting of atoms, with the number of atoms ranging between a few dozen and a few thousand. The atoms in clusters can either be metal or nonmetal atoms. It is found that when clusters are made (for example, by sputtering the surface of a solid), clusters with certain *magic numbers* of atoms occur far more abundantly than for other numbers of atoms. These numbers, which occur for several monovalent metallic elements including sodium, silver, and gold, are 8, 20, 40, 58, 92 and correspond to the electron shells being filled in an external potential. It is predicted that clusters of atoms or cluster compounds in which the number of valence electrons is a magic number should be stable. For example, $Al_{12}C$, which has a total of 40 valence electrons, is predicted to be a stable cluster compound.

cluster compound A type of compound in which a cluster of metal atoms are joined by metal–metal bonds. Cluster compounds are formed by certain transition elements, such as molybdenum and tungsten.

coagulation The irreversible association of particles such as colloids into clusters. *See* flocculation.

coal A brown or black sedimentary deposit that consists mainly of carbon, used as a fuel and as a source of organic chemicals. It is the fossilized remains of decayed plants and algae that chiefly grew in the Carboniferous and Permian periods and were then buried and subjected to high pressures underground. The various types of coal, including ANTHRACITE, BITUMINOUS COAL, and LIGNITE, are classified according to their increasing carbon content.

cobalt A lustrous, silvery-blue, hard ferromagnetic transition metal, the first element of group 9 (formerly subgroup VIIIB) of the periodic table. It occurs in association with nickel, copper, and arsenic in such minerals as cobaltite [(Co,Fe)AsS], skutterudite [$(Co,Ni)As_3$], and erythrite ($Co_3(AsO_4)_2.8H_2O$). It is used in alloys for magnets, high-temperature cutting tools, and electrical heating elements. Cobalt compounds are used in catalysts and some paints. It is an essential dietary mineral, being a component of vitamin B12.

Symbol: Co; m.p. 1495°C; b.p. 2870°C; r.d. 8.9 (20°C); p.n. 27; r.a.m. 58.93320.

cobaltic oxide *See* cobalt(III) oxide.

cobaltous oxide *See* cobalt(II) oxide.

cobalt(II) oxide (cobaltous oxide; CoO) A green powder prepared by the action of heat on cobalt(II) hydroxide in the absence of air. Alternatively it may be prepared by the thermal decomposition of cobalt(II) sulfate, nitrate, or carbonate, also in the absence of air. Cobalt(II) oxide is a basic oxide, reacting with acids to give solutions of cobalt(II) salts. It is stable in air up to temperatures of around 600°C, after which it absorbs oxygen to form tricobalt tetroxide (Co_3O_4). Cobalt(II) oxide can be reduced to cobalt by heating in a stream of carbon monoxide or hydrogen. It is used in the pottery industry and in the production of vitreous enamels.

cobalt(III) oxide (cobaltic oxide; Co_2O_3) A dark gray powder formed by the thermal decomposition of either cobalt(II) nitrate or carbonate in air. If heated in air, cobalt(III) oxide undergoes further oxidation to give tricobalt tetroxide, Co_3O_4.

coherent units A system or subset of UNITS (e.g. SI units) in which the derived units are obtained by multiplying or dividing together base units, with no numerical factor involved.

coinage metals A group of malleable metals forming group 11 (formerly subgroup IB) of the periodic table. They are copper (Cu), silver (Ag), and gold (Au). These metals all have an outer s^1 electronic configuration but they differ from the alkali metals, which also have an outer s^1 configuration, in having inner d-electrons. The coinage metals also have much higher IONIZATION POTENTIALS than the alkali metals and high positive standard ELECTRODE POTENTIALS, and are therefore much more difficult to oxidize. A further significant difference from the alkali metals is the variety of oxidation states observed for the coinage metals. Thus copper in aqueous chemistry is the familiar (hydrated) blue divalent Cu^{2+} ion but colorless copper(I) compounds can be prepared with groups such as CN^- that stabilize low valences, e.g. CuCN. A few compounds of copper(III) have also been prepared. The common form of silver is Ag(I), e.g. $AgNO_3$, with a few Ag(II) compounds stable as solids only. The most common oxidation state of gold is Au(III), although the cyanide ion again stabilizes Au(I) compounds, e.g., $[Au(CN)_4]^{3-}$. The metals all form a large number of coordination compounds, again unlike the alkali metals, and are generally described with the other elements of the appropriate transition series.

coke A dense substance obtained from the carbonization of coal. It is used as a fuel and as a reducing agent.

colligative properties A group of properties of solutions that depends on the number of particles present, rather than the nature of the particles. Colligative properties include:
1. The lowering of vapor pressure.
2. The elevation of boiling point.
3. The lowering of freezing point.
4. Osmotic pressure.

Colligative properties are all based upon empirical observation. The explanation of these closely related phenomena depends on intermolecular forces and the kinetic behavior of the particles, which is qualitatively similar to those used in deriving the kinetic theory of gases.

collimator An arrangement for producing a parallel beam of radiation for use in a spectrometer or other instrument. A system of lenses and slits is utilized.

colloid A heterogeneous system in which the interfaces between phases, though not visibly apparent, are important factors in determining the system's properties. The three important attributes of colloids are:
1. They contain particles, commonly made up of large numbers of molecules, forming the distinctive unit or *dispersed phase*.
2. The particles are distributed in a continuous medium (the *continuous phase*).
3. There is a stabilizing agent, which has an affinity for both the particle and the medium; in many cases the stabilizer is a polar group.

Particles in the dispersed phase typically have diameters in the range 10^{-6}–10^{-4} millimeter. Milk, rubber, and water-based paints are typical examples of colloids. *See also* emulsion; gel; sol.

colorimetric analysis Quantitative analysis in which the concentration of a colored solute is measured by the intensity of the color. The test solution can be compared against standard solutions.

columbium A former name for niobium. It is still sometimes used in metallurgy and mineralogy.

Symbol: Cb.

column chromatography *See* chromatography; gas chromatography.

combustion A reaction with oxygen

with the production of heat and light. The combustion of solids and liquids occurs when they release flammable vapor, which reacts with oxygen in the gas phase. Combustion reactions usually involve a complex sequence of free-radical chain reactions. The light is produced by excited atoms, molecules, or ions. In highly luminous flames it comes from small incandescent particles of carbon.

Sometimes the term is also applied to slow reactions with oxygen, and also to reactions with other gases (for example, certain metals 'burn' in chlorine).

common salt *See* sodium chloride.

complex (coordination compound) A type of compound in which molecules or ions form COORDINATE BONDS with a metal atom or ion. The coordinating species (called *ligands*) have lone pairs of electrons, which they can donate to the metal atom or ion. They are molecules such as ammonia or water, or negative ions such as Cl^- or CN^-. The resulting complex may be neutral or it may be a *complex ion*. For example:

$$Cu^{2+} + 4NH_3 \rightarrow [Cu(NH_3)_4]^{2+}$$
$$Fe^{3+} + 6CN^- \rightarrow [Fe(CN)_6]^{3-}$$
$$Fe^{2+} + 6CN^- \rightarrow [Fe(CN)_6]^{4-}$$

The formation of such coordination complexes is typical of transition metals. Often the complexes contain unpaired electrons and are paramagnetic and colored. *See also* chelate.

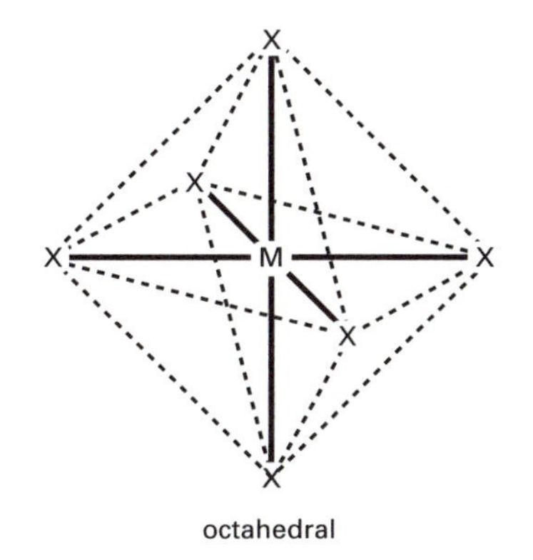

octahedral

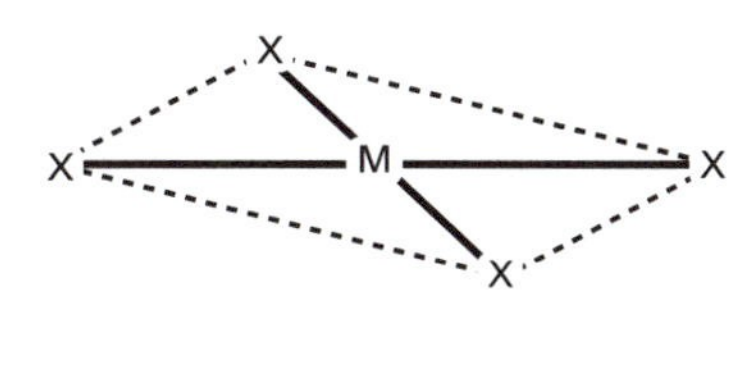

square-planar

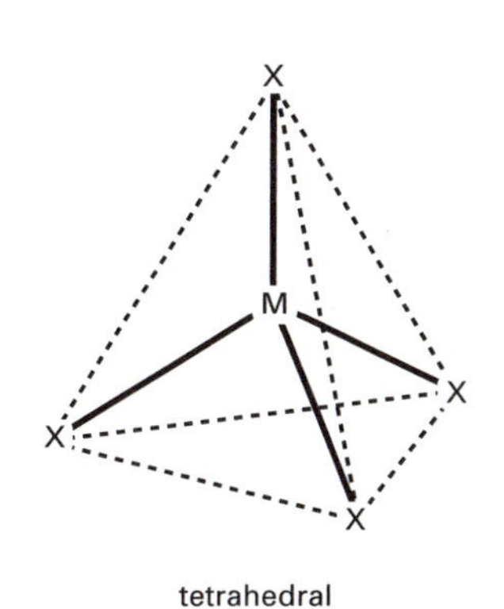

tetrahedral

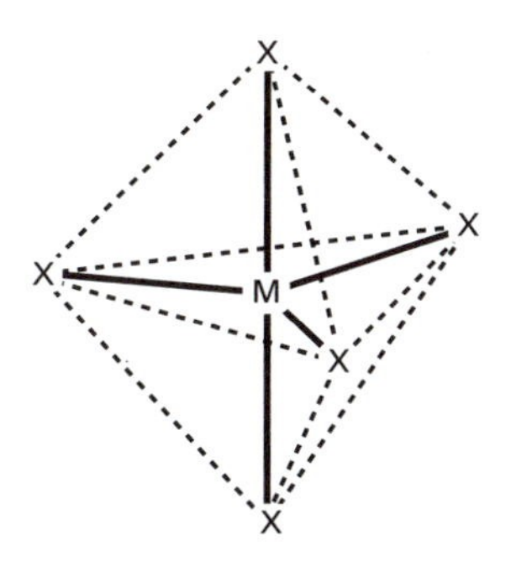

trigonal-bipyramid

Complex: typical shapes of inorganic complexes

complex ion *See* complex.

component One of the separate chemical substances in a mixture in which no chemical reactions are taking place. For example, a mixture of ice and water has one component; a mixture of nitrogen and oxygen has two components. When chemical reactions occur between the substances in a mixture, the number of components is defined as the number of chemical substances present minus the number of equilibrium reactions taking place. Thus, the system: $N_2 + 3H_2 \rightleftharpoons 2NH_3$ is a two-component system. *See also* phase rule.

compound A chemical combination of atoms of different elements to form a substance in which the ratio of combining atoms remains fixed and is specific to that substance. The constituent atoms cannot be separated by physical means; a chemical reaction is required for the compounds to be formed or to be changed. The existence of a compound does not necessarily imply that it is stable. Many compounds have lifetimes of less than a second. *Compare* mixture.

concentrated Denoting a solution in which the amount of solute in the solvent is relatively high. The term is always relative; for example, whereas concentrated sulfuric acid may contain 96% H_2SO_4, concentrated potassium chlorate may contain as little as 10% $KClO_3$. *Compare* dilute.

concentration The amount of substance in a solution per unit quantity of solvent. Molar 'concentration' (symbol c) is the amount of substance per cubic meter. It replaces *molarity*, a term formerly used to describe the amount of substance per cubic decimeter (liter). DENSITY or *mass concentration* (symbol ρ) is kilograms of solute per cubic meter. *Molality* (symbol m) or *molal concentration* is the amount of substance per kilogram of solute.

condensation The conversion of a gas or vapor into a liquid or solid by cooling.

conductiometric titration A titration method in which the electrical conductance of the reaction mixture is continuously measured throughout the addition of the titrant and well beyond the equivalence point. The operation is carried out in a conductance cell, which is part of a resistance bridge circuit. The method depends on the fact that ions have different ionic mobilities, H^+ and OH^- having particularly high values. The method is used in place of traditional end-point determination by indicators, and is especially useful for weak acid–strong base and strong acid–weak base titrations for which color-change titrations are unreliable.

configuration **1.** (electron configuration) The arrangement of electrons about the nucleus of an atom. Configurations are represented by symbols, which contain:

1. An integer, which is the value of the principal quantum number (shell number).
2. A lower-case letter representing the value of the orbital quantum number (l), i.e.

   s means $l = 0$, p means $l = 1$,
   d means $l = 2$, f means $l = 3$.
3. A numerical superscript giving the number of electrons in that particular set; for example, $1s^2$, $2p^3$, $3d^5$.

The ground state electronic configuration (i.e. the most stable or lowest energy state) may then be represented as follows, for example, He, $1s^2$; N, $1s^22s^22p^5$. However, element configurations are commonly abbreviated by using an inert gas to represent the 'core', e.g. Zr has the configuration $[Kr]4d^25s^2$.

2. The arrangement of atoms or groups in a molecule. *See also* atom.

conjugate acid *See* acid; base.

conjugate base *See* acid; base.

conservation of energy, law of Enunciated by Helmholtz in 1847, this law states that in all processes occurring in an isolated system the energy of the system remains constant. The law of course permits energy to be converted from one form to another (including mass, since energy and mass are equivalent).

conservation of mass (matter), law of Formulated by Lavoisier in 1774, this law states that matter cannot be created or destroyed. Thus in a chemical reaction the total mass of the products equals the total mass of the reactants, with 'total mass' including any solids, liquids, and gases – such as air – that participate in the reaction.

constantan A copper-nickel (cupronickel) alloy containing 45% nickel. It has a high electrical resistivity and very low temperature coefficient of resistance and is therefore used in thermocouples and resistors.

constant composition, law of *See* constant proportions; law of.

constant (definite) proportions, law of (constant composition, law of) The principle that the proportion of each element in a compound is fixed or constant. It follows that the composition of a pure chemical compound is independent of its method of preparation. The law was formulated by Proust in 1779 after the analysis of a large number of compounds.

contact process An industrial process for the manufacture of sulfuric acid. Sulfur(IV) oxide and air are passed over a heated catalyst (vanadium(V) oxide or platinum) and sulfur(VI) oxide is produced:

$$2SO_2 + O_2 \rightarrow 2SO_3$$

The sulfur(VI) oxide is dissolved in sulfuric acid:

$$SO_3 + H_2SO_4 \rightarrow H_2S_2O_7$$

The resulting OLEUM is then diluted to give sulfuric acid:

$$H_2S_2O_7 + H_2O \rightarrow 2H_2SO_4$$

continuous phase *See* colloid.

continuous spectrum A spectrum composed of a continuous range of emitted or absorbed radiation. Continuous spectra are produced in the infrared and visible regions by hot solids, liquids, and pressurized gases. *See also* spectrum.

converter *See* Bessemer process.

coordinate bond (dative bond; dipolar bond; semipolar bond) A COVALENT bond in which the bonding pair is visualized as arising from the donation of a LONE PAIR from one species to another species, which behaves as an electron acceptor. The definition includes such examples as the 'donation' of the lone pair of the ammonia molecule to H^+ (an acceptor) to form NH_4^+ or to Cu^{2+} to form $[Cu(NH_3)_4]^{2+}$.

The donor groups are known as Lewis bases and the acceptors are either hydrogen ions or LEWIS ACIDS. Simple combinations, such as $H_3N \rightarrow BF_3$, are known as *adducts*. *See also* complex.

coordination compound *See* complex.

coordination number **1.** The number of atoms, molecules, or ions, surrounding any particular atom, ion, or molecule in a crystal.
2. The number of coordinate bonds formed to a central atom or ion in a complex.

copper A malleable, ductile, golden-red transition metal, the first element of group 11 (formerly subgroup IB) of the periodic table. It occurs naturally principally in sulfides such as chalcopyrite ($CuFeS_2$), chalcocite (Cu_2S), and bornite (Cu_5FeS_4). It is extracted by roasting the ore in a controlled air supply and purified by electrolysis of copper(II) sulfate solution using impure copper as the anode and pure copper as the cathode. Copper is used extensively in electrical conductors and in such alloys as brass and bronze. It is also an excellent roofing material.

Copper(I) compounds are white (except the oxide, which is red), and copper(II) compounds are blue in solution. Copper is unreactive to dilute acids with the exception of nitric acid. Copper(I) compounds are unstable in solution and decompose to copper and copper(II) ions. Both copper(I) and copper(II) ions form complexes, copper(II) ions being identified by the dark blue complex $[Cu(NH_3)_4]^{2+}$ formed with excess ammonia solution.

Symbol: Cu; m.p. 1083.5°C; b.p. 2567°C; r.d. 8.96 (20°C); p.n. 29; r.a.m. 63.546.

copper(II) carbonate ($CuCO_3$) A green crystalline compound that occurs in mineral form as the basic salt in azurite and malachite. *See also* verdigris.

copper(I) chloride (cuprous chloride; CuCl) A white solid, insoluble in water, prepared by heating copper(II) chloride in concentrated hydrochloric acid with excess copper turnings. When the solution is colorless, it is poured into air-free water (or water containing sulfur(IV) oxide) and a white precipitate of copper(I) chloride is obtained. On exposure to air this precipitate turns green due to the formation of basic copper(II) chloride. Copper(I) chloride is essentially covalent in structure. It absorbs carbon monoxide gas and is used in the rubber industry and in organic chemistry.

copper(II) chloride (cupric chloride; $CuCl_2$) A compound prepared by dissolving excess copper(II) oxide or copper(II) carbonate in dilute hydrochloric acid. On crystallization, emerald green crystals of the dihydrate ($CuCl_2.2H_2O$) are obtained. The anhydrous chloride may be prepared as a brown solid by heating copper in excess chlorine. Alternatively the dihydrate may be dehydrated using concentrated sulfuric acid. Dilute solutions of copper(II) chloride are blue, concentrated solutions are green, and solutions in the presence of excess hydrochloric acid are yellow.

copper(II) nitrate (cupric nitrate; $Cu(NO_3)_2$) A compound prepared by dissolving either excess copper(II) oxide or copper(II) carbonate in dilute nitric acid. On crystallization, deep blue crystals of the trihydrate ($Cu(NO_3)_2.3H_2O$) are obtained. The crystals are prismatic in shape and are extremely deliquescent. On heating, copper(II) nitrate decomposes to give copper(II) oxide, nitrogen(IV) oxide, and oxygen. The white anhydrous salt is prepared by adding a solution of dinitrogen pentoxide (N_2O_5) in nitric acid to crystals of the trihydrate.

copper(I) oxide (cuprous oxide; CuO) An insoluble, red, covalent solid powder prepared by the heating of copper with copper(II) oxide or the reduction of an alkaline solution of copper(II) sulfate. Copper(I) oxide is easily reduced by hydrogen when heated; it is oxidized to copper(II) oxide when heated in air. Copper(I) oxide undergoes DISPROPORTIONATION in acid solutions, producing copper(II) ions and copper. The oxide dissolves in concentrated hydrochloric acid due to the formation of the complex ion $[CuCl_2]^-$. It is used in the glass and electronics industries.

copper(II) oxide (cupric oxide; CuO) A black solid prepared by the action of heat on copper(II) nitrate, hydroxide, or carbonate. It is a basic oxide and reacts with dilute acids to form solutions of copper(II) salts. Copper(II) oxide can be reduced to copper by heating in a stream of hydrogen or carbon monoxide. It can also be reduced by mixing with carbon and heating the mixture. Copper(II) oxide is stable up to its melting point, after which it decomposes to give oxygen, copper(I) oxide, and eventually copper.

copper(II) sulfate (cupric sulfate; $CuSO_4$) A compound prepared as the hydrate by the action of dilute sulfuric acid on copper(II) oxide or copper(II) carbonate. On crystallization, blue triclinic crystals of the pentahydrate (*blue vitriol*, $CuSO_4.5H_2O$) are formed. Industrially copper(II) sulfate is prepared by passing air through a hot mixture of dilute sulfuric acid and scrap copper. The solution formed is recycled until the concentration of the copper(II) sulfate is sufficient. Copper(II) sulfate is readily soluble in water. The monohydrate ($CuSO_4.H_2O$) is formed at 100°C and the anhydrous salt at 250°C. Anhydrous copper(II) sulfate is white; it is extremely hygroscopic and turns blue on absorption of water. It decomposes on heating to give copper(II) oxide and sulfur(VI) oxide.

Copper(II) sulfate is used as a wood preservative, as a fungicide in Bordeaux mixture, and in the dyeing and electroplating industries.

coral A form of calcium carbonate that is secreted by various marine animals (such as Anthozoa) for support and living space.

corrosion Reaction of a metal with an acid, oxygen, or other compound with destruction of the surface of the metal. Rusting is a common form of corrosion.

corundum (Al_2O_3) A naturally occurring mineral form of aluminum oxide that sometimes contains small amounts of iron and silicon(IV) oxide. It is found in rocks of igneous and metamorphic origin, and in placer deposits. RUBY and SAPPHIRE are impure crystalline forms. It is extremely hard (only second to diamond) and is used in various polishes, abrasives, and grinding wheels. *See also* emery.

coulomb Symbol: C The SI derived unit of electric charge, equal to the charge transported by a steady electric current of one ampere flowing for one second. In base units, 1 C = 1 A s.

coulombmeter (coulometer; voltameter) A device for determining electric charge or electric current using electrolysis. The mass m of material released in time t is measured, after which this value is used to calculate the charge (Q) and the current (I) from the electrochemical equivalent (z) of the element, using the formula $Q = m/z$ or $I = m/zt$.

coulometer *See* coulombmeter.

Coulson, Charles Alfred (1910–74) British theoretical chemist. Coulson was very influential in promoting the use of quantum mechanics in chemistry, particularly the use of molecular orbital theory. He used this theory to describe conjugated systems such as benzene and to justify the concept of partial valence. Coulson was also able to use molecular orbital theory to predict the properties of many organic molecules. He wrote several books, including *Valence*, the first edition of which was published in 1952. Coulson also contributed to the theoretical understanding of electrons in solids.

coupling A chemical reaction in which two groups or molecules join together.

covalent bond A bond formed by the sharing of an electron pair between two atoms. The covalent bond is conventionally represented as a horizontal line between chemical symbols, thus H–Cl indicates that between the hydrogen atom and the chlorine atom there is an electron pair formed by electrons of opposite spin, implying that the binding forces are strongly localized between the two atoms. Molecules are combinations of atoms bound together by covalent bonds; covalent bonding energies are of the order $10^3 kJmol^{-1}$.

Modern bonding theory treats the electron pairing in terms of the interaction of electron (atomic) ORBITALS and describes the covalent bond in terms of both 'bonding' and 'anti-bonding' molecular orbitals. *See also* coordinate bond; electrovalent bond; polar bond.

covalent carbide *See* carbide.

covalent crystal A crystal in which the atoms are covalently bonded. They are sometimes referred to as giant lattices or MACROMOLECULAR CRYSTALS. The best known completely covalent crystal is diamond.

covalent radius The radius an atom is assumed to have when involved in a covalent bond. For diatomic molecules comprised of the same element (e.g. Cl_2) this is simply half the measured internuclear distance. For molecules comprised of different atoms substitutional methods are used. For example, the internuclear distance of bromine fluoride (BrF) is about 180 picometers (pm). Therefore using 71 pm as the covalent radius of fluorine (from F_2) we get 109 pm as the covalent radius of bromine. This is close to the accepted value of 114 pm.

cream of tartar *See* potassium hydrogentartrate.

critical point The conditions of temperature and pressure under which a liquid being heated in a closed vessel becomes indistinguishable from its gas or vapor phase. At temperatures below the CRITICAL TEMPERATURE (T_c) the substance can be liquefied by applying pressure; at temperatures above T_c this is not possible. For each substance there is one critical point; for example, for carbon dioxide it is at 31.1°C and 7397 kilopascals.

critical pressure The lowest pressure needed to bring about liquefaction of a gas at its CRITICAL TEMPERATURE.

critical temperature The temperature (Tc) below which a gas can be liquefied by applying pressure and above which no amount of pressure is sufficient to bring about liquefaction. Some gases have critical temperatures above room temperature (e.g. carbon dioxide 31.1°C and chlorine, 144°C) and have thus been known in the liquid state for many years. Liquefaction proved much more difficult for those gases (e.g. oxygen, –118°C, and nitrogen, –146°C) that have very low critical temperatures.

critical volume The volume of one mole of a substance at its CRITICAL POINT.

crossed-beam reaction A chemical reaction performed with two molecular beams intersecting at an angle. It is possible to regard one of the beams as the incident (projectile) beam and the other as the target beam. It is possible to control both the incident and target beams of molecules and a considerable amount of information can be obtained about the mechanism and kinetics of the chemical reaction.

cross linkage An atom or short chain joining two longer chains in a polymer.

cryohydrate *See* eutectic.

cryolite *See* sodium hexafluoroaluminate.

cryoscopic constant *See* depression of freezing point.

crystal A solid substance that has a definite geometric shape. A crystal has fixed angles between its faces, which have distinct edges. The crystal will sparkle if the faces are able to reflect light. The fixed angles are caused by the regular arrangements of the atoms, ions, or molecules of the substance in the crystal. The faces and their angles bear a definite relationship to the arrangement of these particles. If broken, a large crystal will form smaller crystals.

crystal-field theory A theory of the properties of metal complexes. Originally it was introduced by Hans Bethe in 1929 to account for the properties of transition elements in ionic crystals. In the theory, the ligands surrounding the metal atom or ion are thought of as negative charges, and the effect of these charges on the energies of the d orbitals of the central metal ion is considered. Crystal-field theory was successful in describing the spectroscopic, magnetic, and other properties of complexes. It has been superseded by *ligand-field theory*, in which the overlap of the d orbitals on the metal ion with orbitals on the ligands is taken into account.

crystal habit The shape of a crystal. The habit depends on the way in which the crystal has grown; i.e. the relative rates of development of different faces and their proportions.

crystalline Describing a substance that has a regular internal arrangement of atoms, ions, or molecules, even though it may not appear visually as geometrically regular crystals. For instance, lead and the rest of the metals are crystalline because they are composed of regular accumulations of tiny invisible crystals. *Compare* amorphous.

crystallite A small rudimentary crystal. The term is often used in mineralogy to describe specimens that contain accumulations of many minute crystals of undetermined chemical composition and crystal structure.

crystallization The process of forming crystals. When a substance cools from the gaseous or liquid state to the solid state, crystallization occurs. Crystals will also form from a solution saturated with a solute.

crystallography The study of the formation, structure, and properties of crystals. *See also* x-ray crystallography.

crystalloid A substance that is not a colloid and which will therefore not pass through a semipermeable membrane. *See* colloid; semipermeable membrane.

crystal structure The particular repeating arrangement of atoms, molecules, or ions, in a crystal. 'Structure' refers to the internal arrangement of particles, not the external appearance.

crystal system A classification of crystals based on the shapes of their UNIT CELL. If the unit cell is a parallelopiped with lengths a, b, and c and the angles between these edges are α (between b and c), β (between a and c), and γ (between a and b), then crystals can be classified as follows:
cubic: $a = b = c$; $\alpha = \beta = \gamma = 90°$
tetragonal: $a = b \neq c$; $\alpha = \beta = \gamma = 90°$
orthorhombic: $a \neq b \neq c$; $\alpha = \beta = \gamma = 90°$
hexagonal: $a = b \neq c$; $\alpha = \beta = 90°$; $\gamma = 120°$
trigonal: $a = b \neq c$; $\alpha = \beta = \gamma \neq 90°$
monoclinic: $a \neq b \neq c$; $\alpha = \gamma = 90° \neq \beta$
triclinic: $a \neq b \neq c$; $\alpha \neq \beta \neq \gamma$
The orthorhombic system is also called the *rhombic* system.

cubic close packing *See* face-centered cubic crystal.

cubic crystal Denoting a crystal in which the UNIT CELL is a cube. In a *simple cubic* crystal the particles are arranged at the corners of a cube. *See also* body-centered cubic crystal; face-centered cubic crystal; crystal system. *See illustration overleaf.*

cupellation A technique used to separate relatively chemically unreactive metals such as silver, gold, platinum, and palladium from impurities by the use of an easily oxidizable metal such as lead. The impure metal is heated in a furnace and a blast of hot air is directed upon it. The lead or other oxidizable metal forms an oxide, leaving the relatively unreactive metal behind.

cupric chloride *See* copper(II) chloride.

cupric nitrate *See* copper(II) nitrate.

cupric oxide *See* copper(II) oxide.

cupric sulfate *See* copper(II) sulfate.

cuprite A red or reddish-black mineral form of copper(I) oxide (Cu_2O). It is a principal ore of copper.

cupronickel A series of durable copper-nickel alloys containing up to 45% nickel. Those containing 20% and 30% nickel are very malleable, can be worked cold or hot, and are very corrosion-resistant. They are used, for example, in condenser tubes in power stations. Cupronickel with 25% nickel is widely used in coinage. *See also* Constantan.

cuprous chloride *See* copper(I) chloride.

cuprous oxide *See* copper(I) oxide.

curie Symbol: Ci A unit of radioactivity, equivalent to the amount of a given radioactive substance that produces 3.7×10^{10} disintegrations per second, approximately equal to the number of disintegrations produced by one gram of radium-226 per second. In SI units it is equal to 37 giga becquerels (gBg).

Curie, Marie (1867–1934) Polish-born French chemist and physicist. Marie Curie

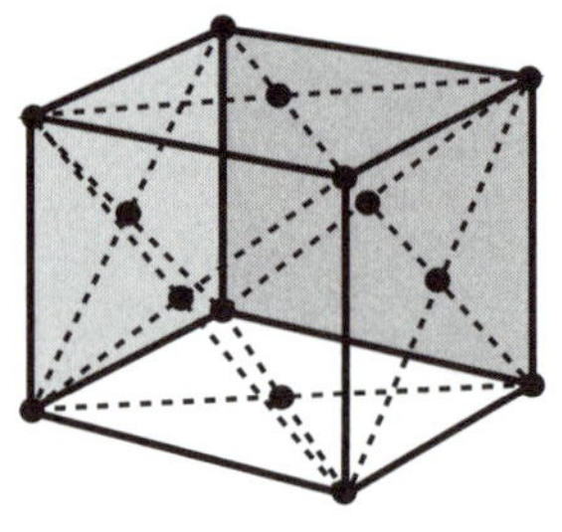

face-centered

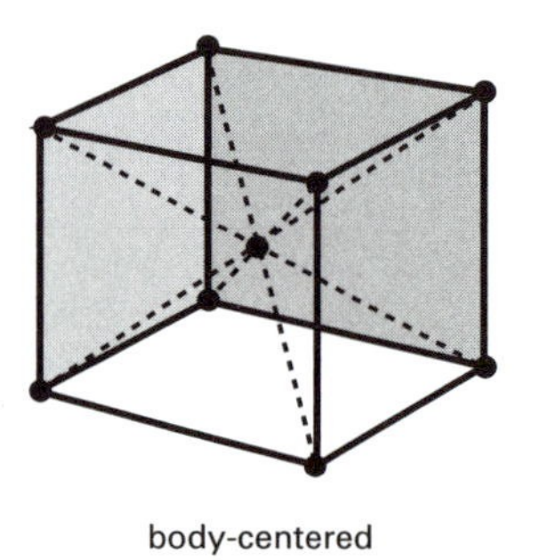

body-centered

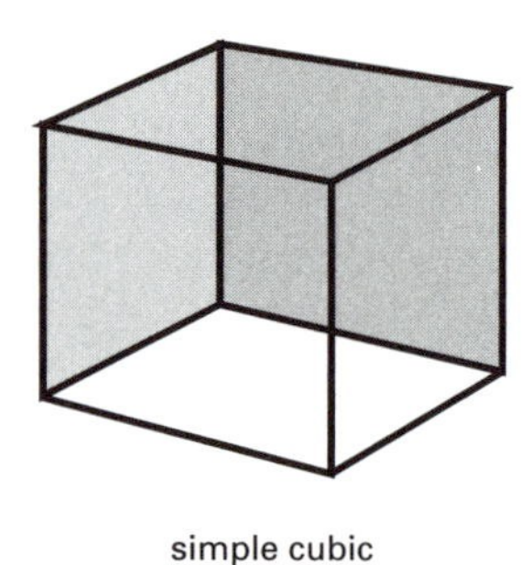

simple cubic

Cubic crystal: types of unit cell

was, together with her husband Pierre, a pioneer in the study of radioactivity. One of her first discoveries was that thorium is radioactive. By conducting a very thorough analysis of a sample of the uranium mineral pitchblende the Curies discovered the radioactive elements polonium and radium in 1898. They shared the 1903 Nobel Prize for physics with Henri Becquerel for their work on radioactivity. By performing experiments with different compounds of uranium under various physical conditions Marie Curie established that only the amount of uranium determined the radioactivity. This established that radioactivity is an atomic property of uranium. Marie Curie was awarded the 1911 Nobel Prize for chemistry for her discovery of polonium and radium.

curium A highly toxic radioactive silvery element of the actinoid series of metals. A transuranic element, it is found naturally on Earth only in very trace amounts in uranium ore. Its discovery in 1944 was based on its synthesis as the result of bombarding a plutonium isotope with helium nuclei. It can accumulate in bone marrow and destroy the body's capacity to produce red blood cells. Curium-244 and curium-242 have been used in thermoelectric power generators.

Symbol: Cm; m.p. 1340±40°C; b.p. ≅ 3550°C; r.d. 13.3 (20°C); p.n. 96; most stable isotope ^{247}Cm (half-life 1.56×10^7 years).

Curl, Robert Floyd Jnr (1933–) American chemist. Curl's initial work was on small clusters of atoms of semiconductors, such as germanium and silicon. In 1984, under the influence of Harry KROTO, he became interested in the possibility of producing long-chain carbon molecules, and persuaded his colleague Richard SMALLEY to deploy the resources of his laboratory towards this end. Although they expected on theoretical grounds to discover linear chain clusters with up to 33 carbon atoms, they in fact came across an unexpected molecule with 60 carbon atoms and with a cage-like structure. The discovery of this new allotrope of carbon, later named BUCKMINSTERFULLERENE, opened up a new branch of materials science. Curl shared the 1996 Nobel Prize for chemistry with Smalley and Kroto.

cyanamide process An industrial process for fixing nitrogen by heating calcium dicarbide in air.

$$CaC_2 + N_2 \rightarrow CaCN_2$$

The product, CALCIUM CYANAMIDE, hydrolyzes to ammonia and can be used as a fertilizer. The process can be expensive if cheap electrical energy is not available to produce CALCIUM DICARBIDE. *See also* nitrogen fixation.

cyanide A salt of hydrogen cyanide, containing the cyanide ion (CN^-). Cyanides are extremely poisonous due to their ability to form coordinate bonds with the iron in red blood cells. *See* hydrocyanic acid.

cyanide process A technique used for the extraction of gold from its ores. After crushing the gold ore to a fine powder it is agitated with a very dilute solution of POTASSIUM CYANIDE. The gold is dissolved by the cyanide to form potassium dicyanoaurate(I) (potassium aurocyanide, $KAu(CN)_2$). This complex is reduced with zinc, filtered, and then melted to obtain pure gold.

cyanoferrate A compound containing the ion $[Fe(CN)_6]^{4-}$ (the *hexacyanoferrate*(II) ion) or the ion $[Fe(CN)_6]^{3-}$ (the hexacyanoferrate(III) ion). These ions are usually encountered as their potassium salts. Potassium hexacyanoferrate(II) ($K_4Fe(CN)_6$, potassium *ferrocyanide*) is a yellow crystalline compound. Potassium hexacyanoferrate(III) ($K_3Fe(CN)_6$, potassium *ferricyanide*) is an orange crystalline compound. Its solution gives a deep blue precipitate with iron(II) ions, and is used as a test for iron(II) (ferrous) compounds. *Prussian blue* is a blue pigment containing hexacyanoferrate ions.

cyanogen (C_2N_2) A highly toxic flammable gas prepared by heating covalent metal hydrides such as mercury(II) cyanide. *See also* pseudohalogens.

cyclic compound A compound containing a ring of atoms. Although the majority of the cyclic compounds are organic, examples are found in inorganic compounds such as the silicates, silicones, and polyphosphates.

cyclization A reaction in which a straight-chain compound is converted into a cyclic compound.

D

dalton *See* atomic mass unit.

Dalton, John (1766–1844) English chemist. Dalton is best remembered as laying the foundations for modern atomic theory. One of his first experiments was to determine the density of water as a function of temperature. In the late 18th and early 19th centuries Dalton investigated properties of gases. This led to the statement of Dalton's law of partial pressures in 1801. Dalton first stated his atomic theory in 1803. He gave a definitive exposition of his views in the book *A New System of Chemical Philosophy*, the first volume of which was published in 1808, with subsequent volumes being published in 1810 and 1827.

Dalton's atomic theory A theory explaining the formation of compounds by elements, first published by John Dalton in 1803. It was the first modern attempt to describe chemical behavior in terms of atoms. The theory was based on certain postulates:

1. All elements are composed of small, indivisible particles, which Dalton called atoms.
2. All atoms of the same element are identical and have the same properties.
3. Atoms can neither be created nor destroyed.
4. Atoms combine to form 'compound atoms' (i.e. molecules) in simple ratios.

Dalton also suggested symbols for the elements. The theory was used to explain the law of CONSERVATION OF MASS and the laws of CHEMICAL COMBINATION.

Dalton's law (of partial pressures) The pressure of a mixture of gases in a particular volume is the sum of the *partial pressures* of each individual constituent gas, with the *partial pressure* of each gas being the pressure that it would exert if it alone occupied the same volume. Dalton's law is strictly true only for ideal gases.

Daniell cell A type of primary cell invented by British chemist John Daniell (1790–1845) in 1836. It consists of two electrodes in different electrolytes separated by a porous pot. The positive electrode is copper immersed in copper(II) sulfate solution and the negative zinc–mercury amalgam electrode is in either dilute sulfuric acid or a zinc sulfate solution. The porous pot prevents mixing of the electrolytes, but allows ions to pass. With sulfuric acid the e.m.f. is about 1.08 volts; with zinc sulfate it is about 1.10 volts.

At the copper electrode copper ions in solution gain electrons from the metal and are deposited as copper atoms:

$$Cu^{2+} + 2e^- \rightarrow Cu$$

The copper electrode thus gains a positive charge. At the zinc electrode, zinc atoms from the electrode lose electrons and dissolve into solution as zinc ions, leaving a net negative charge on the electrode:

$$Zn \rightarrow 2e^- + Zn^{2+}$$

darmstadtium A radioactive metallic element that does not occur naturally on the Earth. It is made either by bombarding a lead target with nickel nuclei or by bombarding a plutonium target with sulfur nuclei. There are several isotopes; the most stable is ^{269}Ds, with a half-life of about 1.7×10^{-4} s. The chemical properties of darmstadtium should be similar to those of platinum.

Symbol Ds; p.n. 110.

dative bond *See* coordinate bond.

Davy, Sir Humphry (1778–1829) English chemist. Davy is best known for the discovery of sodium and potassium and for inventing a safety lamp used in mines. He began his researches in electrochemistry in the early 19th century, soon after the invention of the electric cell. In 1807 he discovered sodium and potassium by electrolysis of fused soda and potash. In 1810 he showed that chlorine, which he named, is an element and did not contain oxygen, as had been thought. This helped destroy the theory that all acids contain oxygen and prompted him to suggest that all acids contain hydrogen.

Davy lamp *See* safety lamp.

d-block elements The TRANSITION ELEMENTS of the first, second, and third long periods of the periodic table, i.e. scandium to zinc, yttrium to cadmium, and lanthanum to mercury. They are so called because in general they have inner d-levels with configurations of the type $(n-1)d^xns^2$ where $x = 1–10$.

deactivation A process in which the reactivity of a substance is lessened, or even totally removed . Usually this deactivation is unwanted, as in the poisoning of catalysts.

de Broglie wave A wave associated with a particle, such as an electron or proton. In 1924, Louis de Broglie (1892–1987) suggested that, since electromagnetic waves can be described as particles (photons), particles of matter could also have wave properties. The wavelength (λ) has the same relationship to momentum (p) as in electromagnetic radiation:

$$\lambda = h/p$$

where h is the Planck constant. *See also* quantum theory.

debye Symbol: D A unit of electric DIPOLE MOMENT equal in SI units to 3.33564 $\times 10^{-30}$ coulomb meter. It is used in expressing the dipole moments of molecules.

Debye, Peter Joseph William (1884–1966) Dutch-born American chemist and physicist. Debye made a number of important contributions to molecular structure and the theory of solids. Debye is perhaps best known for his work with Erich HÜCKEL in 1923 on electrolytes. In his later years he was concerned with scattering of light and sound by liquids and with polymers. He won the 1936 Nobel Prize for chemistry for his work on molecular structure and dipole moments.

Debye–Hückel theory A theory of weak electrolytes. It assumes that the electrolytes are fully dissociated and nonideal behavior arises from electrostatic interactions. It is accurate only for very dilute solutions.

Debye–Scherrer method A method used in X-RAY DIFFRACTION in which a crystal in powder form is exposed to a beam of monochromatic x-rays. Because the crystal is in powder form all possible orientations of the crystal are presented to the x-ray beam. This has the result that the diffracted x-rays form cones concentric with the original beam. The Debye–Scherrer method is particularly useful for determining the lattice type of a crystal and the dimensions of its unit cell. The method was first developed by Peter Debye and Paul Scherrer.

deca- Symbol: da A prefix used with SI units, denoting 10. For example, 1 decameter (dam) = 10 meters (m).

decahydrate A crystalline solid containing ten molecules of water of crystallization per molecule of compound.

decant To pour off the clear liquid above a sediment or heavier, immiscible liquid.

decay **1.** The spontaneous breakdown of a radioactive nuclide. *See* half-life; radioactivity.
2. The transition of excited atoms, ions, molecules, etc., to the ground state.

deci- Symbol: d A prefix used with SI units, denoting 10^{-1}. For example, 1 decimeter (dm) = 10^{-1} meter (m).

decomposition A chemical reaction in which a compound is broken down into compounds with simpler molecules or into elements.

decrepitation A crackling sound heard when certain crystalline solids are heated, usually as a result of loss of water of crystallization.

defect An irregularity in the ordered arrangement of particles in a crystal lattice. There are two main kinds of defects in crystals: *point defects* and *dislocations*. Point defects occur at single lattice points. There are three types. A *vacancy*, also known as a *Schottky defect*, is a missing atom; i.e. a vacant lattice point. An *interstitial* is an atom that is in a position that is not a normal lattice point. If an atom moves off its normal lattice point to an interstitial position the result (vacancy plus interstitial) is called a *Frenkel defect*. All solids above absolute zero have a number of point defects, the concentration of which depends on temperature. Point defects can also be produced by strain or by irradiation.

Dislocations (or *line defects*) are also produced by strain in solids. These defects are irregularities extending over a number of lattice points along a line of atoms.

definite proportions, law of *See* constant proportions.

degassing The removal of adsorbed, dissolved, or absorbed gases from liquids or solids.

degenerate Describing different quantum states that have the same energy. For instance, the five d orbitals in a transition-metal atom all have the same energy but different values of the magnetic quantum number m. Differences in energy occur if a magnetic field is applied or if the arrangement of ligands around the atom is not symmetrical. The degeneracy is then said to be 'lifted'.

degrees absolute *See* absolute temperature.

degrees Kelvin *See* absolute temperature.

degrees of freedom **1.** The independent ways in which particles can take up energy. In a monatomic gas, such as helium or argon, the atoms have three translational degrees of freedom, corresponding to their motion in space in three mutually perpendicular directions (i.e. along x, y, and z coordinates). The mean energy per atom for each degree of freedom is $kT/2$, where k is the Boltzmann constant and T the thermodynamic temperature; the mean energy per atom is thus $3kT/2$.

A diatomic gas, in addition to three translational degrees, also has two rotational degrees of freedom located about the two axes perpendicular to the bond, and one vibrational degree of freedom located along the axis of the bond. The rotational degrees also each contribute $kT/2$ to the average energy. The vibrational degree contributes kT ($kT/2$ for kinetic energy and $kT/2$ for potential energy). Thus, the average energy per molecule for a diatomic molecule is $3kT/2$ (translation) + kT (rotation) + kT (vibration) = $7kT/2$.

Linear triatomic molecules also have two significant rotational degrees of freedom; nonlinear molecules have three. For nonlinear polyatomic molecules, the number of vibrational degrees of freedom is $3N - 6$, where N is the number of atoms in the molecule.

The molar energy of a gas is the average energy per molecule multiplied by the Avogadro constant. For a monatomic gas, for example, it is $3RT/2$.

2. The independent physical quantities (e.g. pressure, temperature, etc.) that define the state of a given system. *See* phase rule.

dehydration **1.** Removal of water from a substance.

2. Removal of the elements of water (i.e. hydrogen and oxygen in a 2:1 ratio) to form a new compound.

deionization A method of removing ions from a solution using ion exchange. The term is commonly applied to the purification of tap water; deionized water has superseded distilled water in many chemical applications. *See* ion exchange.

deliquescent Describing a solid HYGROSCOPIC compound that absorbs so much water from the atmosphere that a solution eventually forms.

delocalization Diffusion of one or more of a molecule's bonding electrons over two or more of the bonds between the molecule's atoms.

delocalized bond (nonlocalized bond) A type of molecular bonding in which the electrons forming the delocalized bond are no longer regarded as remaining between two atoms but instead are spread over several atoms or even the whole molecule.

The electron density of the delocalized bond is spread by means of a delocalized molecular ORBITAL and may be regarded as a series of *pi* bonds extending over several atoms, for example the C–O *pi* bonds in the carbonate ion. *See also* metallic bond; resonance.

delta brass *See* delta metal.

delta metal (delta brass) A strong alloy of copper and zinc, containing also a little iron. Its main use is for making cartridge cases.

dendritic growth Growth of crystals in a branching ('treelike') habit.

denitrification *See* nitrogen cycle.

density (mass density) Symbol: ρ The mass per unit volume of a given substance. It is typically measured in grams per cubic decimeter (g m^{-3}), which is equivalent to grams per liter (g L^{-1}), or in kilograms per cubic meter (kg m^{-3}). *See also* relative density.

density functional theory A method of calculating the electronic structure of molecules using the electron density. The theory was developed by Walter Kohn (1923–) and his colleagues in the mid-1960s and has been used extensively in chemistry and solid state physics since that time.

depolarizer A substance used in a voltaic cell to prevent polarization. For example, hydrogen bubbles forming on the electrode can be removed by an oxidizing agent such as manganese(IV) oxide (MnO_2).

depression of freezing point A COLLIGATIVE PROPERTY of solutions in which the freezing point of a given solvent is lowered by the presence of a solute. The amount of the reduction is proportional to the molal concentration (m) of the solute. The depression depends only on this concentration and is independent of solute composition. The formula is:

$$\Delta t = K_f C_M,$$

where Δt is the lowering of the temperature, K_f is the proportionality constant (also known as the *freezing point constant* or the *cryoscopic constant*), and C_m is the molal concentration. Note that the unit of K_f is the kelvin kilogram $mole^{-1}$ (K kg mol^{-1}). The formula can be applied to the measurement of relative molecular mass with considerable precision. A known mass of pure solvent is slowly frozen, with stirring, in a suitable cold bath and the freezing temperature measured using a Beckmann thermometer. A known mass of solute of known molecular mass is introduced, the solvent thawed out, and the cooling process and measurement repeated. The addition is repeated several times and an average value of K_f for the solvent obtained by plotting Δt against C_m. The whole process is then repeated using the unknown solute and its relative molecular mass determined using the value of K_f previously obtained.

The effect is applied to more precise measurement of relative molecular mass by using a pair of Dewar flasks (pure solvent and solution) and measuring Δt by means of thermocouples. The theoretical explanation for the effect is similar to that for *lowering of vapor pressure*. The freezing point of the solvent is that point at which the curve representing the vapor pressure above the liquid phase intersects the curve representing the vapor pressure above the frozen solvent. The addition of solute depresses the former curve but as the solid phase that separates is always pure solvent (i.e. above the eutectic point), there is no attendant depression of the latter curve. Consequently the point of intersection is depressed, resulting in a lowering of the freezing point. *See also* lowering of vapor pressure. *Compare* elevation of boiling point.

derivative A compound obtained by reaction from another compound. The term is most often used in organic chemistry of compounds that have the same general structure as the parent compound.

derived unit A measurement unit defined in terms of the BASE UNITS of a system of measurement, and not directly from a standard value of the quantity it measures. For example, the newton is a unit of force defined in base SI UNITS as a kilogram meter second^{-2} (kg m s^{-2}).

desalination Any of various techniques for removing the salts (mainly sodium chloride) from seawater to make it fit for drinking, irrigation, use in water-cooled engines, and for making steam for steam turbines. There are various methods. Those based on distillation depend on a cheap source of heat energy, of which solar energy is the most promising, especially in hot climates. Flash evaporation (evaporation under reduced pressure) and freezing to make pure ice are other methods, as are electrodialysis, ion exchange, reverse osmosis, and the use of molecular sieves.

desiccation Removal of moisture from a substance.

desiccator A laboratory apparatus for drying solids or for keeping solids free of moisture. It is a container in which is kept a hygroscopic material (e.g. calcium chloride or silica gel).

destructive distillation The process of heating an organic substance such that it wholly or partially decomposes in the absence of air to produce volatile products, which are subsequently condensed. The destructive distillation of coal was the process for manufacturing coal gas and coal tar. Formerly, methanol was made by the destructive distillation of wood.

detergents A group of substances that improve the cleansing action of solvents, particularly water. The majority of detergents, including *soap*, have the same basic structure. Their molecules have a nonpolar hydrocarbon chain (tail) that lacks an affinity for water molecules and is therefore said to be hydrophobic (water-fearing or repelling). Attached to this tail is a small polar group (head) that has an affinity for water molecules and is said to be hydrophilic (water-loving or attracting).

Detergents reduce the surface tension of water and thus improve its wetting power by anchoring their hydrophilic heads in the water with their hydrophobic tails protruding above it. Consequently the water surface is broken up, enabling the water to spread over the material to be cleaned and penetrate between the material and dirt to which it has been exposed. With the assistance of agitation, the dirt can then be floated off. At the same time the hydrophobic tails of the detergent molecules dissolve in grease and oils. The protruding hydrophilic heads of the molecules repel each other, causing the grease or oil to roll up and form tiny drops, which float off into the water to form an emulsion.

Unlike soaps synthetic detergents, which are often derived from petrochemicals, do not form insoluble scums with hard water. Synthetic detergents are of three types. *Anionic detergents* form ions consisting of a hydrocarbon chain to which is attached either a sulfonate group, $-SO_2-O^-$, or a sulfate group, $-O-SO_2-O^-$. The corresponding metal salts are soluble in water. *Cationic detergents* have organic positive ions of the type RNH_3^+, in which R has a long hydrocarbon chain. *Nonionic* detergents are complex chemical compounds called *ethoxylates*. They owe their detergent properties to the presence of a number of oxygen atoms in one part of the molecule, which are capable of forming hydrogen bonds with the surface water molecules, thus reducing the surface tension of the water.

deuterated compound A compound in which one or more ^1H atoms have been replaced by deuterium (^2H) atoms.

deuteride A compound of deuterium with other elements; i.e. a hydride in which the deuterium isotope is present rather than ^1H. *See* hydride.

deuterium (heavy hydrogen) Symbol: D, ^2H A naturally occurring, stable isotope of hydrogen in which the nucleus contains one proton and one neutron. The atomic mass is thus approximately twice that of ^1H. Chemically it behaves almost identically to hydrogen, forming analogous compounds, although reactions of deuterium compounds are often slower than those of corresponding ^1H compounds. This fact is made use of in KINETICS where the rate of a reaction may depend on transfer of a hydrogen atom (i.e. a kinetic isotope effect).

deuterium oxide (heavy water; (D_2O) A naturally occurring form of water in which deuterium isotopes (^2H) have replaced the far more common isotope of hydrogen (^1H) in the water molecule. Deuterium oxide represents only about 0.003% by mass of the water on Earth, from which it can be separated by either electrolysis or fractional distillation. The chief use of deuterium oxide is to slow down fast neutrons in nuclear fission reactors designed to use unenriched uranium (^{238}U) as fuel.

deuteron The nucleus of the deuterium atom, $^2H^+$ or D^+. It is used to bombard other nuclei in nuclear accelerators.

Dewar flask (vacuum flask) A double-walled container of thin glass with the space between the walls evacuated and sealed to stop conduction and convection of energy through it. The glass is often silvered to reduce radiation.

dextro-form *See* optical activity.

dextrorotatory *See* optical activity.

D-form *See* optical activity.

diagonal relationship There is a general trend in the periodic table for electronegativity to increase from left to right along a period and to decrease down a group. Thus a move of 'one across and one down', i.e. a diagonal move in the table, gives rise to effects that tend to cancel each other. There is a similar general trend and combined effect for size, which decreases along a period but increases down a group. These diagonal relationships give rise to similarities in chemical properties, which are particularly noticeable for the following diagonal pairs. Li–Mg; Be–Al; B–Si. Li–Mg: 1. both have carbonates that give CO_2 on heating; 2. both burn in air to give the normal oxide only; 3. both form a nitride; 4. both form hydrated chlorides that hydrolyze slowly. Be–Al: 1. both give hydrogen with alkalis; 2. both give water-insoluble hydroxides that dissolve in alkali; 3. both form complex ions of the type MCl_3^-; 4. both have covalently bridged chlorides. B–Si: 1. both form acidic oxides of a giant covalent-molecule type with high melting points; 2. both form low-stability hydrides that ignite in air; 3. both form readily hydrolyzable chlorides that fume in air; 4. both have amorphous and crystalline forms and both form glasses with basic oxides, such as Na_2O.

diamagnetism *See* magnetism.

diamond An allotrope of carbon and the hardest naturally occurring substance known. It is used for jewelry and, industrially, for cutting and drilling equipment. In diamond, each carbon atom is surrounded by four equally spaced carbon atoms arranged tetrahedrally. The carbon atoms form a three-dimensional network with each carbon-carbon bond equal to 0.154 nm and set at an angle of 109.5° with its neighbors. Millions of atoms are covalently bonded in diamonds to form a giant molecular structure, the great strength of which results from the strong covalent bonds between carbon atoms. Diamonds can be formed synthetically from graphite under extreme temperature and pressure in the presence of a catalyst. Although small, such diamonds are of adequate size for many industrial uses. Natural diamonds are thought to form deep in the Earth's crust and are often associated with ancient volcanic vents when discovered. Major deposits occur in Africa, Australia, Brazil, Canada, and Russia. *See also* carbon.

diatomaceous earth *See* diatomite.

diatomic Describing a molecule that consists of two atoms. Hydrogen (H_2), oxygen (O_2), nitrogen (N_2), and the halogens are examples of elements that form idatomic molecules. Sodium chloride (NaCl) is another example of a diatomic molecule.

diatomite (diatomaceous earth; kieselguhr) A whitish powdered mineral consisting mainly of SILICON(IV) OXIDE derived from the shells of diatoms. It is used to make fireproof cements and as an absorbent in the manufacture of dynamite.

dibasic acid An acid that has two active protons, such as sulfuric acid. Dibasic acids can give rise to two series of salts. For example, sulfuric acid (H_2SO_4) forms sulfates (SO_4^{2-}) and hydrogensulfates (HSO_4^-).

diboron trioxide *See* boron oxide.

dicarbide *See* carbide.

dichlorine oxide (chlorine monoxide; chlorine(I) oxide; Cl_2O) An orange gas made by passing chlorine over mercury(II) oxide. It is a strong oxidizing agent and dissolves in water to give chloric(I) acid.

dichromate(VI) *See* chromium; potassium dichromate.

dielectric constant (relative permittivity) A quantity that characterizes how a medium reduces the electric field strength associated with a distribution of electric charges. If two point charges Q_1 and Q_2 are a distance d apart in a medium with a *permittivity* ε this means that the force F between the charges is given by $F = (1/4\pi\varepsilon)(Q_1Q_2/d^2)$.
The dielectric constant (relative permittivity) ε_r of the medium is given by $\varepsilon_r = \varepsilon/\varepsilon_0$, where ε_0 is the permittivity of free space. The dielectric constant of air is very slightly greater than 1, while that of water is about 80. The value of the dielectric constant of a medium has important physical and chemical consequences, particularly for ions in solutions.

diffusion Movement of a gas, liquid, or solid as a result of the random thermal motion of its particles (atoms or molecules). A drop of ink in water, for example, will slowly spread throughout the liquid. Diffusion in solids takes place very slowly at normal temperatures. *See also* Graham's law.

dihydrate A crystalline compound having two molecules of water of crystallization per molecule of compound.

diiodine hexachloride (iodine trichloride; I_2Cl_6) A yellow crystalline solid made by reacting excess chlorine with iodine. It is a strong oxidizing agent. At 70°C it dissociates into iodine monochloride and chlorine.

dilead(II) lead(IV) oxide (red lead; Pb_3O_4) A powder made by heating lead(II) oxide or lead(II) carbonate hydroxide at 400°C. It is black when hot and red or orange when cold. On strong heating it decomposes to lead(II) oxide and oxygen. Dilead(II) lead (IV) oxide is used as a pigment and in glass making. It is not stoichiometric and tends to have less oxygen than denoted by its formula.

diluent A solvent that is added to reduce the strength of a solution.

dilute Denoting a solution in which the amount of solute is low relative the amount of solvent. The term is always relative and includes dilution at trace level as well as the common term *bench dilute acid*, which usually means a 2M solution. *Compare* concentrated.

dimensionless units Units defined in terms of the ratio of two comparable quantities expressed in like units, which therefore reduce to one. For example, the radian and steradian, both SI derived units, are dimensionless because they are expressed as ratios of meter per meter and square meter per square meter, respectively. *See* SI units.

dimer A compound (or molecule) formed by combination or association of two molecules of a monomer. For instance, aluminum chloride ($AlCl_3$) is a dimer (Al_2Cl_6) in the vapor phase.

dimorphism *See* polymorphism.

dinitrogen oxide (nitrous oxide; N_2O) A colorless gas with a faintly sweet odor and taste. It is appreciably soluble in water but more soluble in ethanol. It is prepared commercially by carefully heating ammonium nitrate:

$$NH_4NO_3(s) \rightarrow N_2O(g) + 2H_2O(g)$$

Dinitrogen oxide is fairly easily decomposed on heating to temperatures above 520°C, giving nitrogen and oxygen. The gas is used as a mild anesthetic in medicine and dentistry, being marketed in small steel cylinders. It is sometimes called *laughing gas* because it induces a feeling of elation. It is also used as an aerosol propellent.

dinitrogen tetroxide (N_2O_4) A colorless gas that becomes a pale yellow liquid below 21°C and solidifies below –11°C. On heating, the gas dissociates to nitrogen dioxide molecules:

$$N_2O_4(g) \rightarrow 2NO_2(g)$$

This dissociation is complete at 140°C. Liquid dinitrogen tetroxide has good solvent properties and is used as a nitrating agent.

diphosphane (diphosphine; P_2H_4) A yellow liquid that can be condensed out from PHOSPHINE in a freezing mixture. It ignites spontaneously in air.

diphosphine *See* diphosphane.

dipolar bond *See* coordinate bond.

dipole–dipole interactions The interaction of two molecules resulting from their DIPOLE MOMENTS. The strength of this interaction depends on the size of the dipole moments of the molecules and their distance apart. The dipole moments that can interact include permanent dipole moments in molecules such as water, induced dipole moments caused by the presence of polar molecules in close proximity, thus giving rise to *induced dipole–dipole interactions*, and the charge separation associated with VAN DER WAALS FORCES.

dipole moment Symbols: μ, p A quantitative measure of polarity in either a bond (*bond moment*) or a molecule as a whole (*molecular dipole moment*). The unit is the debye (equivalent to $3.335\ 64 \times 10^{-30}$ coulomb meter). Molecules such as HF, H_2O, and NH_3, possess dipole moments; CCl_4, N_2, and PF_5 do not.

The molecular dipole moment can be estimated by vector addition of individual bond moments if the bond angles are known. The possession of a dipole moment permits direct interaction with electric fields or interaction with the electric component of radiation.

Dirac, Paul Adrien Maurice (1902–84) English physicist. Dirac was one of the founders of quantum mechanics. In a remarkable series of papers in the second half of the 1920s he formulated quantum mechanics in a general way that incorporated the matrix mechanics of Werner HEISENBERG and the wave mechanics of Erwin SCHRÖDINGER as special cases. He shared the 1933 Nobel Prize for physics with Schrödinger.

dislocation *See* defect.

disodium hydrogen orthophosphate *See* disodium hydrogen phosphate(V).

disodium hydrogen phosphate(V) (disodium hydrogen orthophosphate; Na_2HPO_4) A white solid prepared by titrating phosphoric acid with sodium hydroxide solution using phenolphthalein as the indicator. On evaporation the solution yields efflorescent monoclinic crystals of the dodecahydrate, $Na_2HPO_4.12H_2O$. The effloresced salt contains $7H_2O$. Disodium hydrogen phosphate is used in the textile industry.

disodium oxide *See* sodium monoxide.

disodium tetraborate decahydrate (borax; $Na_2B_4O_7.10H_2O$) A white crystalline solid, sparingly soluble in cold water but readily soluble in hot water. It occurs naturally as salt deposits in dry lake beds, especially in California. It is an important industrial material, being used in the manufacture of enamels and heat-resistant glass, as a paper glaze, and as a source of borium compounds. It is also used in laundry products and as a mild antiseptic.

In solution hydrolysis occurs:

$$B_4O_7^{2-} + 7H_2O = 2OH^- + 4H_3BO_3$$

See also borax-bead test.

dispersed phase *See* colloid.

dispersing agent A compound used to help produce EMULSIONS or dispersions of IMMISCIBLE liquids, such as water and oil.

dispersion force *See* van der Waals force.

displacement pump A commonly used device for transporting liquids and gases around chemical plants. It works on the principle of the bicycle pump: a piston raises the pressure of the fluid and, when it is high enough, a valve opens and the fluid is discharged through an outlet pipe. As the piston moves back the pressure falls and the cycle continues. Displacement pumps can be used to generate very high pressures (e.g. in the synthesis of ammonia) but because of the system of valves, they are more expensive than other types of pump. *Compare* centrifugal pump.

displacement reaction A chemical reaction in which an atom or group displaces another atom or group from a molecule. A common example is the displacement of hydrogen from acids by metals:

$$Zn + 2HCl \rightarrow ZnCl_2 + H_2$$

See also double decomposition.

disproportionation A chemical reaction in which there is simultaneous oxidation and reduction of the same compound. An example is the reaction of copper(I) chloride to copper and copper(II) chloride:

$$2CuCl \rightarrow Cu + CuCl_2$$

in which there is simultaneous oxidation (to Cu(II)) and reduction (to Cu(0)).

Another example is the reaction of chlorine with water:

$$Cl_2 + H_2O \rightarrow 2H^+ + Cl^- + ClO^-$$

in which there is reduction to Cl^- and oxidation to ClO^-.

dissociation Breakdown of a molecule into two molecules, atoms, radicals, or ions. Often the reaction is reversible, as in the ionic dissociation of weak acids in water:

$$HCN + H_2O \rightleftharpoons H_3O^+ + CN^-$$

See also dissociation constant.

dissociation constant Symbol: K The EQUILIBRIUM CONSTANT of a reversible dissociation reaction. For example, the dissociation constant of a reaction:

$$AB \rightleftharpoons A + B$$

is given by:

$$K = [A][B]/[AB]$$

where the brackets denote concentration (activity).

Often the degree of dissociation is used – the fraction (α) of the original compound that has dissociated at equilibrium. For an original amount of AB of n moles in a volume V, the dissociation constant is given by:

$$K = \alpha^2 n/(1 - \alpha)V$$

Note that this expression is for dissociation into two molecules.

Acid dissociation constants (or *acidity constants*, symbol: K_a) are dissociation constants for the dissociation into ions in solution:

$$HA + H_2O \rightleftharpoons H_3O^+ + A^-$$

If the concentration of water is taken as unity, the acidity constant is given by:

$$K_a = [H_3O^+][A^-]/[HA]$$

The acidity constant is a measure of the strength of the acid. Base dissociation constants (symbol K_b) are similarly defined. The expression:

$$K = \alpha^2 n/(1 - \alpha)V$$

applied to an acid is known as *Ostwald's dilution law*. In particular if α is small (a weak acid) then $K = \alpha^2 n/V$, or $\alpha = C\sqrt{V}$, where C is a constant. The degree of dissociation is then proportional to the square root of the dilution.

distillation The process of boiling a liquid and condensing the vapor. Distillation is used to purify liquids or to separate components of a liquid mixture. *See also* fractional distillation; steam distillation; vacuum distillation.

distilled water Water that has been purified by distillation in order to separate it from dissolved solids or other substances. Laboratory-grade water is usually distilled several times.

disulfur dichloride (sulfur monochloride; S_2Cl_2) A red fuming liquid with a strong smell. It is prepared by passing chlorine over molten sulfur and is used to harden (vulcanize) rubber.

dithionate A salt of dithionic acid. Dithionates are reducing agents.

dithionic acid (hyposulfuric acid; $H_2S_2O_6$) A strong sulfur oxyacid that decomposes slowly on standing or heating.

dithionite (sulfinate) A salt of dithionous acid. Dithionites are powerful reducing agents.

dithionous acid (sulfinic acid; hyposulfurous acid; $H_2S_2O_4$) An unstable sulfur oxyacid that is known only in solution.

divalent (bivalent) Having a valence of two.

***dl*-form** *See* optical activity.

Döbereiner's triads Groups of three chemically similar elements in which the central member, when placed in order of increasing relative atomic mass, has a relative atomic mass approximately equal to the average of the outer two. Other chemical and physical properties of the central member also lie between those of the first and last members of the triad. German chemist Johann Döbereiner (1780–1849) noted this relationship in 1817; each triad is now recognized as comprised of consecutive members of a group of the periodic table; e.g. calcium, strontium, and barium; chlorine, bromine, and iodine.

dolomite (pearl spar) A typically white, pink, or colorless mineral, $CaCO_3.MgCO_3$, used as an ore of magnesium and for the lining of open-hearth steel furnaces and Bessemer converters.

donor **1.** The atom, ion, or molecule that provides the pair of electrons in forming a covalent bond.
2. The impurity atoms used in doping semiconductors.

doping The incorporation of impurities within the crystal lattice of a solid so as to alter its physical properties. For instance, silicon, when doped with boron, becomes semiconducting.

double bond A covalent bond between two atoms that includes two pairs of electrons, one pair being the single bond equivalent (the sigma bond) and the other forming an additional bond, the pi bond (π bond). It is conventionally represented by two lines, for example $H_2C=O$. *See* multiple bond; orbital.

double decomposition (metathesis) A chemical technique for making an insoluble salt by mixing two soluble salts. The ions from the two reactants 'change partners' to form a new soluble compound and an insoluble compound, which is precipitated. For example, mixing solutions of potassium iodide (KI) and lead(II) nitrate ($Pb(NO_3)_2$) forms a solution of potassium nitrate (KNO_3) and a yellow precipitate of lead(II) iodide (PbI_2).

double salt When equivalent quantities of certain salts are mixed in aqueous solution and the solution evaporated, a salt having two different anions or cations may form, e.g. $FeSO_4.(NH_4)_2SO_4.6H_2O$. In aqueous solution the salt behaves as a mixture of the two individuals. These salts are called double salts to distinguish them from complex salts, which yield complex ions in solution. *See also* alum.

Downs process An electrolytic process for making sodium and chlorine from fused sodium chloride. The process takes place in a *Downs cell*, which has a centrally located graphite anode surrounded by a cylindrical steel cathode. A coaxial steel grid keeps the electrodes and their products separate. Chlorine gas is collected via a cone-shaped dome above the anode, while molten sodium floats on top of the denser electrolyte to be siphoned off via a collecting pipe.

Dow process An industrial method whereby magnesium is extracted from seawater via the precipitation of magnesium hydroxide ($Mg(OH)_2$) when calcium hydroxide ($Ca(OH)_2$) is added, followed by SOLVATION of the precipitated hydroxide by hydrochloric acid (HCl).

dry cell A voltaic cell in which the electrolyte is typically in the form of a jelly or paste and the metal container holding the electrolyte forms the negative electrode. Dry cells are extensively used for flashlights, toys, and other portable applications.

dry ice *See* carbon dioxide.

dubnium A radioactive synthetic transactinide element first made at Dubna, a town near Moscow, in 1967 by bombarding americium ions with neon ions. It was almost simultaneously made in the U.S.A. by bombarding californium nuclei with nitrogen nuclei.

Symbol: Db; m.p., b.p., and r.d. unknown, p.n. 105; most stable isotope ^{262}Db (half-life 34 s).

Dulong and Petit's law The law that states that the molar heat capacity of a solid element is approximately equal to $3R$, where R is the GAS CONSTANT (25 J K^{-1} mol^{-1}). The law applies only to elements with simple crystal structures at normal temperatures. At lower temperatures the molar heat capacity falls with decreasing temperature. *Molar heat capacity* was formerly called *atomic heat* – the product of the relative atomic mass of a substance and its specific heat capacity.

The law was derived by French chemists Pierre Dulong (1785–1838) and Alexis Petit (1791–1820) in 1819.

Duralumin (*Trademark*) A strong lightweight aluminum alloy containing 3–4% copper, with small amounts of magnesium, manganese, and sometimes silicon. It is widely used in aircraft bodies and similar applications requiring strength combined with lightness.

dysprosium A very soft malleable silvery-white element of the lanthanoid series of metals. It occurs in association with other lanthanoids. It is used in lasers and compact disks, and as a neutron absorber in nuclear reactors; it is also a constituent of certain magnetic alloys.

Symbol: Dy; m.p. 1412°C; b.p. 2562°C; r.d. 8.55 (20°C); p.n. 66; most common isotope 164 Dy; r.a.m. 162.50.

E

ebullioscopic constant *See* elevation of boiling point.

ebullition The boiling or bubbling of a liquid.

Edison cell *See* nickel–iron accumulator.

edta Abbreviation for ethylenediaminetetraacetic acid, a tetrabasic carboxylic acid. It is important in inorganic chemistry because the carboxylate ion can act as a ligand in forming complexes. Edta is a hexadentate ligand, able to coordinate to a metal ion at the four COO^- groups and the two nitrogen atoms. It is able to form octahedral complexes with certain metal ions.

effervescence The evolution of gas in the form of bubbles in a liquid.

efficiency Symbol: η A measure used for processes of energy transfer; the ratio of the useful energy produced by a system or device to the energy input. For a reversible heat engine, the idealized maximum efficiency is given by

$$\eta = (T_1 - T_2)/T_1$$

where T_1 is the temperature of the heat as sourced and T_2 is the temperature of the heat given out at the end of the process. *See also* Carnot's principle; Carnot cycle.

efflorescence The process in which a crystalline hydrated solid loses water of crystallization to the air. A powdery deposit is gradually formed on the surface of the compound.

Eigen, Manfred (1927–) German chemist. Eigen was a pioneer of the study of very fast chemical reactions. Starting in 1954, he studied such reactions by relaxation techniques in which very quick changes in the pressure, temperature, and electric field were applied to the system and the consequences studied using spectroscopy. He shared the 1967 Nobel Prize for chemistry with Ronald NORRISH and George PORTER for this work. Subsequently Eigen used relaxation techniques to study very fast biochemical reactions.

eigenfunction A function that is a solution of an EIGENVALUE equation. An example of an eigenfunction is the WAVEFUNCTION for a quantum mechanical system described by the SCHRÖDINGER EQUATION.

eigenvalue One of the set of possible values of numbers that make up the solu-

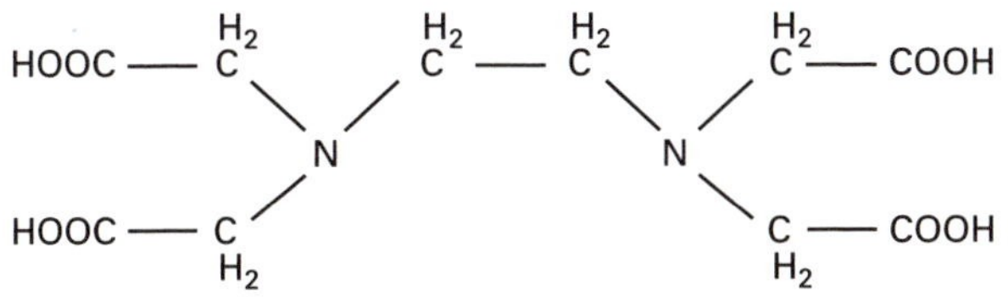

Edta

tions of an *eigenvalue equation*. An eigenvalue equation has the general form: $\hat{O}\psi = E\psi$, where $\hat{O}$ is an *operator* which operates on the EIGENFUNCTION ψ and E is a number which multiplies the same eigenfunction. The solution of an eigenvalue equation consists in finding the sets of eigenfunctions ψ and eigenvalues E that satisfy the eigenvalue equation. The fundamental equations of quantum mechanics such as the SCHRÖDINGER EQUATION are eigenvalue equations. In the case of the Schrödinger equation the eigenfunctions are the WAVEFUNCTIONS for the quantum states of the system and the eigenvalues are the values of the quantized energy levels associated with these quantum states.

einsteinium A radioactive transuranic element of the actinoid series, not found naturally on Earth and originally discovered in the radioactive fallout of the first large hydrogen bomb. It can be produced in milligram quantities by bombarding ^{239}Pu with neutrons to give ^{253}Es (half-life 20.47 days). Several other short-lived isotopes have also been synthesized.

Symbol: Es; m.p. 860 ± 30°C; b.p. and r.d. unknown; p.n. 99; most stable isotope ^{254}Es (half-life 276 days).

electrochemical equivalent Symbol: *z* The mass of an element released from a solution of its ions when a current of one ampere flows for one second during electrolysis (i.e. by one coulomb of electricity).

electrochemical series (activity series; electromotive series) A series giving the activities of metals for reactions that involve ions in solution. In decreasing order of activity, the series for the most commercially important metals is

K, Ba, Ca, Na, Mg, Al, Mn,
Zn, Cr, Fe, Cd, Co, Ni, Sn,
Pb, H, Cu, Hg, Ag, Pt, Au

Any member of the series will displace ions of a lower member from solution. For example, zinc metal will displace Cu^{2+} ions:

$$Zn(s) + Cu^{2+}(aq) \rightarrow Zn^{2+}(aq) + Cu(s)$$

Zinc has a greater tendency than copper to form positive ions in solution. Similarly, metals above hydrogen displace hydrogen from acids:

$$Zn + 2HCl \rightarrow ZnCl_2 + H_2$$

The series is based on ELECTRODE POTENTIALS, which measure the tendency of elements to form positive ions. The series is one of increasing electrode potential for half cells of the type $M^{n+}|M$. Thus, copper ($E^{\ominus}$ for $Cu^{2+}|Cu$ = + 0.34 V) is lower than zinc ($E^{\ominus}$ for $Zn^{2+}|Zn$ = –0.76V). The hydrogen half cell by definition has the value $E^{\ominus} = 0$.

electrochemistry The study of the formation and behavior of ions in solutions. It includes electrolysis and the generation of electricity by chemical reactions in cells.

electrochromatography *See* electrophoresis.

electrode Any part of an electrical device or system that emits or collects electrons or other charge carriers. An electrode may also be used to deflect charged particles by the action of the electrostatic field that it produces. *See also* half cell.

electrodeposition (electroplating) The process of depositing a layer of solid (metal) on an electrode by means of electrolysis. Positive ions in solution gain electrons at the cathode and are deposited as atoms. Copper, for instance, can be deposited on a metal cathode from an acidified copper sulfate solution. Electrodeposition allows metals to be plated with other, more decorative or corrosion-resistant metals, e.g. chromium on steel.

electrode potential (reduction potential) Symbol: *E* A measure of the tendency of an element to form ions in solution. For example, a metal in a solution containing M^+ ions may dissolve in the solution as M^+ ions; the metal then has an excess of electrons and the solution an excess of positive ions. Thus, the metal becomes negative with respect to the solution. Alternatively, the positive ions may gain electrons from the metal and be deposited as metal atoms. In this case, the metal becomes positively charged with respect to the solution. In either case, a potential difference is developed between solid and solution, and an equilibrium state will be reached at which further reaction is prevented. The equilibrium value of this potential difference would give an indication of the tendency to form aqueous ions.

It is not, however, possible to measure this value for an isolated half cell, because any measurement requires a circuit, which sets up another half cell in the solution. To overcome this limitation, electrode potentials are defined by comparison with a HYDROGEN ELECTRODE (a type of half cell), which is connected to the half cell under investigation by a salt bridge. The e.m.f. of the full cell can then be measured.

In referring to a given half cell the more reduced form is written on the right for a half-cell reaction. For the half cell $Cu^{2+}|Cu$, the half-cell reaction is a reduction:

$$Cu^{2+}(aq) + 2e^- \rightarrow Cu$$

The cell formed in comparison with a hydrogen electrode is:

$$Pt(s)H_2(g)|H^+(aq)|Cu^{2+}(aq)|Cu$$

The e.m.f. of this cell is +0.34 volt (V) measured under standard conditions. Thus, the standard electrode potential (symbol: $E^{\ominus}$) is +0.34 V for the half cell $Cu^{2+}|Cu$. The standard conditions are 1.0 molar solutions of all ionic species, standard pressure, and a temperature of 298 K.

Half cells can also be formed by a solution of two different ions (e.g. Fe^{2+} and Fe^{3+}). In such cases, a platinum electrode, a half cell employing a platinum plate, is used under standard conditions.

electrodialysis A method of removing ions from water by selective flow through membranes under the influence of an electric field. In the simplest arrangement a cell is divided into three compartments by two semipermeable membranes, one permeable to positive ions and the other permeable to negative ions. Electrodes are placed in the outer compartments of the cell – the positive electrode next to the membrane that allows negative ions to pass, and the negative electrode next to the membrane that allows positive ions to pass. In this arrangement, positive and negative ions pass through the membranes, leaving deionized water in the center compartment of the cell. In practice, an array of alternating membranes is used. Electrodialysis is used to make drinking water in areas where the available water supply is brackish. *See also* desalination.

electrolysis The production of chemical change by passing electric charge through certain conducting liquids known as electrolytes. Electrolytes may be solutions or molten salts, etc. The current is conducted by the migration of ions – positive ones (cations) to the cathode (negative electrode), and negative ones (anions) to the anode (positive electrode). Reactions take place at the electrodes via the transfer of electrons to or from them.

In the electrolysis of water (containing a small amount of acid to make it conduct adequately) hydrogen gas is given off at the cathode and oxygen is evolved at the anode. At the cathode the reaction is:

$$H^+ + e^- \rightarrow H$$
$$2H \rightarrow H_2(g)$$

At the anode:

$$OH^- \rightarrow e^- + OH$$
$$2OH \rightarrow H_2O + O$$
$$2O \rightarrow O_2(g)$$

In certain cases the electrode material may dissolve. For instance, in the electrolysis of copper(II) sulfate solution with copper electrodes, copper atoms of the anode dissolve as copper ions:

$$Cu \rightarrow 2e^- + Cu^{2+}$$

See Faraday's laws.

electrolyte A liquid containing positive and negative ions that conducts electricity by the flow of those charges. Electrolytes can be solutions of acids or metal salts ('ionic compounds'), usually in water. Alternatively they may be molten ionic compounds, which also allow ions to move freely. Liquid metals in which conduction is by free electrons rather than ions are not classified as electrolytes. *See also* electrolysis.

electrolytic Relating to the behavior or reactions of ions in solution.

electrolytic cell *See* cell; electrolysis.

ELECTROMAGNETIC SPECTRUM
(note: the figures are only approximate)

Radiation	*Wavelength (m)*	*Frequency (Hz)*
gamma radiation	-10^{-10}	$10^{19}-$
x-rays	$10^{-12}-10^{-9}$	$10^{17}-10^{20}$
ultraviolet radiation	$10^{-9}-10^{-7}$	$10^{15}-10^{17}$
visible radiation	$10^{-7}-10^{-6}$	$10^{14}-10^{15}$
infrared radiation	$10^{-6}-10^{-4}$	$10^{12}-10^{14}$
microwaves	$10^{-4}-1$	$10^{9}-10^{13}$
radio waves	$1-$	-10^{9}

electrolytic corrosion Corrosion by electrochemical reaction, e.g. rusting.

electrolytic refining A method of purifying metals by electrolysis. Copper is purified by making the impure metal the anode in an electrolytic cell containing an acidified copper sulfate electrolyte. The cathode is a thin strip of pure copper. The copper at the anode dissolves as Cu^{2+} ions and pure copper is deposited on the cathode. In this particular process, gold and silver are obtained as by-products, deposited as an *anode sludge* on the bottom of the cell.

electromagnetic radiation Energy waves propagated by electric and magnetic fields that oscillate at right angles to one another as they travel through space. Electromagnetic radiation forms a whole *electromagnetic spectrum*, depending on frequency, ranging from high-frequency radio waves to low-frequency gamma rays.

Electromagnetic radiation can be thought of as waves (*electromagnetic waves*) or as streams of photons. The frequency (ν) and wavelength (λ) are related by

$$\lambda\nu = c$$

where c is the speed of light. The energy carried depends on the frequency.

electromagnetic spectrum *See* electromagnetic radiation.

electromagnetic waves *See* electromagnetic radiation.

electromotive force (e.m.f.) The greatest potential difference that can be supplied by a source of electrical energy. The unit is the volt (V).

electromotive series *See* electrochemical series.

electron An elementary particle of negative charge ($-1.602\ 192 \times 10^{-19}$ coulomb) and rest mass $9.109\ 558 \times 10^{-31}$ kilogram. Electrons are present in all atoms in shells around the nucleus.

electron affinity Symbol: A The energy released when an atom (or molecule or group) gains an electron in the gas phase to form a negative ion. It is thus the energy of:

$$A + e^- \rightarrow A^-$$

A positive value of A (often in electronvolts) indicates that heat is given out. Often the molar enthalpy is given for this process of electron attachment (ΔH). Here the units are joules per mole ($J\ mol^{-1}$), and, by the usual convention, a negative value indicates that energy is released.

electron configuration *See* configuration.

electron-deficient compounds Compounds in which the number of electrons available for bonding is insufficient for the bonds to consist of conventional two-electron covalent bonds. In diborane, B_2H_6, for example, each boron atom has two terminal hydrogen atoms bound by conventional electron-pair bonds and in addition two hydrogen atoms bridging the boron atoms (B–H–B). In each bridge there are only two electrons for the bonding orbital. *See also* boron hydride; multicenter bond.

electron diffraction A technique used to help determine the structure of substances, principally the shapes of molecules in the gaseous phase. A beam of electrons directed through a gas at low pressure produces a series of concentric rings on a photographic plate. The dimensions of these rings are related to the interatomic distances in the molecules. *See also* x-ray diffraction.

electron donor *See* reduction.

electronegative Describing an atom or molecule that attracts electrons, forming negative ions. Examples of electronegative elements include the halogens (fluorine, chlorine etc.), which readily form negative ions, e.g. F^-, Cl^-, etc. *See also* electronegativity.

electronegativity A measure of the tendency of an atom in a molecule to attract electrons to itself. Elements to the right-hand side of the periodic table are strongly electronegative (values from 2.5 to 4); those on the left-hand side have low electronegativities (0.8–1.5) and are sometimes called ELECTROPOSITIVE elements. Different electronegativities of atoms in the same molecule give rise to polar bonds and sometimes to polar molecules.

As the concept of electronegativity is not precisely defined it cannot be precisely measured and several electronegativity scales exist. Although the actual values between these scales differ, they are in good relative agreement. *See also* electron affinity; ionization potential.

electronic energy level *See* energy level.

electronic spectra of molecules The spectra associated with transitions between the electronic states of molecules. These transitions correspond to the visible or ultraviolet regions of the electromagnetic spectrum. There are changes in vibrational and rotational energy when electronic transitions occur. This has the effect that there are spectral bands associated with changes in vibrational motion, with these bands having fine structure due to changes in rotational motion. Because electronic transitions are associated with changes in vibrational motion the corresponding spectra are sometimes called *vibrational spectra.* In contrast to the cases of rotational and vibrational spectroscopy, all molecules can give rise to electronic spectra. The electronic spectra of molecules are used to obtain information about energy levels in molecules, interatomic distances, dissociation energies of molecules, and force constants of chemical bonds.

electronic transition The demotion or promotion of an electron between electronic energy levels in an atom or molecule.

electron pair Two electrons in one orbital with opposing spins (*spin paired*), such as the electrons in a covalent bond or lone pair.

electron spin *See* spin.

electron spin resonance (ESR) A technique similar to that used in NUCLEAR MAGNETIC RESONANCE, but applied to unpaired electrons in a molecule rather than to the nuclei. It is a powerful means by which to study free radicals and transition-metal complexes.

electron-transfer reaction A chemical reaction that involves the transfer, addition, or removal of electrons. Many electron-transfer reactions involve complexes of transition metals. The rates of such reactions vary enormously and can be explained in terms of the way in which molecules of the solvent that start off solvating the reactants rearrange so as to solvate the products.

electronvolt Symbol: eV A unit of energy equal in SI derived units to 1.602192×10^{-19} joule. It is defined as the energy required to move an electron charge across a potential difference of one volt. It is often used to measure the kinetic energies of elementary particles or ions, or the ionization potentials of molecules.

electrophoresis The application of an electric field between two electrodes located at either side of a solution in order to cause charged particles of a colloid to move through the solution. The technique is used to separate and identify colloidal substances such as carbohydrates, proteins, and nucleic acids. Various experimental arrangements are used. One simple technique uses a strip of adsorbent paper soaked in a buffer solution with electrodes placed at two points on the paper. This technique is sometimes called *electrochromatography*. In *gel electrophoresis*, used to separate DNA fragments, the solution is replaced by a layer of gel.

electroplating The process of coating a solid surface with a layer of metal by means of electrolysis (i.e. by electrodeposition).

electropositive Describing an atom or molecule that tends to lose electrons, forming positive ions. Examples of electropositive elements include the alkali metals (lithium, sodium, etc.), which readily form positive ions, e.g. Li^+, Na^+, etc. *See also* electronegative.

electrovalent bond (ionic bond) A binding force between the ions in compounds in which the ions are formed by complete transfer of electrons from one element to another element or radical. For example, Na + Cl becomes $Na^+ + Cl^-$. The electrovalent bond arises from the excess of the net attractive force between the ions of opposite charge over the net repulsive force between ions of like charge. The magnitude of electrovalent interactions is of the order 10^2–10^3 kJ mol^{-1} and electrovalent compounds are generally solids with rigid lattices of closely packed ions. The strengths of electrovalent bonds vary with the reciprocal of the inter-ionic distances and are discussed in terms of lattice energies.

electrum **1.** A naturally occurring or artifically created alloy of gold and silver (containing up to 45% silver) that resembles pure gold in appearance.
2. A NICKEL-SILVER containing 52% copper, 26% nickel, and 22% zinc.

element A substance that cannot be chemically decomposed into more simple substances. The atoms of an element all have the same number of protons (given by the element's proton number) and thus the same number of electrons, which determines the element's chemical activity.

At present there are 114 reported chemical elements, although research is continuing to synthesize new ones. The elements from hydrogen (proton number 1) to uranium (92) all occur naturally on Earth, with the exception of technetium (43), which is produced artificially by particle bombardment. Neptunium (93), plutonium (94), americium (95) and curium (96) also occur naturally in very small quantities in uranium ores. Technetium and elements with proton numbers higher than 84 (polonium) are radioactive. Radioactive isotopes also exist for other elements, either naturally in small amounts or synthetically via particle bombardment. The elements with proton number higher than 92 are the *transuranic elements*. The transuranics are all synthesized. Thus, neptunium and plutonium can be made by neutron bombardment of uranium nuclei while the other transuranics are made by high-energy collision processes between nuclei. The higher proton number elements have been detected only in extremely minute quantities – in some cases, only a few atoms have been produced.

There has been some controversy about the naming of the higher synthetic elements because of disputes about their discovery. A long-standing problem concerned element-104 (rutherfordium) which was formerly also known by its Russian name of *kurchatovium*. More recent confusion has been caused by differences between names suggested by the International Union of Pure and Applied Chemistry (IUPAC) and the names suggested by the American Chemical Union (ACU).
The original IUPAC names (1994) were:

mendelevium (Md, 101)
nobelium (No, 102)
lawrencium (Lr, 103)
dubnium (Db, 104)
joliotium (Jl, 105)
rutherfordium (Rf, 106)
bohrium (Bh, 107)
hahnium (Hn, 108)
meitnerium (Mt, 109)

The ACU names were:

mendelevium (Md, 101)
nobelium (No, 102)
lawrencium (Lr, 103)
rutherfordium (Rf, 104)
hahnium (Ha, 105)
seaborgium (Sg, 106)
nielsbohrium (Ns, 107)
hassium (Hs, 108)
meitnerium (Mt, 109)

A compromise list of names was adopted by IUPAC in 1997:

mendelevium (Md, 101)
nobelium (No, 102)
lawrencium (Lr, 103)
rutherfordium (Rf, 104)
dubnium (Db, 105)
seaborgium (Sg, 106)
bohrium (Bh, 107)
hassium (Hs, 108)
meitnerium (Mt, 109)

These are the names used in this dictionary. Dubnium is named after Dubna in Russia, where much work has been done on synthetic elements. Hassium is the Latin name for the German state of Hesse, where the German synthetic element research group is located.

In addition to the above elements, five other elements have been reported with proton numbers 110, 111, 112, 114, and 116. Element 110 has been named darmstadtium (Ds, 110). So far the balance of these elements have not been named and have their temporary systematic IUPAC names (*see* unun-).

These are: unununium (Uuu, 111), ununbium (Uub, 112), ununquadium (Uuq, 114) and ununhexium (Uuh, 116).

elevation of boiling point A colligative property of solutions in which the boiling point of a solution is raised relative to that of the pure solvent. The elevation is directly proportional to the number of solute molecules introduced rather than to any specific aspect of the solute composition. The proportionality constant, k_B, is called the *boiling-point elevation constant* or sometimes the *ebullioscopic constant.* The relationship is

$$\Delta t = k_B C_m$$

where Δt is the rise in boiling point and C_m is the molal concentration; the units of k_B are kelvins kilograms moles^{-1} (K kg mol^{-1}). The property permits the measurement of the relative molecular mass of involatile solutes. *Compare* depression of freezing point.

Elinvar (*Trademark*) A steel alloy of chromium, iron, and nickel containing some tungsten and manganese. It is used for making hairsprings for clocks and watches because its elasticity does not vary with temperature.

elixir of life *See* alchemy.

elution The removal of an adsorbed substance in a chromatography column or ion-exchange column using a solvent (*eluent*), giving a solution called the *eluate.* The CHROMATOGRAPHY column can selectively adsorb one or more components from the mixture. To ensure efficient recovery of these components graded elution is used. The eluent is changed in a regular manner starting with a nonpolar solvent and gradually replacing it by a more polar one. This will wash the strongly polar components from the column.

emery A mineral of corundum (aluminum oxide, Al_2O_3) containing some magnetitie (Fe_3O_4), hematite (Fe_2O_3), or spinel ($MgAl_2O_4$). It is extremely hard and is used as an abrasive and polishing material.

e.m.f. *See* electromotive force.

emission spectrum *See* spectrum.

empirical formula The formula of a compound showing the simplest ratio of the atoms present. The empirical formula is the formula obtained by experimental analysis of a compound. It can be related to a molecular formula only if the relative molecular mass is known. For example P_2O_5 is the empirical formula of phosphorus(V) oxide although its molecular formula is P_4O_{10}. *Compare* molecular formula; structural formula.

emulsion A colloid in which a liquid phase (small droplets with a diameter range 10^{-5}–10^{-7} centimeter) is dispersed or suspended in a liquid medium. Emulsions are classed as lyophobic (solvent-repelling and generally unstable) or lyophilic (solvent-attracting and generally stable).

en *See* ethylenediamine.

enantiomer (enantiomorph) A compound whose structure is not superimposable on its mirror image: one of any pair of optical isomers. *See also* isomerism; optical activity.

enantiomorph *See* enantiomer.

enantiotropy The existence of different stable allotropes of an element at different temperatures; sulfur, for example, exhibits enantiotropy. The phase diagram for an enantiotropic element has a point at which all the allotropes can coexist in a stable equilibrium. At temperatures above or below this point, one of the allotropes will be more stable than the other(s). *See also* allotropy; monotropy.

enclosure compound *See* clathrate.

endothermic Describing a process in which heat is absorbed (i.e. heat flows from outside the system, or the temperature falls). The dissolving of a salt in water, for instance, is often an endothermic process. *Compare* exothermic.

end point *See* equivalence point; volumetric analysis.

energy Symbol: W A property of a system; a measure of its capacity to do work. Energy and work have the same SI derived unit: the joule (J). It is convenient to divide energy into *kinetic energy* (energy of motion) and *potential energy* ('stored' energy). Names are given to many different forms of energy depending on their source, i.e. chemical, electrical, nuclear, etc.; the only real difference lies in the system under discussion. For example, chemical energy is the kinetic and potential energies of electrons in a chemical compound.

energy level One of the discrete energies that an atom or molecule, for instance, can have according to quantum theory. Thus in an ATOM there are certain definite shells and orbitals that the electrons can be in, corresponding to definite *electronic energy levels* of the atom. Similarly, a vibrating or rotating molecule can have discrete vibrational and rotational energy levels.

energy profile A diagram that traces the changes in the energy of a system during the course of a reaction. Energy profiles are obtained by plotting the potential energy of the reacting particles against the *reaction coordinate*, i.e. the pathway for which the energy is a minimum. To obtain the reaction coordinate the energy of the total interacting system is plotted against position for the molecules.

enthalpy Symbol: H The sum of the internal energy (U) and the product of pressure (p) and volume (V) of a system:

$$H = U + pV$$

In a chemical reaction carried out at constant pressure, the change in enthalpy measured is the internal energy change plus the work done by the volume change:

$$\Delta H = \Delta U + p\Delta V$$

In SI units enthalpy is measured in joules (J) or kilojoules (kJ).

entropy Symbol: S In any system that undergoes a reversible change, the change of entropy is defined as the heat Q absorbed divided by the thermodynamic temperature T:

$$\Delta S = \Delta Q/T$$

A given system is said to have a certain entropy, although absolute entropies are seldom used: it is rather the change in entropy that is important, particularly as the entropy of a system measures the availability of energy to do work.

In any real (i.e. irreversible) change in a closed system entropy always increases. Although the total energy of the system has not changed, according to the first law of THERMODYNAMICS, the available energy is less as a consequence of the second law of thermodynamics.

The concept of entropy has been widened to take in the general idea of disorder – the higher the entropy, the more disordered the system. For instance, a chemical reaction involving polymerization may well have a decrease in entropy because there is a change to a more ordered system. The 'thermal' definition of entropy is actually a special case of this idea of disorder, because it explains how transferred energy is distributed among particles of matter.

Epsom salt *See* magnesium sulfate.

equation *See* chemical equation.

equation of state An equation that interrelates the pressure, temperature, and volume of a system, such as a gas. The equation for an ideal gas and the VAN DER WAALS EQUATION are examples. *See* gas laws.

equilibrium In a reversible chemical reaction:

$$A + B \rightleftharpoons C + D$$

The reactants are forming the products:

$$A + B \rightarrow C + D$$

which also react to give the original reactants:

$$C + D \rightarrow A + B$$

The concentrations of A, B, C, and D change with time until a state is reached at which both reactions are taking place at the same rate. The concentrations (or pressures) of the components are then constant – the system is said to be in a state of *chemical equilibrium*. Note that the equilibrium is a dynamic one; the reactions still take place but at equal rates. The relative proportions of the components determine the 'position' of the equilibrium, which may be changed by changing the conditions (e.g. temperature or pressure).

equilibrium constant Symbol: Kc, Kp In a chemical equilibrium of the type

$$xA + yB \rightleftharpoons zC + wD$$

The expression:

$$[A]^x[B]^y/[C]^z[D]^w$$

where the square brackets indicate concentrations, is a constant (K_c) when the system is at equilibrium. K_c is the equilibrium constant of the given reaction; its units depend on the stoichiometry of the reaction. For gas reactions, pressures are often used instead of concentration. The equilibrium constant is then K_p, where $K_p = K_c^n$. Here n is the number of moles of product minus the number of moles of reactant; for instance, in

$$3H_2 + N_2 \rightleftharpoons 2NH_3$$

n is $2 - (1 + 3) = -2$.

equipartition of energy The principle that the total energy of a molecule is, on average, equally distributed among the available DEGREES OF FREEDOM. It is only approximately true in most cases.

equivalence point The point in a titration at which the reactants have been added in equivalent proportions, so that there is no excess of either. It differs slightly from the *end point*, which is the observed point of complete reaction, because of the effect of the indicator, errors, etc.

equivalent proportions, law of (law of reciprocal proportions) When two chemical elements both form compounds with a third element, a compound of the first two elements will contain each in the same relative proportions in which they exist in compounds with the third element. For example, the mass ratio of carbon to hydrogen in methane (CH_4) is 12:4; the mass ratio of oxygen to hydrogen in water (H_2O) is 16:2. In carbon monoxide (CO), the ratio of carbon to oxygen is 12:16.

equivalent weight A measure of 'combining power' formerly used in calculations for chemical reactions. The equivalent weight of an element is the number of grams that could combine with or displace one gram of hydrogen (or 8 grams of oxygen or 35.5 grams of chlorine). It is the relative atomic mass (atomic weight) divided by the valence. For a compound the equivalent weight depends on the reaction considered. An acid, for instance, in acid-base reactions has an equivalent weight equal to its molecular weight divided by the number of acidic hydrogen atoms.

erbium A soft, malleable, silvery element of the lanthanoid series of metals. It occurs in association with other lanthanoids in such minerals as apatite, gadolinite, monazite, and bastnaesite. It does not oxidize as easily as other lanthanoids. It is used to make vanadium alloys more workable, to make pink pigments for paints and ceramics, and as a signal amplifier in fiber-optic cables.

Symbol: Er; m.p. 1529°C; b.p. 2863°C; r.d. 9.066 (25°C); p.n. 68; most common isotope ^{166}Er; r.a.m. 167.26.

Erlenmeyer flask A conical glass laboratory flask with a narrow neck and wide bottom.

ESR *See* electron spin resonance.

ethanoate (acetate; CH_3COO^-) A salt or ester of ETHANOIC ACID (acetic acid).

ethanoic acid (acetic acid; CH_3COOH) A colorless viscous liquid or glassy solid organic acid with a pungent odor. It is the acid in vinegar. Below 16.6°C it solidifies to a glassy solid known as *glacial ethanoic acid*. It is made by the oxidation of ethanol or butane, or by the bacterial fermentation of beer or wine, and is used as a food preservative and for making polymers.

ethanol (ethyl alcohol; alcohol; C_2H_5OH) A colorless volatile liquid alcohol. Ethanol occurs in intoxicating drinks, as the result of the anaerobic fermentation of simple sugars:

$$C_6H_{12}O_6 \rightarrow 2C_2H_5OH + 2CO_2$$

Apart from its use in drinks, alcohol is used as a solvent and to form ethanal (acetaldehyde). Formerly, the main source was by fermentation of molasses, but now catalytic hydration of ethene is used to manufacture industrial ethanol.

ethene (ethylene; C_2H_4) A gaseous alkene. Ethene is not normally present in the gaseous fraction of crude oil but can be obtained from heavier fractions by catalytic cracking. This is the principal industrial source. The compound is important as a starting material in the organic-chemicals industry (e.g. in the manufacture of ethanol) and as the starting material for the production of polyethene. *See also* Zeise's salt.

ethoxylate *See* detergents.

ethyl alcohol *See* ethanol.

ethylene *See* ethene.

ethylenediamine An organic compound, $H_2NCH_2CH_2NH_2$. It is important in inorganic chemistry because it functions as a bidentantate ligand, coordinating to a metal ion by the lone pairs on the two nitrogen atoms. In the names of complexes it is given the abbreviation *en*.

ethylenediaminetetraacetic acid *See* edta.

ethyne (acetylene; C_2H_2) An unstable colorless, gaseous, organic compound. Traditionally ethyne was made by the action of water on calcium dicarbide, and the gas has found use in oxy-acetylene welding torches, since its combustion with oxygen produces a flame of very high temperature. It is also important in the organic chemicals industry for the production of vinyl compounds.

ethynide *See* carbide.

eudiometer An apparatus for the volumetric analysis of gases. It often consists of a sealable graduated glass tube housing wires that are used to spark a reaction between gas mixtures.

europium A silvery-white element of the lanthanoid series of metals. It occurs in association with other lanthanoids in such minerals as monazite, bastnaesite, gadolinite, and samarskite. It is highly reactive and will ignite in air when heated above 150°C. Its main use is in a mixture of europium and yttrium oxides widely employed as the red phosphor in television screens. The metal is used in some superconducting alloys.

Symbol: Eu; m.p. 822°C; b.p. 1597°C; r.d. 5.23 (25°C); p.n. 63; most common isotope ^{153}Eu; r.a.m. 151.965.

eutectic A mixture of two substances in such proportions that no other mixture of the same substances has a lower freezing point. If a solution containing the proportions of the eutectic is cooled no solid is deposited until the freezing point of the eutectic is reached, when two solid phases are simultaneously deposited in the same proportions as the mixture; these two solid phases form the eutectic. If one of the substances is water the eutectic can also be referred to as a *cryohydrate*.

evaporation **1.** A change of state from liquid to gas (or vapor). Evaporation can take place at any temperature, although the rate increases with temperature. Evaporation occurs because some molecules in the liquid, if they are near the surface and moving in the right direction, have enough energy to escape into the gas phase. Because these are the molecules with higher kinetic energies, the liquid is cooled as a result.
2. A change from solid to vapor, especially occurring at high temperatures close to the melting point of the solid. Thin films of metal can be evaporated onto a surface in this way.

exa- Symbol: E A prefix used with SI units, denoting 10^{18}.

excitation The process by which an atom, molecule, etc. is elevated from its ground to an excited state.

excitation energy The energy required to change an atom, molecule, etc. from one quantum state to a state with a higher energy. The excitation energy (sometimes called *excitation potential*) is the difference between two energy levels of the system.

excitation potential *See* excitation energy.

excited state The state of an atom, molecule, or other system when it has an energy greater than that of its GROUND STATE.

exclusion principle *See* atom.

exothermic Denoting a chemical reaction in which heat is evolved (i.e. heat flows from the system or the temperature rises). Combustion is an example of an exothermic process. *Compare* endothermic.

explosive A substance or mixture that can rapidly decompose upon detonation, producing large amounts of heat and gases. The three most important classes of explosives are:
1. *Propellants*, which burn steadily, and are used as rocket fuels,
2. *Initiators*, which are very sensitive and are used in small amounts to detonate less sensitive explosives, and
3. *High explosives*, which need an initiator but are very powerful.

Eyring, Henry (1901–81) American chemist. Eyring was a prolific chemist whose main contribution was to the theory of chemical reactions. In the late 1920s and early 1930s he calculated a number of potential energy surfaces in terms of quantum mechanics. He then expressed the temperature dependence of the rate of chemical reactions in terms of the parameters of potential energy surfaces. Eyring gave an account of this work in the book *The Theory of Rate Processes* (1941) which he wrote with Samuel Glasstone and Keith Laidler.

F

face-centered cubic crystal (f.c.c.) (cubic close packing) A close-packed crystal structure made up of layers in which each atom, ion, or molecule is surrounded by six others arranged hexagonally. The layers are packed one on top of the other, with the second layer fitting into the recesses left in the first layer, and the third layer fitting into yet a different set of holes formed by the second layer. If the layers are designated A, B, C, the packing is ABCABC. This is in contrast to HEXAGONAL CLOSE PACKED CRYSTAL, in which the layers are arranged ABABAB. Copper and aluminum have face-centered cubic crystal structures. *See* close packing.

fac-isomer *See* isomerism.

Fahrenheit scale A temperature scale in which the temperature of pure melting ice at standard pressure is fixed at 32° and the temperature of pure boiling water at standard pressure is fixed at 212°. The scale is not used for scientific purposes. To convert from degrees Fahrenheit (°F) to degrees Celsius (°C) the formula

$$°F = 1.8(°C) + 32$$

is used. *See also* Celsius scale; temperature scale.

Fajan's rules Rules that deal with variations in the degree of covalent character in ionic compounds in terms of polarization effects. Decrease in size and increase in charge for cations is regarded as making them more polarizing, while for anions an increase in both charge and size makes them more polarizable. Thus covalent character is said to increase with increasing polarizing power of the cation and/or increasing polarizability of the anion. These were originally stated as rules one and two. Fajan's third rule states that covalent character is greater for cations with a nonrare gas configuration than for those with a complete octet, charge and size being approximately equal. For example, Cu^+ is more polarizing than Na^+ (ions of approximately the same size) because the d electrons around Cu^+ do not shield the nucleus as effectively as the complete octet in Na^+.

farad Symbol: F The SI derived unit of capacitance. When the plates of a capacitor are charged by one coulomb and there is a potential difference of one volt between them, the capacitor is said to have a capacitance of one farad. One farad = 1 coulomb volt^{-1} (CV^{-1}).

Faraday, Michael (1791–1867) English physicist and chemist. Faraday was responsible for a remarkably large number of important discoveries in chemistry and physics. His early work was mostly devoted to chemistry. In 1820 he prepared chlorides of carbon. In 1835 he discovered benzene. In the 1830s Faraday discovered what are now known as FARADAY'S LAWS of electrolysis. He also introduced the terms anode, cathode, cation, anion, electrode, and electrolyte. Faraday made fundamental contributions to the study of electricity and magnetism. He constructed the first electric motor and discovered electromagnetic induction. He also developed the concept of electric and magnetic fields. His investigations of matter in magnetic fields led to the discovery of diamagnetism and paramagnetism. In 1845 he discovered *Faraday rotation*, i.e. the rotation of the plane of polarization when polarized light passes through matter in a magnetic field.

faraday Symbol: *F* A non-SI unit of electric charge equal to the charge required to discharge one mole of a singly-charged ion. In SI units one faraday is equal to 96.48670 kilocoulombs (kC). *See* Faraday's laws.

Faraday constant *See* Faraday's laws.

Faraday's laws (of electrolysis) Two laws resulting from the work of Michael Faraday on electrolysis:

1. The amount of chemical change produced is proportional to the electric charge passed.
2. The amount of chemical change produced by a given charge depends on the ion concerned.

More strictly it is proportional to the relative ionic mass of the ion, and the charge on it. Thus the charge Q needed to deposit m grams of an ion of relative ionic mass M carrying a charge Z is given by:

$$Q = FmZ/M$$

F, the *Faraday constant*, has a value of one faraday, i.e. 9.648670×10^4 coulombs.

f-block elements The block of elements in the periodic table that have two outer s electrons and an incomplete penultimate subshell of f electrons. The f block consists of the lanthanoids (cerium to lutetium) and the actinoids (thorium to lawrencium).

f.c.c. *See* face-centered cubic crystal.

feldspar (felspar) A member of a large group of very abundant crystalline igneous rocks, mainly silicates of aluminum with calcium, potassium, and sodium. The alkali feldspars, e.g. orthoclase, microcline, have mainly potassium with a little sodium and no calcium; the plagioclase feldspars, e.g. albite, anorthite, labradorite, vary from sodium-only to calcium-only, with little or no potassium. Feldspars are used in making ceramics, enamels, and glazes.

femto- Symbol: f A prefix used with SI units denoting 10^{-15}. For example, 1 femtometer (fm) = 10^{-15} meter (m).

femtochemistry The study of atomic and molecular processes, particularly chemical reactions, that take place on timescales of about a femtosecond (10^{-15}s). Femtochemistry became possible with the construction of lasers that could be pulsed in femtoseconds. Femtochemistry can be regarded as very high speed 'photography' of chemical reactions. It has enabled entities that exist only for a short time, such as activated complexes, to be studied. The detailed mechanisms of many chemical reactions have been investigated in this way, notably by the Egyptian-born American chemist Ahmed Zewail.

fermi A non-SI unit of length equal to one femtometer (10^{-15} meter) in SI units. It was formerly used in atomic and nuclear physics.

fermium A highly radioactive transuranic element of the actinoid series, not found naturally on Earth. It was discovered in the fallout from the first hydrogen bomb. It can be produced in very small quantities by bombarding ^{239}Pu with neutrons to give ^{253}Fm (half-life 3 days). Several other short-lived isotopes have also been synthesized.

Symbol: Fm; m.p., b.p. and r.d. unknown; p.n. 100; most stable isotope ^{257}Fm (half-life 100.5 days).

ferric chloride *See* iron(III) chloride.

ferric oxide *See* iron(III) oxide.

ferric sulfate *See* iron(III) sulfate.

ferricyanide *See* cyanoferrate.

ferrite A nonconducting, magnetic, metal oxide ceramic having the general formula $MO.Fe_2O_3$, where M is a divalent transition metal such as cobalt, nickel, or zinc. Ferrites are used to make powerful magnets for use in high-frequency electronic devices such as radar sets.

ferrocene ($Fe(C_5H_5)_2$) An orange crystalline solid. It is an example of a sandwich compound, in which an iron(II) ion is coordinated to two cyclopentadienyl ions. The bonding involves overlap of d orbitals on the iron with the pi electrons in the cyclopentadienyl rings. The compound melts at 173°C, is soluble in organic solvents, and can undergo substitution reactions on the rings. It can be oxidized to the blue cation $(C_5H_5)_2Fe^+$. The systematic name is di-π-cyclopentadienyl iron(II).

ferrocyanide *See* cyanoferrate.

ferromagnetism *See* magnetism.

ferrosoferric oxide *See* triiron tetroxide.

ferrous chloride *See* iron(II) chloride.

ferrous oxide *See* iron(II) oxide.

ferrous sulfate *See* iron(II) sulfate.

fertilizer A substance added to soil to increase its fertility. Artificial fertilizers generally consist of a mixture of chemicals designed to chiefly supply three main nutrients: nitrogen (N), phosphorus (P), and potassium (K), and are consequently known as PKN fertilizers. Typical chemicals used include ammonium nitrate, ammonium phosphate, ammonium sulfate, and potassium carbonate.

Fick's law A law describing the diffusion that occurs when solutions of different concentrations come into contact, with molecules moving from regions of higher concentration to regions of lower concentration. Fick's law states that the rate of diffusion dn/dt, called the *diffusive flux* and denoted J, across an area A is given by: $dn/dt = J = -DA\partial c/\partial x$, where D is a constant called the *diffusion constant*, $\partial c/\partial x$ is the concentration gradient of the solute and dn/dt is the amount of solute crossing the area A per unit time. D is constant for a specific solute and solvent at a specific temperature. Fick's law was formulated by the German physiologist Adolf Eugen Fick (1829–1901) in 1855.

filler A solid material used to modify the physical properties or reduce the cost of synthetic compounds, such as rubbers, plastics, paints, and resins. Slate powder, glass fiber, mica, and cotton are all used as fillers.

filter *See* filtration.

filter pump A type of vacuum pump in which a jet of water forced through a nozzle carries air molecules out of the system. Filter pumps cannot produce pressures below the vapor pressure of water. They are used in the laboratory for vacuum filtration, distillation, and similar techniques requiring a low-grade vacuum.

filtrate *See* filtration.

filtration The process of removing suspended particles from a fluid by passing or forcing the fluid through a porous material (the *filter*). The fluid that passes through the filter is the *filtrate*. In laboratory filtration, filter paper or sintered glass is commonly used.

fine structure Closely spaced lines seen at high resolution in a spectral line or band. Fine structure may be caused by vibration of the molecules or by electron spin. *Hyperfine structure*, seen at very high resolution, is caused by the atomic nucleus affecting the possible energy levels of the atom.

fireclay A refractory clay containing high proportions of alumina (aluminum oxide) and silica (silicon(IV) oxide), used for making firebricks and furnace linings.

firedamp Methane and other combustible gases occurring in coal mines. Explosions or fires can occur if quantities of these gases accumulate. The deoxygenated, suffocating air left after an explosion or fire, consisting chiefly of carbon dioxide and nitrogen, is called *blackdamp*. *Afterdamp* is the poisonous carbon monoxide formed.

first-order reaction A reaction in which the rate of reaction is proportional to the concentration of one of the reacting substances. The concentration of this reacting substance is raised to the power one; i.e. rate = k[A]. For example, the decomposition of hydrogen peroxide is a first-order reaction,

$$\text{rate} = k[H_2O_2]$$

Similarly the rate of decay of radioactive material is a first-order reaction,

$$\text{rate} = k[\text{radioactive material}]$$

For a first-order reaction, the time for a definite fraction of the reactant to be consumed is independent of the original concentration. The units of k, the RATE CONSTANT, are s^{-1}.

Fischer, Ernst Otto (1918–94) German inorganic chemist. Fischer is noted for his work on inorganic complexes. In 1951 two chemists, T. Kealy and P. Pauson, were attempting to join two five-carbon (cyclopentadiene) rings together and discovered a compound $C_5H_5FeC_5H_5$, which they proposed had an iron atom joined to a carbon atom on each ring. Fischer, on reflection, considered such a structure inadequate for he failed to see how it could provide sufficient stability with its carbon–iron–carbon bonds. The British chemist Geoffrey WILKINSON suggested a more novel structure in which the iron atom was sandwiched between two parallel rings and thus formed bonds with the electrons in the rings, rather than with individual carbon atoms. Compounds of this type are called SANDWICH COMPOUNDS. By careful x-ray analysis Fischer confirmed the proposed structure of FERROCENE, as the compound was called, and for this work shared the Nobel Prize for chemistry with Wilkinson in 1973.

fixation of nitrogen *See* nitrogen fixation.

fixing, photographic A stage in the development of exposed photographic film or paper in which unexposed silver halides in the emulsion are dissolved away. The fixing bath typically contains a solution of ammonium or sodium thiosulfate(IV), and is employed after the initial development (chemical reduction) of the latent silver image.

FLAME TEST COLORS

Element	*Flame color*
barium	green
calcium	brick red
lithium	crimson
potassium	pale lilac
sodium	yellow
strontium	red

flame test A preliminary test in qualitative analysis in which a small sample of a chemical is introduced into a nonluminous Bunsen-burner flame on a clean platinum wire. The flame vaporizes part of the sample and excites some of the atoms, which emit light at wavelengths characteristic of the metallic elements in the sample. Thus the observation of certain colors in the flame indicates the presence of these elements. The same principles are applied in the modern instrumental method of spectrographic analysis.

flare stack A special chimney in an oil refinery, rig, or other chemical plant at the top of which unwanted gases are burnt.

flash photolysis A technique for investigating free radicals in gases. The gas is held at low pressure in a long glass or quartz tube, and an absorption spectrum taken using a beam of light passing down the tube. The gas can be subjected to a very brief intense flash of light from a lamp outside the tube, producing free radicals, which are identified by their spectra. Measurements of the intensity of spectral lines with time can be made using an oscilloscope, and the kinetics of very fast reactions can thus be investigated.

flash point The lowest temperature at which vapor given off by a flammable liquid will be sufficient to momentarily sustain a flame in the presence of a spark.

flocculation The combining of the particles of a finely divided precipitate, such as a colloid, into *flocculent* clumps that sink and are easier to filter off. *See* coagulation.

flocculent *See* flocculation.

flotation *See* froth flotation.

fluid A state of matter that is not a solid, i.e. a liquid or a gas. All fluids can flow, and the resistance they pose to such flow is called VISCOSITY.

fluidization The suspension of a finely-divided solid, e.g. a catalyst, in an upward-flowing liquid or gas. This suspension mimics many properties of liquids, including allowing the solids to 'float' in it and achieve a uniform temperature with the fluid medium. Fluidized beds so prepared are used industrially to carry out catalytic reactions and to minimize the production of pollutants from coal-fired furnaces.

fluorescein A fluorescent dye used as an absorption indicator and water flow marker.

fluorescence The absorption of energy by atoms, molecules, etc., followed by immediate emission of electromagnetic radiation as the particles make transitions to lower energy states. *Compare* phosphorescence.

fluoridation The introduction of small quantities of fluoride compounds, e.g. sodium fluoride (NaF), into the water supply as a public-health measure to reduce the incidence of tooth decay.

fluoride *See* halide.

fluorination *See* halogenation.

fluorine A slightly greenish-yellow, highly reactive, poisonous, gaseous element belonging to the halogens; i.e. group 17 (formerly VIIA) of the periodic table, of which it is the first element. It occurs notably as fluorite (CaF_2) and cryolite (Na_3AlF_3) but traces of fluorine are also widely distributed with other minerals. It is slightly more abundant than chlorine, accounting for about 0.065% of the Earth's crust. Fluorine's high reactivity delayed its isolation until 1886, and means it is never found as a free element. Fluorine is now prepared by electrolysis of a molten mixture of potassium fluoride and hydrogen fluoride electrolytes, using copper or steel apparatus. Its preparation by chemical methods is impossible.

Fluorine compounds are used in the steel industry, glass and enamels, uranium processing, aluminum production, and in a range of fine organic chemicals. Fluorine is the most reactive and most electronegative element known; in fact it reacts directly with almost all the elements, including the rare gas xenon. As the most powerful oxidizing agent known, it has the pronounced characteristic of bringing out the higher oxidation states when combined with other elements. Ionic fluorides contain the small F^- ion, which is very similar in reactivity to the O^{2-} ion. With the more electronegative elements, fluorine forms an extensive range of compounds, e.g. SF_6, NF_3, and PF_5, in which the bonding is essentially covalent.

Fluorine reacts explosively with hydrogen, even in the dark, to form hydrogen fluoride, HF, which polymerizes as a result of hydrogen bonding and has a boiling point very much *higher* than that of HCl (HF b.p. 19°C; HCl b.p. –85°C). Unlike the other halogens, fluorine does not form higher oxides or oxyacids; oxygen difluoride in fact reacts with water to give hydrogen fluoride.

Fluorine exhibits a strong electron withdrawing effect on adjacent bonds, thus CF_3COOH is a strong acid whereas CH_3COOH is not. Although the coordination number seldom exceeds one there are cases in which fluorine bridging occurs, e.g. $(SbF_5)_n$. The element is sufficiently reactive to combine directly with the rare gas xenon at 400°C to form XeF_2 and XeF_4. At 600°C and high pressure it will even form XeF_6.

Fluorine and hydrogen fluoride are extremely dangerous and should be used only in purpose-built apparatus; gloves and face shields should be used when working with hydrofluoric acid, and accidental exposure should be treated as a hospital emergency.

Symbol: F; m.p. –219.62°C; b.p. –188.14°C; d. 1.696 kg m^{-3} (0°C); p.n. 9; r.a.m. 18.99840.32.

fluorite (fluorspar) *See* calcium fluoride.

fluorite structure *See* calcium-fluoride structure.

fluorspar *See* calcium fluoride.

flux 1. A substance used to keep metal surfaces free of oxide in soldering. *See* solder.
2. A substance used in smelting metals to react with silicates and other impurities and form a low-melting slag.

fluxional molecule A molecule in which the constituent atoms change their relative positions so quickly at room temperature that the normal concept of structure is inadequate; i.e. no specific structure exists for longer than about 10^{-2} second and the relative positions become indistinguishable. For example ClF_3 at –60°C has a distinct 'T' shape but at room temperature the fluorine atoms are visualized as moving rapidly over the surface of the chlorine atom in a state of exchange and are effectively identical.

formula (chemical formula) A representation of a chemical compound using symbols for the atoms and subscript numbers to show the numbers of atoms present. *See* empirical formula; general formula; molecular formula; structural formula.

fraction *See* fractional distillation.

fractional crystallization Crystallization of one component from a mixture in solution. When two or more substances are present in a liquid (or in solution), on gradual cooling to a lower temperature one substance will preferentially form crystals, leaving the other substance in the liquid (or dissolved) state. Fractional crystallization can thus be used to purify or separate substances.

fractional distillation (fractionation) Distillation in which a mixture of liquids is separated into components (*fractions*) depending on their boiling points. A common technique is to use a long vertical column (*fractionating column*) containing an inert material (e.g. glass beads) attached to the vessel containing the mixture. There is a decreasing temperature gradient from the bottom to the top of the column. Vapor rises in the column until it reaches a part at which it condenses back to liquid, and runs back down the column. A steady state is reached at which the more volatile components are concentrated near the top of the column and the less volatile components are near the bottom. It is possible to draw off fractions at points on the column. On an industrial scale, large towers containing perforated trays are used for extracting fractions in petroleum refining.

fractionating column *See* fractional distillation.

fractionation *See* fractional distillation.

francium A radioactive element of the alkali-metal group, the seventh element of group 1 (formerly IA) of the periodic table. It is found on Earth only as a short-lived product of radioactive decay, occurring in uranium ores in minute quantities. At least 22 isotopes are known.

Symbol: Fr; m.p. 27°C; b.p. 677°C; r.d. 2.4°; p.n. 87; most stable isotope ^{223}Fr (half-life 21.8 minutes).

Frasch process *See* sulfur.

free electron An electron that can move from one atom or molecule to another under the influence of an applied electric field. Movement of free electrons in a conductor constitutes an electric current. Free electrons are also involved in conducting heat and in METALLIC BONDS.

free energy A measure of the ability of a system to do useful work. *See* Gibbs function; Helmholtz function.

free radical An atom or group of atoms with a single unpaired electron. Free radicals are produced by breaking a covalent bond; for example:

$$CH_3Cl \rightarrow CH_3\bullet + Cl\bullet$$

They are often formed in light-induced reactions and as a result of chemical decomposition at high temperatures. Free radicals are extremely reactive and can be stabilized and isolated only under special conditions.

freezing The process by which a liquid is converted into a solid by cooling; the reverse of melting.

freezing mixtures Two or more substances mixed together to produce a temperature below the freezing point of water (0°C). A mixture of sodium chloride and ice in water, for example, can yield a temperature as low as –20°C.

freezing point The temperature at which a liquid is in equilibrium with its solid phase at standard pressure and below which the liquid freezes or solidifies. This temperature is always the same for a particular pure liquid and is numerically equal to the melting point of the solid.

freezing point constant *See* depression of freezing point.

French chalk *See* talc.

Frenkel defect *see* defect.

frontier orbital theory A theory of the chemical and spectroscopic properties of molecules that emphasizes the *frontier orbitals*, i.e. the *highest occupied molecular orbital* (HOMO) and the *lowest unoccupied molecular orbital* (LUMO). The theory was developed by the Japanese chemist Kenichi Fukui (1919–98) in the 1950s. Frontier orbital theory has sometimes been used in conjunction with the WOODWARD–HOFFMANN RULES.

froth flotation (flotation) An industrial technique for separating the required parts of ores from unwanted gangue. The mineral is pulverized and mixed with water, to which a frothing agent is added. The mixture is aerated, after which particles of one constituent are carried to the surface by bubbles of air, which adhere to it preferentially. The froth is then skimmed off. Various additives are also used to modify the surface properties of the mineral particles (e.g. to increase the adherence of air bubbles).

fuel cell A type of cell in which fuel is converted directly into electricity. In one form, hydrogen gas and oxygen gas are fed to the surfaces of two separately compartmentalized porous nickel electrodes immersed in a potassium hydroxide solution. The oxygen reacts to form hydroxyl (OH^-) ions, which it releases into the solution, leaving a positive charge on one electrode. The hydrogen reacts with the OH^- ions in the solution to form water, giving up electrons to leave a negative charge on the other electrode. Large fuel cells can generate tens of amperes. Usually the e.m.f. is about 0.9 volt and the efficiency around 60%.

fullerene *See* buckminsterfullerene.

fullerite *See* buckminsterfullerene.

fuller's earth A natural clay, the predominate mineral in bentonite, used as an absorbent and decolorizing agent, and as a drilling mud and industrial catalyst.

fuming sulfuric acid *See* oleum.

fundamental units The BASE UNITS of a measurement system, especially those that deal with force and motion. In most systems these are the units for length, mass, and time, plus at least one electrical unit. The fundamental units in the SI are considered to be the meter, the kilogram, the second, and the ampere.

fused Describing a solid that has been melted and solidified into a single mass. Fused silica, for example, is produced by melting sand.

fused ring *See* ring.

fusion Melting.

G

gadolinium A ductile malleable silvery element of the lanthanoid series of metals. It occurs in association with other lanthanoids in such minerals as gaolinite, bastnaesite, monazite, and xenotime. It is also a product of nuclear fission and can be found in association with uranium ores. Gadolinium is used in alloys and magnets, and in electronic components designed to withstand very high temperatures. Two isotopes, gadolinium-155 and gadolinium-157, are the best neutron absorbers known and thus find use in the nuclear industry.

Symbol: Gd; m.p. 1313°C; b.p. 3266°C; r.d. 7.9 (25°C); p.n. 64; most common isotope ^{158}Gd; r.a.m. 157.25.

galena (lead glance) A mineral form of lead(II) sulfide (PbS), usually occurring as grayish-blue cubes or octahedrons. It is the principal ore of lead.

gallium A soft silvery low-melting metallic element belonging to group 13 (formerly IIIA) of the periodic table. It is found in minute quantities in several ores, including sphalerite (ZnS), bauxite ($Al_2O_3.H_2O$), and kaolinite ($Al_2(OH)_4$-$S_{12}O_5$. Gallium is used in low-melting alloys, high-temperature thermometers, and as a doping impurity in semiconductors. Gallium arsenide is a semiconductor used in light-emitting diodes and in microwave apparatus.

Symbol: Ga; m.p. 29.78°C; b.p. 2403°C; r.d. 5.907 (solid at 20°C), 6.114 (liquid); p.n. 31; most common isotope $^{69}6a$; r.a.m. 69.723.

galvanic cell *See* cell.

galvanizing The process by which steel is coated with zinc by dipping it in a bath of molten zinc, or by electrodeposition. The zinc is preferentially attacked by carbonic acid (H_2CO_3), thus protecting the steel by forming zinc carbonates.

gamma radiation A form of ELECTROMAGNETIC RADIATION emitted by changes in the nuclei of atoms. Gamma rays have very high frequencies and thus short wavelength. Gamma radiation shows particle properties much more often than wave properties. The radiant energy (W) of a gamma photon, given by

$$W = hv$$

where h is the Planck constant and v the velocity, can be very high. A gamma photon of 10^{24} hertz (Hz) has an energy of 6.6×10^{-10} joule (J).

gamma rays Streams of gamma radiation.

gangue The waste rock or other undesirable material occurring in an ore.

gas The state of matter in which forces of attraction between the particles of a substance are small. The particles have freedom of movement and gases therefore have no fixed shape or volume. The atoms and molecules of a gas are in a continual state of motion and are continually colliding with each other and with the walls of any containing vessel. These collisions create the pressure of a gas.

gas chromatography A technique widely used for the separation and analysis of mixtures. Gas chromatography employs a column packed with either a solid stationary phase (*gas-solid chromatography* or *GSC*) or a solid coated with a non-volatile liquid (*gas-liquid chromatography* or *GLC*). The whole column is placed in a thermostatically controlled heating jacket. A volatile sample is introduced into the column using a syringe, and an unreactive carrier gas, such as nitrogen, passed through it. The components of the sample will be carried along in this mobile phase. However, some of the components will cling more readily to the stationary phase than others, either because they become attached to the solid surface or because they dissolve in the liquid. The time taken for different components to pass through the column is characteristic and can be used to identify them. The emergent sample is passed through a detector, which registers the presence of the different components in the carrier gas.

Two types of detector are in common use: the katharometer, which measures changes in thermal conductivity, and the flame-ionization detector, which turns the volatile components into ions and registers the change in electrical conductivity.

gas constant (universal gas constant) Symbol: R The universal constant 8.31451 J mol^{-1} K^{-1} appearing in the equation of state for an ideal gas. *See* gas laws.

gaseous diffusion separation A technique for the separation of gases that relies on their slightly different atomic masses. For example, the isotopes of uranium (^{235}U and ^{238}U) can be separated by first preparing gaseous uranium hexafluoride (UF_6) from uranium ore. If this is then made to pass through a series of fine pores, the molecules containing the lighter isotope of uranium will pass through more quickly. As the gases pass through successive 'diaphragms', the proportion of lighter molecules increases. A separation of over 99% is thus attainable. Gaseous diffusion separation has also been applied to the separation of the isotopes of hydrogen.

gas equation *See* gas laws.

gas laws Laws relating the temperature, pressure, and volume of a fixed mass of gas. The main gas laws are BOYLE'S LAW and CHARLES' LAW. The laws are not obeyed exactly by any real gas, but many common gases obey them under certain conditions, particularly at high temperatures and low pressures. A gas that would obey the laws over all pressures (p) and temperatures (T) is a *perfect* or *ideal gas*.

Boyle's and Charles' laws can be combined into a *gas equation* of state for ideal gases:

$$pV_{\mathrm{m}} = RT$$

where V_{m} is the molar volume and R is the molar gas constant. For n moles of gas, then

$$pV = nRT$$

All real gases deviate to some extent from the gas laws, which are applicable only to idealized systems of particles of negligible volume with no intermolecular forces. There are several modified equations of state that give a better description of the behavior of real gases, the best known being the VAN DER WAALS EQUATION.

gas-liquid chromatography *See* gas chromatography.

gas-solid chromatography *See* gas chromatography.

gauss Symbol: G The unit of magnetic flux density in the c.g.s. system. It is equal to 10^{-4} tesla (T) in SI units.

Gay-Lussac, Joseph-Louis (1778–1850) French chemist and physicist. Gay-Lussac is best remembered for his work on gases. In 1802 he formulated the law sometimes called Charles' law (after the work of Jacques Charles in 1787) or Gay-Lussac's law that, if the pressure remains constant, gases expand equally for a given increase in temperature. In 1805 he found that two volumes of hydrogen combine with one volume of oxygen to produce water. In 1808 he formulated what is known as *Gay-Lussac's law of combining volumes*, which states that gases combine in simple proportions by volume, with the volumes of the products being related to the volume of the reactants. He was an early investigator of boron, iodine, and the cyanides. He also invented the *Gay-Lussac tower* for recovering nitrogen oxides produced in the manufacture of sulphuric acid.

Gay-Lussac's law **1.** Gases react in volumes that are in simple ratios to each other and to their products, if they are gases, assuming that all volumes are measured at the same temperature and pressure.
2. *See* Charles' law.

gel A lyophilic colloid that is normally stable but may be induced to coagulate partially under certain conditions (e.g. lower temperatures). It is these conditions that produce a pseudo-solid or easily deformable jellylike mass called a gel, in which intertwining particles enclose the whole dispersing medium. Gels may be further subdivided into elastic gels (e.g. gelatin) and rigid gels (e.g. silica gel).

gel electrophoresis *see* electrophoresis.

gel filtration *See* molecular sieve.

gem positions Positions in a molecule on the same atom. For example, 1,1-dichloroethane (CH_3CHCl_2) is a gem dihalide.

general formula A representation of the chemical formula common to a group of compounds. For example, ROH (where R is a hydrocarbon group) is the general formula for an alcohol. *See also* empirical formula; molecular formula; structural formula.

geochemistry The study of the chemistry of the Earth. This includes the chemical composition of the Earth, the abundance of the elements and their isotopes, the atomic structure of minerals, and chemical processes that occur at the high temperatures and pressures inside the Earth.

geometrical isomerism *See* isomerism.

germanium A hard, brittle, grayish-white metalloid element belonging to group 14 (formerly IVA) of the periodic table. It is found in sulfide ores such as argyrodite ($4Ag_2S.GeS_2$) and in zinc ores and coal. Most germanium is recovered during zinc or copper refining as a by-product. Germanium was extensively used in early semiconductor devices but has now been largely superseded by silicon. It is used as an alloying agent and catalyst, and in phosphors and infrared equipment.

Symbol: Ge; m.p. 937.45°C; b.p. 2830°C; r.d. 5.323 (20°C); p.n. 32; most common isotope ^{74}Ge; r.a.m. 72.61.

germanium(IV) oxide (GeO_2) A compound made by strongly heating germanium in air or by hydrolyzing germanium(IV) chloride. It occurs in two forms. One is slightly soluble in water forming an acidic solution; the other is insoluble in water. Germanium oxide dissolves in alkalis to form the anion GeO_3^{2-}.

German silver *See* nickel-silver.

Gibbs, Josiah Willard (1839–1903) American mathematician and physicist. Between 1873 and 1878 Gibbs founded the subject of chemical thermodynamics in a series of lengthy papers. In one of these papers entitled *On the Equilibrium of Heterogeneous Substances*, he put forward the phase rule concerning physical systems with different phases. His work on chemical thermodynamics also included the concepts of the Gibbs free energy and chemical potential. In addition, he contributed to work on the thermodynamics of surfaces. Gibbs was one of the pioneers of statistical mechanics. He gave a classic exposition of the subject in the book *Elementary Principles of Statistical Mechanics* (1902).

Gibbs free energy *See* Gibbs function.

Gibbs function (Gibbs free energy) Symbol: G A thermodynamic function defined by

$$G = H - TS$$

where H is the enthalpy, T the thermodynamic temperature, S the entropy, and G the energy absorbed or liberated. It is useful for specifying the conditions of chemical equilibrium for reactions for constant temperature and pressure (G is a minimum). It is named for U.S. physicist and mathematician Josiah Willard Gibbs (1839–1903), who proposed it in 1877. *See also* free energy.

giga- Symbol: G A prefix used with SI units, denoting 10^9. For example, 1 gigahertz (GHz) = 10^9 hertz (Hz).

glass A hard transparent material made by heating calcium oxide (lime, CaO), sodium carbonate (Na_2CO_3), and sand (silicon(IV) oxide, SiO_2). This produces a calcium silicate (Ca_2SiO_4) – the normal type of glass, called *soda glass*. Special types of glass can be obtained by incorporating boron oxide (B_2O_3) in the glass (borosilicate glass, used for laboratory apparatus) or by including metals other than calcium, e.g. lead or barium.

Glass is an amorphous substance, in the sense that there is no long-range ordering of the atoms on a lattice. It can be regarded as a supercooled liquid, which has not crystallized. Solids with similar noncrystalline structures, e.g. obsidian, are also called *glasses*.

Glauber's salt *See* sodium sulfate.

GLC Gas-liquid chromatography. *See* gas chromatography.

gold A ductile malleable lustrous yellow transition metal, the third element of group 11 (formerly subgroup IB) of the periodic table. Gold is largely unreactive and is therefore found native, often in veins in igneous rock or in placer deposits. Native gold almost always contains 2–20% silver. Gold is used in jewelry, often alloyed with copper, and in electronics and colored glass. The purity of gold is traditionally stated in karats (or carats), with 24 karats indicating pure gold. Thus, a ring stated to be 18 karat gold would consist of an alloy of 18 parts gold and six parts other metals.

Symbol: Au; m.p. 1064.43°C; b.p. 2807°C; r.d. 19.320 (20°C); p.n. 79; most common isotope ^{197}Au; r.a.m. 196.96654.

gold(III) chloride (gold trichloride; auric chloride; $AuCl_3$) A compound prepared by dissolving gold in aqua regia. The bright yellow crystals (chloroauric acid, $HAuCl_4$) produced on evaporation are heated to form dark red crystals of gold(III) chloride. The chloride decomposes easily (at 175°C) to give gold(I) chloride (AuCl) and chlorine; at higher temperatures it decomposes further to give gold and chlorine. Gold(III) chloride is used in photography. It exists as a dimer, Au_2Cl_6.

Goldschmidt, Victor Moritz (1888–1947) Swiss-born Norwegian chemist. Goldschmidt is frequently described as the founder of geochemistry. He pioneered the application of chemical thermodynamics to processes in geology. After the development of x-ray crystallography by the Braggs, he and his colleagues worked out the structure of over 200 crystals. This led him to produce tables of atomic and ionic radii. Goldschmidt also devoted a lot of attention to the problem of determining the cosmic abundance of the chemical elements. He summarized his work in the book *Geochemistry* (1954), which was published posthumously.

Goldschmidt process A process for extracting certain metals from their oxides by reduction with aluminum, named for German chemist Hans Goldschmidt (1861–1923), who discovered it. *See* thermite.

gold trichloride *See* gold(III) chloride.

Graham's law (of diffusion) The principle that gases diffuse at a rate that is inversely proportional to the square root of their density. It was discovered in 1829 by the Scottish chemist Thomas Graham (1805–69). Light molecules, in other words, diffuse faster than heavy molecules. The principle is used in the separation of isotopes.

grain A crystal in a metal that has been prevented from attaining its regular geometrical form.

gram (gramme) Symbol: g A metric unit of mass defined as 10^{-3} kilogram. It was the fundamental unit of mass in the c.g.s. system.

granulation A process for enlarging particles to improve the flow properties of solid reactants and products in industrial chemical processes. The larger a particle, and the freer from fine materials in a solid, the more easily it will flow. Dry granulation produces pellets from dry materials, which are crushed into the desired size. Wet granulation involves the addition of a liquid to the material, and the resulting paste is extruded and dried before cutting into the required size.

graphite An allotrope of CARBON. The atoms are arranged in layers as a series of flat, hexagonal rings. Graphite is a good conductor of heat and electricity. The layers cleave easily, making graphite useful as a solid lubricant.

gravimetric analysis A method of quantitative analysis in which the final analytical measurement is made by determination of mass. There are many variations in the method but in essence they all consist of:

1. taking a sample whose mass has been accurately determined into solution;
2. precipitating a known compound by a quantitative reaction;
3. performing digestion and coagulation procedures;
4. filtering and washing;
5. drying and determining mass as a pure compound.

Filtration is a key element in the method and a variety of special filter papers and sinter-glass filters are available.

gray Symbol: Gy The SI derived unit of absorbed energy dose per unit mass resulting from the passage of ionizing radiation through living tissue. One gray is an energy absorption of one joule per kilogram of mass.

gray cast iron *See* cast iron.

ground state The state of lowest energy of an atom (or ion, molecule, etc.). For example, the hydrogen atom in the ground state has its single electron in the 1s orbital. If energy is provided (e.g. by electromagnetic radiation, electron impact, high temperature, etc.) it is possible to form hydrogen atoms in an EXCITED STATE (for example, with the electron in the 2s orbital).

group In the periodic table, a series of chemically similar elements that have similar electronic configurations. A group is thus a column of the periodic table. For example, the alkali metals, all of which have outer s^1 configurations, belong to group 1. *See also* periodic table.

group 0 elements *See* rare gases.

group 1 elements *See* alkali metals.

group 2 elements *See* alkaline-earth metals.

group 3-12 elements *See* transition elements.

group 13 elements A group of elements in the periodic table consisting of the elements boron (B), aluminum (Al), gallium (Ga), indium (In), and thallium (Tl). These elements were formerly classified as group III elements and belonged to the IIIA subgroup. The IIIB subgroup consisted of the elements scandium (Sc), yttrium (Y), lanthanum (La), and actinium (Ac).

The group 13 elements all have three outer electrons (s^2p^1) with no partly filled inner shells. The group is the first group of elements in the p-block. As with all the groups there is an increase in metallic character as the group is descended.

The boron atom is both small and has a high ionization potential, so bonds to boron are largely covalent with only small degrees of polarization. In fact boron is classed as a metalloid. It has volatile hydrides and a weakly acidic oxide. As the group is descended the ionization potentials tend to decrease and the atomic radii increase, leading to increasingly polar interactions and the formation of distinct M^{3+} ions. The increase in metallic character is clearly illustrated by the hydroxides: boric acid $B(OH)_3$ is acidic; aluminum and gallium hydroxides are amphoteric, dissolving in acids to give Al^{3+} and Ga^{3+} and in bases to give aluminates and gallates; the hydroxides of indium and thallium are distinctly basic. As the elements get heavier the bond energies with other elements become generally smaller. The monovalent state (removal of the p electron only) becomes progressively stable. For example gallium(I) is known in the gas phase and in coordination complexes (sometimes as '$GaCl_2$'). Monovalent indium halides are known, for example InX, and thallium(I) is distinctly more stable than thallium(III).

group 14 elements A group of elements in the periodic table consisting of the elements carbon (C), silicon (Si), germanium (Ge), tin (Sn), and lead (Pb). These elements were formerly classified as group IV elements and belonged to the IVA subgroup. The IVB subgroup consisted of the elements titanium (Ti), zirconium (Zr), and hafnium (Hf), which are transition metals.

The group 14 elements all have electronic structures with four outer electrons (s^2p^2) and no partly filled inner shells. The group shows the common trend toward metallic character with the heavier elements; thus carbon is a typical nonmetal, silicon and germanium are metalloids, and tin and lead are characteristically metallic. As observed with other groups, the first member of the group is quite different from the rest. The carbon atom is smaller and has a higher ionization potential, both favouring predominance of covalence, but additional factors are:

1. The widespread nature of extensive catenation in carbon compounds.
2. The possibility of strong overlap of p orbitals or *double bonds*
3. The unavailability of d orbitals in carbon compounds.

Significant differences are:

1. Both oxides of carbon are gaseous, CO and CO_2; the other elements form solid oxides, e.g. SiO_2, Pb_3O_4.
2. The heavier elements readily expand their coordination number, e.g. SiF_6^{2-} and $SnCl_6^{2-}$.

group 15 elements A group of elements in the periodic table consisting of the elements nitrogen (N), phosphorus (P), arsenic (As), antimony (Sb), and bismuth (Bi). These elements were formerly classified as group V elements and belonged to the VA subgroup. The VB subgroup consisted of the elements vanadium (V), niobium (Nb), and tantalum (Ta), which are transition elements.

The group 15 elements all have electronic configurations corresponding to outer s^2p^3 electrons with no vacancies in inner levels. The ionization potentials are high and the lighter members of the group, N and P, are distinctly electronegative and nonmetallic in character. Arsenic and antimony are metalloids and bismuth is weakly metallic (Bi_2O_3 dissolves in acids to give $Bi(OH)_3$).

Nitrogen is dissimilar from the other members of the group in that it forms stable multiple bonds, it is limited to coordination number 4, e.g. NH_4^+, it is sufficiently electronegative to form hydrogen bonds and the nitride ion N^{3-}, and in that its oxides are irregular with the rest of the group.

Group members display a remarkable number of allotropes, showing the trend from nonmetallic forms through to metallic forms. Thus nitrogen has only the diatomic form; phosphorus has a highly reactive form P_1 (brown), and a tetrahedral form P_4 (white), forms based on broken tetrahedra P_n (red and violet), and a hexagonal layer-type lattice (black). Arsenic and antimony have the As_4 and Sb_4 forms, which are less stable than the layer-type lattice form; bismuth has the layer-lattice form only.

group 16 elements (chalcogens) A group of elements of the periodic table consisting of the elements oxygen (O), sulfur (S), selenium (Se), tellurium (Te), and polonium (Po). These elements were formerly classified as group VI elements and belonged to the VIA subgroup. The VIB subgroup consisted of the elements chromium (Cr), molybdenum (Mo), and tungsten (W), which are transition elements.

The electronic configurations of the group 6 elements are all s^2p^4 with no vacant inner orbitals. These configurations are all just two electrons short of a rare-gas structure. Consequently they have high electron affinities and are almost entirely nonmetallic in character.

The group shows the normal property of a trend toward metallic character as it is descended. Selenium, tellurium, and polonium have 'metallic' allotropes, and polonium has generally metalloid-type properties where it has been possible to study these (Po is very rare). All the elements combine with a large number of other elements, both metallic and nonmetallic, but in contrast to compounds of the halogens they are more generally insoluble in water, and even when soluble do not ionize readily.

Oxygen behaves like the first members of other groups in displaying great differences from the rest of its group. For example, oxygen forms hydrogen bonds whereas sulfur and the others do not; oxygen cannot expand its outer shell and positive oxidation states are uncommon (OF_2, O_2F_2, $O_2^+PtF_6^-$). Oxygen is paramagnetic. Excluding oxygen, group 16 shows similar trends with increasing size to those in group 15. For example, with increasing size there is decreasing stability of H_2X, greater tendency to form complexes such as $SeBr_6^{2-}$, decreasing stability of high formal positive oxidation states, and marginal metallic properties (e.g. Po forms a hydroxide).

group 17 elements A group of elements in the periodic table consisting of the elements fluorine (F), chlorine (Cl), bromine (Br), iodine (I), and astatine (At). Formerly they belonged to group VII and were classified in the VIIA subgroup. The VIIB subgroup consisted of the elements manganese (Mn), technetium (Te), and rhenium (Re), which are transition elements. *See* halogens.

group 18 elements *See* rare gases.

group theory A branch of mathematics used to analyze symmetry in a systematic way. A group consists of a set of elements A, B, C, etc. and a rule for forming the product of these elements, such that certain conditions are satisfied:

1. Every product AB of two elements A and B is also an element of the set.
2. Any three elements A, B, C, must satisfy A(BC) = (AB)C.
3. There is an element of the set, denoted I, which is the *identity element*, i.e. for all A in the set AI = IA =A.
4. For each element A of the set there is an *inverse*, denoted A^{-1}, which also belongs to the set which satisfies the relation $AA^{-1} = A^{-1} = A = I$.

In the case of symmetry operations the product is the successive performance of these operations. If a group has a finite number of elements it is said to be a *discrete group*. The *point groups*, which arise from the rotations and reflections of bodies such as isolated molecules are discrete groups.

GSC Gas-solid chromatography. *See* gas chromatography.

gunmetal *See* bronze.

gunpowder A finely powdered mixture of sulfur, charcoal, and potassium nitrate, used as an explosive.

gypsum *See* calcium sulfate.

gyromagnetic ratio The ratio of the magnetic moment of a system to the angular momentum of that system. The gyromagnetic ratio is a useful quantity in the theory of electrons and atomic nuclei.

H

Haber, Fritz (1868–1934) German physical chemist. Haber is noted for his discovery of the industrial process for synthesizing ammonia from nitrogen and hydrogen. The need at the time was for nitrogen compounds for use as fertilizers – most plants cannot utilize free nitrogen from the air, and need 'fixed' nitrogen. The main source was deposits of nitrate salts in Chile, but these would have a limited life. Haber, in an attempt to solve this problem, began investigating the reaction:

$$N_2 + 3H_2 \rightleftharpoons 2NH_3$$

Under normal conditions the yield is very low. Haber (1907–09) showed that practical yields could be achieved at high temperatures (250°C) and pressures (250 atmospheres) using a catalyst (iron is the catalyst now used). The process was developed industrially by Carl BOSCH around 1913 and is still the main method for the fixation of nitrogen. Haber received the Nobel Prize for chemistry for this work in 1918.

Haber process An important industrial process for the manufacture of ammonia, which is used for fertilizers and for making nitric acid. The reaction is the equilibrium:

$$N_2 + 3H_2 \rightleftharpoons 3NH_3$$

The nitrogen used is obtained by fractional distillation of liquid air and the hydrogen by the oxidation of hydrocarbons (from natural gas). The nitrogen and hydrogen are purified and mixed in the correct proportions. The equilibrium amount of ammonia is favoured by low temperatures, but in practice the reaction rate would be too slow to be economic. An optimum temperature of about 450°C is therefore used, along with a catalyst. High pressure also favors the reaction and a pressure of about 250 atmospheres is used. The catalyst is iron with small amounts of potassium and aluminum oxides present. The yield is about 15%. Liquid ammonia is condensed out of the gas equilibrium mixture at –50°C.

habit *See* crystal habit.

hafnium A bright silvery transition metal, the third element of group 4 (formerly subgroup IVB) of the periodic table. It is found in zirconium ores. Hafnium is difficult to work and can burn in air. It is used in control rods for nuclear reactors and in certain specialized alloys and ceramics.

Symbol: Hf; m.p. 2230°C; b.p. 5197°C; r.d. 13.31 (20°C); p.n. 72; most common isotope ^{180}Hf; r.a.m. 178.49.

hahnium Symbol: Ha or Hn A name formerly suggested for element-105, now known as dubnium (Db), and also for element-108, now known as hassium (Hs). *See* element.

half cell An electrode in contact with a solution of ions. In general there will be an e.m.f. set up between electrode and solution by transfer of electrons to or from the electrode. The e.m.f. of a half cell cannot be measured directly, since setting up a circuit results in the formation of another half cell. *See* electrode potential.

half-life (half-life period; Symbol: $t_{1/2}$) The time taken for half the nuclei of a sample of a radioactive nuclide to decay. The half-life of a nuclide is a measure of its stability. (Stable nuclei can be thought of as having infinitely long half-lives.) If N_0 is the original number of nuclei, the number remaining at the end of one half-life is $N_0/2$, at the end of two half-lives is $N_0/4$, etc.

half sandwich compound *See* sandwich compound.

halide A compound containing a halogen. The inorganic halides of electropositive elements contain ions of the type Na^+Cl^- (sodium chloride) or K^+Br^- (potassium bromide). Transition metal halides often have some covalent bonding. Nonmetal halides are covalent compounds, which are usually volatile. Examples are tetrachloromethane (CCl_4) and silicon tetrachloride ($SiCl_4$). Halides are named as bromides, chlorides, fluorides, or iodides.

halite (rock salt) A naturally occurring mineral form of sodium chloride, NaCl. It forms colorless or white crystals when pure, but is often colored by impurities.

halogenation A reaction in which a HALOGEN atom is introduced into a generally organic molecule. Halogenations are specified as *chlorinations*, *brominations*, *fluorinations*, etc., according to the element involved. There are several methods.

1. Direct reaction with the element using high temperature or ultraviolet radiation:
$$CH_4 + Cl_2 \rightarrow CH_3Cl + HCl$$
2. Addition to a double bond:
$$H_2C{:}CH_2 + HCl \rightarrow C_2H_5Cl$$
3. Reaction of a hydroxyl group with a halogenating agent, such as PCl_3:
$$C_2H_5OH \rightarrow C_2H_5Cl + OH^-$$
4. In aromatic compounds direct substitution can occur using aluminum chloride as a catalyst:
$$2C_6H_6 + Cl_2 \rightarrow 2C_6H_5Cl$$
5. Alternatively in aromatic compounds, the chlorine can be introduced by reacting the diazonium ion with copper(I) chloride:
$$C_6H_5N_2^+ + Cl^- \rightarrow C_6H_5Cl + N_2$$

halogens A group of elements (group 17, formerly VIIA, of the periodic table) consisting of fluorine, chlorine, bromine, iodine, and the short-lived element astatine. The halogens all have outer valence shells that are one electron short of a rare-gas configuration. Because of this, the halogens are characterized by high electron affinities and high electronegativities, fluorine being the most electronegative element known. The high electronegativities of the halogens favor the formation of both uninegative ions, X^-, particularly combined with electropositive elements (e.g. NaCl, KBr, CsI, CaF_2), and the single covalent bond –X with elements of moderate to high electronegativity (e.g. HCl, SiF_4, SF_4, $BrCH_3$, Cl_2O). The halogens all form diatomic molecules, X_2, which are characterized by their high reactivities with a wide range of other elements and compounds. As a group, the elements increase in both size and polarizability as the proton number increases and there is an attendant decrease in electronegativity and reactivity. This means that there is no chemistry of positive species apart from a few cationic iodine compounds. The heavier halogens Cl, Br, and I all form oxo-species with formal positive oxidation numbers +1 and +5, chlorine and iodine also forming +3 and +7 species (e.g. HOCl, $HBrO_3$, I_2O_5, HIO_4).

The elements decrease in oxidizing power in the order $F_2>Cl_2>Br_2>I_2$ and the ions X^- may be arranged in order of increasing reducing power $F^-<Cl^-<Br^-<I^{nd\ ash;}$. Thus any halogen will displace the elements below it from their salts in solution, for example
$$Cl_2 + 2Br^- \rightarrow Br_2 + 2Cl^-$$
A wide range of organic halides is formed in which the C–F bond is characteristically resistant to chemical attack; the C–Cl bond is also fairly stable, particularly in aryl compounds but the alkyl halogen compounds become increasingly susceptible to nucleophilic attack and are generally more reactive.

hammer mill A device used in the chemical industry for crushing and grinding solid materials at high speeds to a specified size. The impact between the particles, grinding plates, and grinding hammers pulverizes the particles. Hammer mills can be used for a greater variety of soft material than other types of grinding equipment. *Compare* ball mill.

hardness (of water) Hard water is water that will not readily form a lather with soap owing to the presence of dissolved calcium, iron, and magnesium compounds. Such compounds can react with soap to produce insoluble salts, which collect as solid scum. The effectiveness of the cleansing solution is thus reduced. Hardness of water is of two types, temporary hardness and permanent hardness. Only temporary hardness can be removed by boiling the water.

Hardness of water is usually expressed by assuming that all the hardness is due to dissolved calcium carbonate, which is present as ions. It can be estimated by titration with a standard soap solution or with edta. *See also* permanent hardness; detergents; temporary hardness; water softening.

hard vacuum *See* vacuum.

hard water *See* hardness.

hassium A transactinide element that is formed artificially.

Symbol: Hs; p.n. 108; most stable isotope ^{265}Hs (half-life 2×10^{-3}s).

hausmannite *See* manganese.

heat Energy transferred as a result of a temperature difference. The term is often loosely used to mean internal energy (i.e. the total kinetic and potential energy of the particles). It is common in chemistry to define such quantities as *heat of combustion*, *heat of neutralization*, etc. These are in fact molar enthalpies for the change, given the symbol $\Delta H_M^{\ominus}$. The superscript symbol denotes standard conditions, while the subscript M indicates that the enthalpy change is for one mole. The unit is usually the kilojoule per mole (kJ mol^{-1}). By convention, ΔH is negative for an exothermic reaction. Molar enthalpy changes stated for chemical reactions are changes for standard conditions, which are defined as 298 K (25°C) and 101325 Pa (1 atmosphere). Thus, the standard molar enthalpy of reaction is the enthalpy change for reaction of substances under these conditions producing reactants under the same conditions. The substances involved must be in their normal equilibrium physical states under these conditions (e.g. carbon as graphite, water as the liquid, etc.). Note that the measured enthalpy change will not usually be the standard change. In addition, it is common to specify the entity involved. For instance $\Delta H_f^{\ominus}(H_2O)$ is the standard molar enthalpy of formation for one mole of H_2O species.

heat engine (thermodynamic engine) A device for converting heat energy into work. Heat engines operate by transferring energy from a high-temperature source to a low-temperature sink. The theoretical operation of heat engines is useful in the theory of thermodynamics. *See* Carnot cycle.

heat exchangers Devices that enable the heat from a hot fluid to be transferred to a cool fluid without allowing the two fluids to come into contact. The normal arrangement is for one of the fluids to flow in a coiled tube through a jacket containing the second fluid. Both the cooling and heating effect may be of benefit in conserving the energy used in a chemical plant and in controlling a process.

heat of atomization The energy required in dissociating one mole of a substance into atoms. *See* heat.

heat of combustion The energy liberated when one mole of a substance burns in excess oxygen. *See* heat.

heat of crystallization The energy liberated when one mole of a substance crystallizes from a saturated solution of that substance.

heat of dissociation The energy required to dissociate one mole of a substance into its constituent elements.

heat of formation The energy change when one mole of a substance is formed from its elements. *See* heat.

heat of neutralization The energy liberated when one mole of an acid or base is neutralized.

heat of reaction The energy change when molar amounts of given substances react completely. *See* heat.

heat of solution The energy change when one mole of a substance is dissolved in a given solvent to infinite dilution (in practice, to form a dilute solution).

heavy hydrogen *See* deuterium.

heavy-metal pollution *See* pollution.

heavy water *See* deuterium oxide.

hecto- Symbol: h A prefix used with SI units, denoting 10^2. For example, 1 hectometer (hm) = 10^2 meters (m).

Heisenberg, Werner Karl (1901–76) German physicist. Heisenberg was one of the founders of quantum mechanics. In 1925, along with Max BORN and Pascual Jordan, he formulated *matrix mechanics.* In 1926 he studied the spectrum of molecular hydrogen and predicted the existence of ortho- and para-hydrogen from the splitting of the spectral lines. His most famous contribution is the HEISENBERG UNCERTAINTY PRINCIPLE, put forward in 1927.

Heisenberg uncertainty principle The impossibility of making simultaneous measurements of both the position and the momentum of a subatomic particle (e.g. an electron) with unlimited accuracy. The uncertainty arises because, in order to detect the particle, radiation has to be 'bounced' off it, and this process itself disrupts the particle's position. Heisenberg's uncertainty principle is not a consequence of 'experimental error'. It represents a fundamental limit to objective scientific observation, and arises from the wave–particle duality of particles and radiation. In one direction, the uncertainty in position Δx and momentum Δp are related by $\Delta x \Delta p \sim h/4\pi$, where h is the Planck constant. It is named for the German physicist Werner Heisenberg, who discovered it in 1927.

Heitler, Walter (1904–81) German-born physicist. Heitler is most famous for a classic paper he wrote with Friz London in 1927 in which they showed that the chemical bond in the hydrogen molecule could be described by quantum mechanics. Heitler extended this work to more complicated molecules and was a pioneer of the use of group theory in quantum mechanics. He wrote a classic book entitled *The Quantum Theory of Radiation*, the third edition of which was published in 1954.

helicate *See* supramolecular chemistry.

helium An inert colorless odorless monatomic gas, the first member of the rare gases; i.e. group 18 (formerly VIIIA or 0) of the periodic table. Helium has the electronic configuration $1s^2$ and consists of a nucleus of two protons and two neutrons (equivalent to an alpha particle) with two extra-nuclear electrons. It has an extremely high ionization potential and is completely resistant to chemical attack of any sort. Helium is the second most abundant element in the universe, the primary process in the Sun and other stars being the nuclear fusion of hydrogen to give helium. Nevertheless the gas accounts for only $5.2 \times 10^{-4}\%$ of the Earth's atmosphere although some natural gas deposits contain up to 7%.

Helium is recovered commercially by the fractional distillation of natural gas and also forms part of ammonia plant tail gas if natural gas is used as a feedstock. Its applications are in fields in which inertness is required and where cheaper alternatives, such as nitrogen, would prove too reactive. Examples include high-temperature metallurgy, powder technology, and use as a coolant in nuclear reactors. Helium is also favoured over nitrogen for diluting oxygen in breathing mixtures designed for deep-sea diving due to its lower solubility in blood and as a pressurizer for liquefied gas fuels in rockets (due to its total inertness). It is also used as an ideal gas for balloons (no fire risk) and for low-temperature physics research.

Helium is unusual in that it is the only known substance for which there is no triple point (i.e., no combination of pressure and temperature at which all three phases can co-exist in equilibrium). This is because the interatomic forces, which normally participate in the formation of solids, are so weak in helium that they are of the same order as the zero-point energy. At 2.186 K helium undergoes a transition from liquid helium I to liquid helium II, the latter being a true liquid but exhibiting superconductivity and an immeasurably low viscosity (SUPERFLUIDITY). This low viscosity allows the liquid to spread in layers a few atoms thick, described by some as an action like 'flowing uphill'.

Helium has two naturally occurring isotopes, of which helium-4 (^4He) is the most common. Helium-3 (^3He) is formed in nuclear reactions and by the decay of tritium. This isotope also undergoes a phase change at temperatures close to absolute zero.

Symbol: He; m.p. 0.95 K (under pressure); b.p. 4.216 K; r.d. 0.178; p.n. 2; most common isotope ^4He; r.a.m. 4.002602.

Helmholtz free energy *See* Helmholtz function.

Helmholtz function (Helmholtz free energy) Symbol: *F* A thermodynamic function defined by

$$F = U - TS$$

where U is the internal energy, T the thermodynamic temperature, and S the entropy. It is a measure of the ability of a system to do useful work in an isothermal process. *See also* free energy.

hematite A reddish-brown, gray, or black mineral form of iron(III) oxide (Fe_2O_3), found in igneous, metamorphic, and sedimentary rocks. It is the principal ore of iron and is also used as a pigment.

hemihydrate A crystalline compound having one molecule of water of crystallization per two molecules of compound (e.g. $2CaSO_4.H_2O$).

hemimorphite *See* calamine.

henry Symbol: H The SI derived unit of inductance, equal to the inductance of a closed circuit that has a magnetic flux of one weber per ampere of current in the circuit. 1 H = 1 Wb A^{-1}. It is named for U.S. physicist Joseph Henry (1797–1878).

Henry's law The concentration (C) of a gas in solution is proportional to the partial pressure (p) of that gas in equilibrium with the solution, i.e. $p = kC$, where k is a proportionality constant. The relationship is similar in form to that for Raoult's law, which deals with ideal solutions.

A consequence of Henry's law is that the 'volume solubility'of a gas is independent of pressure. It was discovered in 1801 by British chemist William Henry (1774–1836).

heptahydrate A crystalline hydrated compound containing one molecule of compound per seven molecules of water of crystallization.

heptavalent (septivalent) Having a valence of seven.

hertz Symbol: Hz The SI derived unit of frequency, defined as one cycle per second (s^{-1}). Note that the hertz is used for regularly repeated processes, such as vibrations or wave motions. It is named for the German physicist Heinrich Hertz (1857–94).

Herzberg, Gerhard (1904–99) German-born Canadian spectroscopist. Herzberg investigated and interpreted the spectra of many molecules. He wrote a series of classic, definitive books on atomic and molecular spectra between the 1930s and the 1970s. Herzberg was awarded the Nobel Prize for chemistry in 1971.

Hess's law (law of constant heat summation) A derivative of the first law of thermodynamics. It states that the total heat change for a given chemical reaction involving alternative series of steps is independent of the route taken. It is named for the Swiss-born Russian chemist Germain

Hess (1802–50), who stated it in 1840.

heterogeneous Relating to more than one phase. A heterogeneous mixture, for instance, contains two or more distinct phases.

heterogeneous catalyst *See* catalyst.

heterolysis *See* heterolytic fission.

heterolytic fission (heterolysis) The breaking of a covalent bond so that both electrons of the bond remain with one fragment. A positive ion and a negative ion are produced:

$$RX \rightarrow R^+ + X^-$$

Compare homolytic fission.

hexacyanoferrate *See* cyanoferrate.

hexagonal close-packed crystal A close-packed crystal structure in which layers of close-packed atoms, ions, or molecules are stacked in an ABABAB arrangement. In this arrangement, the B layer fits into the recesses left in the A layer, with the third layer then fitting in the recesses in the second layer that are directly above the atoms, etc. in the first layer. Zinc and magnesium have hexagonal close-packed structures. *See* close packing. *Compare* face-centered cubic crystal.

hexagonal crystal *See* crystal system.

Hinshelwood, Sir Cyril Norman (1897–1967) British chemist. Hinshelwood was a leading figure in the study of the kinetics and mechanisms of chemical reactions. He wrote a classic book on this topic entitled *The Kinetics of Chemical Change in Gaseous Systems* (1926). In 1956 he shared the Nobel Prize for chemistry with Nikolay SEMENOV for his work on the mechanisms of chain reactions. Hinshelwood worked extensively on chemical reactions in bacteria, summing up his research in the book *The Chemical Kinetics of the Bacterial Cell* (1954).

holmium A soft malleable silvery element of the lanthanoid series of metals. It occurs in association with other lanthanoids in such minerals as euxenite, gadolinite, monazite, samarskite, and xenotime. It has a few applications in electronics due to the highly magnetic nature of its compounds.

Symbol: Ho; m.p. 1474°C; b.p. 2695°C; r.d. 8.795 (25°C); p.n. 67; most common isotope ^{165}Ho; r.a.m. 164.93032.

HOMO *See* frontier orbital theory.

homogeneous Relating to a single phase. A homogeneous mixture, for instance, consists of only one phase.

homogeneous catalyst *See* catalyst.

homolysis *See* homolytic fission.

homolytic fission (homolysis) The breaking of a covalent bond so that one electron from the bond is left on each fragment. Two free radicals result:

$$RR' \rightarrow R\bullet + R'\bullet$$

Compare heterolytic fission.

hornblende Any of a group of dark-colored rock-forming minerals consisting of complex silicates of calcium, sodium, aluminum, iron, and magnesium. It is a major component of granite and other igneous and metamorphic rocks.

host–guest chemistry A branch of supramolecular chemistry in which a molecular structure acts as a 'host' to hold a 'guest' ion or molecule. The guest may be coordinated to the host or may be trapped by its structure. *See also* supramolecular chemistry.

HPLC High-performance liquid chromatography; a sensitive analytical technique, similar to gas-liquid chromatography but using a liquid carrier. The carrier is specifically choosen for the particular substance to be detected.

Humphreys series *See* hydrogen atom spectrum.

Hund's rule A rule that states that the

electronic configuration in degenerate orbitals will have the minimum number of paired electrons. It is named for the German physicist Friedrich Hund (1896–1993), who stated it in 1925.

hybrid orbital *See* orbital.

hydrate A compound coordinated with water molecules. When water is bound up in a compound it is known as the WATER OF CRYSTALLIZATION.

hydration The solvation of species such as ions in water.

hydraulic cement *See* cement.

hydrazine (N_2H_4) A colorless liquid that can be prepared by the oxidation of ammonia with sodium chlorate(I) or by the gas phase reaction of ammonia with chlorine. Hydrazine is a weak base, forming salts (e.g. $N_2H_4.HCl$) with strong acids. It is a powerful reducing agent, reducing salts of the noble metals to their respective metals. Anhydrous hydrazine ignites spontaneously in oxygen and reacts violently with oxidizing agents. The aqueous solution, *hydrazine hydrate*, has been used as a fuel for jet engines and for rockets.

hydrazine hydrate *See* hydrazine.

hydrazoic acid (hydrogen azide; HN_3) A colorless liquid with a nauseating smell. It is highly poisonous and explodes in the presence of oxygen and oxidizing agents. It can be made by distilling a mixture of sodium azide (NaN_3) and a dilute acid. It is usually used as an aqueous solution. The salts of hydrazoic acid (azides), especially lead azide ($Pb(N_3)_2$), are used in detonators because they explode when given a mechanical shock.

hydride A compound of hydrogen. Ionic hydrides are formed with highly electropositive elements and contain the H^- ion (hydride ion). Nonmetals form covalent hydrides, as in methane (CH_4) or silane (SiH_4). The boron hydrides are electron-deficient covalent compounds. Many transition metals absorb hydrogen to form interstitial hydrides.

hydrobromic acid (HBr) A colorless liquid produced by dissolving HYDROGEN BROMIDE in water. It shows the typical properties of a strong acid and is a strong reducing agent. A convenient way of producing hydrobromic acid is to bubble hydrogen sulfide through bromine water. Although it is not as strong as hydrochloric acid it dissociates extensively in water and is a good proton donor.

hydrocarbon Any compound containing only the elements carbon and hydrogen. Methane and ethane are examples.

hydrochloric acid (HCl) A colorless fuming liquid made by dissolving HYDROGEN CHLORIDE to water:

$$HCl(g) + H_2O1(l). \rightarrow H_3O^+(aq) + Cl^-(aq)$$

Dissociation into ions is extensive and hydrochloric acid shows the typical properties of a strong acid. It reacts with carbonates to give carbon dioxide and yields hydrogen when reacted with all but the most unreactive metals. Hydrochloric acid is used in the manufacture of dyes, drugs, and photographic materials. It is also used to pickle metals, i.e. clean their surfaces prior to electroplating. It donates protons with ease and is the strongest of the hydrohalic acids. The concentrated acid is oxidized to chlorine by such agents as potassium manganate(VII) and manganese(IV) oxide.

hydrocyanic acid (prussic acid; HCN) A highly poisonous weak acid formed when hydrogen cyanide gas dissolves in water. Its salts are cyanides. Hydrogen cyanide is used in making acrylic plastics.

hydrofluoric acid (HF) A colorless liquid produced by dissolving HYDROGEN FLUORIDE in water. It is a weak acid, but will dissolve most silicates and hence can be used to etch glass. As the interatomic distance in HF is relatively small, the H–F BOND ENERGY is very high and hydrogen fluoride is not a good proton donor. It does, however, form hydrogen bonds.

hydrogen A colorless odorless gaseous element traditionally placed in group 1 (formerly IA) of the periodic table. Although it has some similarities to both the alkali metals (group 1) and the halogens (group 17), hydrogen is now not normally classified in any particular periodic group. It is the most abundant element in the universe and the ninth most abundant element in the Earth's crust and atmosphere (by mass). It occurs in nature principally as a constituent of water, compounds formed by organisms, and hydrocarbon deposits (coal, oil, and natural gas); traces of molecular hydrogen are found in some natural gases and in the upper atmosphere.

The gas may be prepared in the laboratory by the reaction of dilute hydrochloric acid with a metal that lies above hydrogen in the electromotive series, magnesium and zinc being commonly used:

$$Zn + 2HCl \rightarrow H_2 + ZnCl_2$$

Reactions of the amphoteric metals zinc and aluminum with dilute aqueous alkali, to form zincates or aluminates, are also convenient means of obtaining laboratory hydrogen. Although electrolysis of dilute mineral acids may be used as well, care must be taken to avoid mixing the hydrogen released at the cathode with the oxygen released at the anode, because such mixtures are potentially explosive. Industrially, hydrogen is obtained as a by-product of the electrolytic cells used in the production of sodium hydroxide, i.e. by reacting the Na/Hg amalgam (produced when sodium liberated from the brine solution alloys with the mercury cathode) with water, or by the water-gas route in which steam is decomposed by hot coke.

The main use of hydrogen is as a chemical feedstock for the manufacture of ammonia and a range of organic compounds. Small-scale uses include reducing atmospheres for metallurgy, hydrogenation of edible oils, and pharmaceutical manufacturing. Hydrogen also has significant potential as a fuel, especially using fuel cell technology.

The hydrogen atom consists of a proton (positive charge) with one extranuclear electron (s^1). Hydrogen is thus the simplest of all atoms, giving it a unique position among the elements. The chemistry of hydrogen depends on one of three processes:

1. Loss of the electron to form H^+.
2. Gain of an electron to form H^-.
3. Sharing of electrons by covalent bond formation, as in H_2 or HCl.

The hydrogen atom with the 1s electron removed would have an extremely small ionic radius and the positive hydrogen ion in fact occurs only in association with other species, as in H^+NH_3 or H^+FH. The ion commonly written H^+ in solution is actually the solvated proton (hydroxonium or hydronium) H_3O^+, which is formed by ionization of acids, e.g.:

$$H_2O + HCl \rightarrow H_3O^+ + Cl^-$$

It is believed that the lifetime of any one H_3O^+ ion is extremely short, as the protons appear to undergo very rapid exchange between water molecules.

Hydrogen also forms a number of compounds in which it is regarded as gaining an electron and becoming H^- (the hydride ion). It can form ionic HYDRIDES only with the most electropositive elements (i.e., those in groups 1 and 2). The ionic nature of these compounds is indicated by the fact that their melts are good conductors and that hydrogen is liberated at the anode during their electrolysis. These hydrides are prepared by heating the element in a stream of hydrogen:

$$H_2 + 2M \rightarrow 2MH$$

Examples of hydrides with covalent bonds are CH_4, NH_3, and HCl. These compounds are generally low-boiling. General methods of preparation are:

1. The hydrolysis of the appropriate '-ide' compound, e.g. silicides give silane, nitrides give ammonia, sulfides give H_2S.
2. Reduction of a chloride; for example:

$$SiCl_4 + LiAlH_4 \rightarrow SiH_4 + LiCl + AlCl_3$$

A third kind of hydride is that of metallic hydrides formed between hydrogen and many transition metals. Palladium in particular is renowned for its ability to absorb hydrogen as PdH_x, where x takes values up to 1.8. Titanium and zirconium behave similarly, but the exact nature of the compounds formed and the type of bond involved remains uncertain. There are changes in the magnetic properties of the metal, indicating some electron interaction, and in the lattice dimensions of the compound, but discrete phases of the type MH or MH_2 have not been isolated. These compounds are sometimes referred to as the *interstitial hydrides*.

Atomic nuclei possess the property of 'spin' and for diatomic molecules there exists the possibility of having the spins of adjacent nuclei aligned (ortho) or opposed (para). Because of the small mass of hydrogen, these forms are more important in hydrogen molecules than in other diatomic molecules. The two forms are in equilibrium, with parahydrogen dominant at low temperatures. However the equilibrium ratio rises to 3 parts orthohydrogen to 1 part parahydrogen at room temperatures. Although chemically identical, the melting point and boiling point of the para form are both about 0.1°C lower than the 3:1 equilibrium mixture.

Natural hydrogen in molecular or combined forms contains about one part in 2000 of DEUTERIUM, symbol D, an isotope of hydrogen that contains one proton and one neutron in its nucleus. The artificially created radioactive isotope TRITIUM, symbol T, has one proton and two neutrons. Although the effect of isotopes on chemical properties is normally small, in the case of hydrogen the difference in mass number leads to a lowering of some reaction rates, a phenomenon known as the 'isotope effect'.

Hydrogen also exhibits two less common forms of bonding. Boron hydrides form a wide variety of compounds in which the hydrogen acts as a bridging species involving 'three-center two-electron' bonds. Such species are said to be 'electron deficient' because they do not have sufficient electrons for conventional two-electron covalent bonds. The second, less common, form is that of the coordinated hydrides in which the H^- ion acts as a ligand bound to a transition metal atom.

Symbol: H; m.p. 14.01 K; b.p. 20.28 K; r.d. 0.089 88; p.n. 1; most common isotope ^1H; r.a.m. 1.0079.

hydrogenation The reaction of a compound with hydrogen. An example is the hydrogenation of nitrogen to form ammonia in the Haber process. In organic chemistry, hydrogenation refers to the addition of hydrogen to multiple bonds, usually with the aid of a catalyst. Unsaturated natural liquid vegetable oils can be hydrogenated to form saturated semisolid fats – a reaction used in making types of margarine and cooking oils. Consumption of such *hydrogenated oils* is now thought to be associated with increased risk of cardiovascular disease. *See* Bergius process.

hydrogen atom spectrum The spectrum of the hydrogen atom is characterized by several series of sharp spectral lines described by simple laws. The general law for these series of lines is:

$$1/\lambda = R\,(1/n_1^2 - 1/n_2^2),$$

where λ is the wavelength associated with a spectral line, R is the RYDBERG CONSTANT and n_1 and n_2 are integers, with $n_2 \geq n_1$.

The first of these series to be discovered was the *Balmer series* in which $n_1 = 2$, $n_2 = 3,4,5,\ldots$ This series is in the visible region and was discovered by the Swiss mathematician and physicist Johann Jakob Balmer (1825–98) in 1885. The series in which $n_1 = 1$ is the *Lyman series* which lies in the ultraviolet region. This series was discovered by Theodore Lyman (1874–1954). The Lyman series is a conspicuous feature of the spectrum of the Sun.

In the *Paschen series* ($n_1 = 3$), the *Brackett series* ($n_1 = 4$), the *Pfund series* ($n_1 = 5$) and the *Humphreys series* ($n_1 = 6$) the spectral lines occur in the infrared region.

The explanation for these regular series lies in the existence of discrete, quantized energy levels. In 1913 Niels Bohr was able to derive the formula for these series in terms of the ad hoc quantum assumptions of the BOHR THEORY. In the mid-1920s the formula was derived in a deductive way from quantum mechanics. In the wave mechanics formulation of quantum mechanics it is possible to derive the formula because the Schrödinger equation can be solved exactly for the hydrogen atom.

hydrogen azide *See* hydrazoic acid.

hydrogen bond (H-bonding) An intermolecular bond between molecules in which hydrogen is bound to a strongly electronegative element. Bond polarization by the electronegative element X leads to a positive charge on hydrogen $X^{\delta-}–H^{\delta+}$; this hydrogen can then interact directly with electronegative elements of adjacent molecules. The hydrogen bond is represented as a dotted line:

$$X^{\delta-} - H^{\delta+} \ldots\ldots X^{\delta-} - H^{\delta+} \ldots$$

The length of a hydrogen bond is characteristically 0.15–0.2 nm. Hydrogen bonding may lead to the formation of dimers (for example, in carboxylic acids) and is used to explain the anomalously high boiling points of H_2O and HF.

hydrogen bromide (HBr) A colorless sharp-smelling gas that is very soluble in water, giving HYDROBROMIC ACID. It is produced by direct combination of hydrogen and bromine in the presence of a platinum catalyst or by the reaction of phosphorus tribromide with water. Hydrogen bromide is rather inactive chemically. It will not conduct electricity in the liquid state, indicating that it is a molecular compound.

hydrogencarbonate (bicarbonate) A salt containing the ion $^{-}HCO_3$.

hydrogen chloride (HCl) A colorless gas that has a strong irritating odor and fumes strongly in moist air. It is prepared by the action of concentrated sulfuric acid on sodium chloride. The gas is made industrially by burning a stream of hydrogen in chlorine. It is not particularly reactive but will form dense white clouds of ammonium chloride when mixed with ammonia. It is very soluble in water and ionizes almost completely to give HYDROCHLORIC ACID. It will also dissolve in many nonaqueous solvents, including toluene, but will not ionize, indicating that the resulting solution will show no acidic properties. Hydrogen chloride is used in the manufacture of organic chlorine compounds, such as polyvinyl chloride (PVC).

Unlike the other hydrogen halides, hydrogen chloride will not dissociate on heating, indicating a strong H–Cl bond.

hydrogen cyanide *See* hydrocyanic acid.

hydrogen electrode (standard hydrogen half cell) A type of half cell based on hydrogen, and assigned zero electrode potential, so that other elements may be compared with it. The hydrogen is bubbled over a platinum electrode, coated in 'platinum black', in a 1 M acid solution. Hydrogen is adsorbed on the platinum black, which has a high surface area, enabling the equilibrium

$$H(g) \rightleftharpoons H^{+}(aq) + e^{-}$$

to be set up. The platinum is inert and has no tendency to form platinum ions in solution. *See also* electrode potential.

hydrogen fluoride (HF) A colorless liquid produced by the reaction of concentrated sulfuric acid with calcium fluoride:

$$CaF_2(s) + H_2SO_4(aq) \rightarrow CaSO_4(aq) + 2HF(l)$$

It produces toxic corrosive fumes and dissolves readily in water to give HYDROFLUORIC ACID.

Hydrogen fluoride is atypical of the hydrogen halides inasmuch as the individual H–F units are associated into much larger units, forming zigzag chains and rings. This effect is caused by hydrogen bonds that form between the hydrogen and the highly electronegative fluoride ions. Hydrogen fluoride is used extensively as a catalyst in the petroleum industry.

hydrogen ion A positively charged hydrogen atom, H^+, i.e. a proton. Hydrogen ions are produced by all acids in water, in which they are hydrated to hydroxonium (hydronium) ions, H_3O^+. *See* acid; pH.

hydrogen molecule ion (H_2^+) The simplest type of molecule. It consists of two hydrogen nuclei and one electron. In the BORN–OPPENHEIMER APPROXIMATION in which the nuclei are regarded as being fixed the SCHRÖDINGER EQUATION for the hydrogen molecule ion can be solved exactly. This enables ideas and approximation techniques concerned with chemical bonding to be tested quantitatively.

hydrogen peroxide (H_2O_2) A colorless syrupy liquid, usually used in solution in water. Although it is stable when pure, on contact with bases such as manganese(IV) oxide it gives off oxygen, the manganese(IV) oxide acting as a catalyst:

$$2H_2O_2 \rightarrow 2H_2O + O_2$$

Hydrogen peroxide can act as an oxidizing agent, converting iron(II) ions to iron(III) ions, or as a reducing agent with potassium manganate(VII). It is used as a bleach and mild antiseptic, and as an oxidant in rocket fuel. The strength of H_2O_2 solutions is usually given as *volume strength*, i.e. the volume (in cubic decimeters) of oxygen released at STP by the decomposition of one cubic decimeter of the solution.

hydrogensulfate (bisulfate; HSO_4^-) An acidic salt of sulfuric acid (H_2SO_4), in which only one of the acid's hydrogen atoms has been replaced by a metal. An example is sodium hydrogensulfate, $NaHSO_4$.

hydrogen sulfide (sulfuretted hydrogen; H_2S) A colorless very poisonous gas with an odor of bad eggs. Hydrogen sulfide is prepared by reacting hydrochloric acid with iron(II) sulfide. Its presence can be determined by mixing with lead nitrate solution, with which H_2S gives a black precipitate. Its aqueous solution is weakly acidic. Hydrogen sulfide reduces iron(III) chloride to iron(II) chloride, forming hydrochloric acid and a yellow precipitate of sulfur. Hydrogen sulfide precipitates metals from insoluble sulfides in acid solution, and is used in qualitative analysis. It burns with a blue flame in oxygen to form sulfur(IV) oxide and water. Natural gas contains some hydrogen sulfide, which is removed before supply to the consumer.

hydrogensulfite (bisulfite; HSO_3^-) An acidic salt of sulfurous acid (H_2SO_3), in which only one of the acid's hydrogen atoms has been replaced by a metal. An example is sodium hydrogensulfite, $NaHSO_3$.

hydrohalic Describing acids formed by *hydrogen halides*, e.g. HF, HCl, HBr, when dissolved in water.

hydroiodic acid *See* iodine.

hydrolysis A reaction between a compound and water, e.g.:
Salts of weak acids

$$Na_2CO_3 + 2H_2O \rightarrow 2NaOH + H_2CO_3$$

and certain inorganic halides

$$SiCl_4 + 4H_2O \rightarrow Si(OH)_4 + 4HCl$$

hydronium *See* acid; hydrogen ion.

hydrophilic *See* lyophilic.

hydrophobic *See* lyophobic.

hydrosol A colloid in aqueous solution.

hydroxide A compound containing the ion OH^- or the group –OH.

hydroxonium ion *See* acid; hydrogen ion.

hydroxyl group A group (–OH) containing hydrogen and oxygen, characteristic of alcohols and some hydroxides. It should not be confused with the hydroxide ion (OH^-).

hygroscopic Describing a substance that absorbs moisture from the atmosphere. *See also* deliquescent.

hyperfine structure *See* fine structure.

hypo *See* sodium thiosulfate(IV).

hypobromous acid *See* bromic(I) acid.

hypochlorous acid *See* chloric(I) acid.

hypophosphorous acid *See* phosphinic acid.

hyposulfuric acid *See* dithionic acid.

hyposulfurous acid *See* dithionous acid.

I

Iceland spar A pure transparent nearly flawless mineral form of calcite (calcium carbonate, $CaCO_3$), noted for its property of birefringence (double refraction). It comes from Eskifjördhur in Iceland and has been used in optics.

icosahedron A polyhedron bounded by 20 triangular faces. The icosahedron is one of the main structural units of boron and its compounds. Icosohedral symmetry also occurs in certain CLUSTER COMPOUNDS.

ideal gas (perfect gas) *See* gas laws; kinetic theory.

ideal-gas scale *See* absolute temperature.

ideal solution *See* Raoult's law.

immiscible Describing two or more liquids that will not mix, such as oil and water. After being shaken together and left to stand they form separate layers.

indicator A compound that reversibly changes color depending on the pH of the solution in which it is dissolved. The visual observation of this change is therefore a guide to the pH of the solution and it follows that careful choice of indicators permits a wide range of end points to be detected in acid–base titrations.

Redox titrations require either specific indicators, which detect one of the components of the reaction (e.g. starch for iodine, potassium thiocyanate for Fe^{3+}), or true redox indicators in which the transition potential of the indicator between oxidized and reduced forms is important. The transition potential of a redox indicator is analogous to the transition pH in acid–base systems.

Complexometric titrations require indicators that complex with metal ions and change color between the free state and the complex state. *See also* absorption indicator.

indium A soft malleable ductile silvery metallic element belonging to group 13 (formerly IIIA) of the periodic table. A rare element, it is found in minute quantities, primarily in zinc ores. It is used in alloys, in several electronic devices, and in electroplating.

Symbol: In; m.p. 155.17°C; b.p. 2080°C; r.d. 7.31 (25°C); p.n. 49; most common isotope ^{115}In; r.a.m. 114.818.

inert gases *See* rare gases.

infrared (IR) Electromagnetic radiation with longer wavelengths than visible radiation. The wavelength range is approximately 0.7 μm to 1 mm. Many materials transparent to visible light are opaque to infrared, including glass. Rock salt, quartz, germanium, or polyethene prisms and lenses are suitable for use with infrared. Infrared radiation is produced by movement of charges on the molecular scale; i.e. by vibrational or rotational motion of molecules. *Infrared spectroscopy* is of particular importance in organic chemistry and absorption spectra are used extensively in identifying compounds. Certain bonds between pairs of atoms (C–C, C=C, C=O, etc.) have characteristic vibrational frequencies, which correspond to bands in the infrared spectrum. Infrared spectra are thus used in finding the structures of new organic compounds by indicating the presence of certain groups. They are also used to 'fingerprint' and thus identify known compounds. At shorter wavelengths, infrared absorption corresponds to transitions between rotational energy levels, and can be used to find the dimensions of molecules (by their moment of inertia).

inhibitor A substance that slows down the rate of reaction. Hydrogen sulfide, hydrogen cyanide, mercury salts, and arsenic compounds readily inhibit heterogeneous catalysts by adsorption. For example, arsenic compounds inhibit platinum catalysts in the oxidation of sulfur(IV) oxide to sulfur(VI) oxide. Inhibitors must not be confused with negative catalysts; inhibitors do not change the pathway of a reaction.

inorganic chemistry The branch of chemistry concerned with elements other than carbon and with the preparation, properties, and reactions of their compounds. Certain simple carbon compounds are treated in inorganic chemistry, including the oxides, carbon disulfide, the halides, hydrogen cyanide, and salts, such as the cyanides, cyanates, carbonates, and hydrogencarbonates.

insoluble Describing a compound that has a very low solubility (in a specified solvent).

instrumentation The measurement of the conditions and the control of processes within a chemical plant. The instruments can be classified into three groups: those for current information using mercury thermometers, weighing scales, and pressure gauges; those for recording viscosity, fluid flow, pressure, and temperature; and those instruments that control and maintain the desired conditions including pH and the flow of materials.

intercallation compound A compound that has a structure based on layers and in which there are layers of a different character interleaved in the basic structural units. For example, the micas phlogopite ($KMg_3(OH)_2Si_3AlO_{10}$) and muscovite ($KAl_2(OH)_2Si_3AlO_{10}$) are formed by interleaving K^+ ions replacing a quarter of the silicon layers in talc and pyrophyllite respectively. *See also* lamellar compound.

interhalogen *See* bromine.

intermediate **1.** A compound that requires further chemical treatment to produce a finished industrial product such as a dye or pharmaceutical chemical.
2. A transient chemical entity in a complex reaction. *See also* precursor.

intermediate bonding Describing a form of covalent bond that also has an ionic or electrovalent character. *See* polar bond.

intermolecular forces Forces of attraction between molecules rather than forces within the molecule (chemical bonding). If these intermolecular forces are weak the material will be gaseous and as their strength progressively increases materials become progressively liquids and solids. The intermolecular forces are divided into H-bonding forces and van der Waals forces, and the major component is the electrostatic interaction of dipoles. *See* hydrogen bond; van der Waals force.

internal energy Symbol: *U* The energy of a system that is the total of the kinetic and potential energies of its constituent particles (e.g. atoms and molecules). If the temperature of a substance is raised, by transferring energy to it, the internal energy increases (the particles move faster). Similarly, work done on or by a system results in an increase or decrease in the internal energy. The relationship between heat, work, and internal energy is given by the first law of thermodynamics. Sometimes the internal energy of a system is loosely spoken of as 'heat' or 'heat energy'. Strictly, this is incorrect; heat is the transfer of energy as a result of a temperature difference.

internal resistance Resistance of a source of electricity. In the case of a cell, when a current is supplied, the potential difference (p.d.) between the terminals is lower than the e.m.f. The difference (i.e.

e.m.f. – p.d.) is proportional to the current supplied. The internal resistance (r) is given by:

$$r = (E - V)/I$$

where E is the e.m.f., V the potential difference between the terminals, and I the current.

interstitial *See* defect.

interstitial carbide *See* carbide.

interstitial compound A crystalline compound in which atoms of a nonmetal (e.g. carbon, hydrogen, or boron) occupy interstitial positions in the crystal lattice of a metal (usually a transition metal). Interstitial compounds are often nonstoichiometric. Their physical properties are often similar to those of metals; e.g. they have a metallic luster and are electrical conductors.

interstitial hydrides *See* hydrogen.

iodate *See* iodine.

iodic acid *See* iodic(V) acid.

iodic(V) acid (iodic acid; HIO_3) A colorless deliquescent crystalline solid produced by the reaction of concentrated nitric acid with iodine. Iodic(V) acid is a strong oxidizing agent. It will liberate iodine from solutions containing iodide ions and it reacts vigorously with organic materials, often producing flames. It dissociates extensively in water and hence is a strong acid.

iodic(VII) acid (periodic acid; H_5IO_6) A white crystalline solid made by low-temperature electrolysis of concentrated iodic(V) acid. It exists in a number of forms, the most common of which is paraiodic(VII) acid. Iodic(VII) acid is a powerful oxidizing agent. It is a weak acid, which – by cautious heating under vacuum – can be converted to *dimesoiodic(VII) acid* ($H_4I_2O_9$), and *metaiodic(VII) acid* (HIO_4). All three will oxidize manganese to manganate(VII), reactions that can be used to determine the presence of small amounts of manganese in steel.

iodide *See* halide.

iodine A dark-violet volatile solid element belonging to the halogens i.e. group 17 (formerly VIIA) of the periodic table, of which it is the fourth element. It occurs in seawater and is concentrated by various marine organisms in the form of iodides. Significant deposits also occur in the form of *iodates*, i.e. salts containing iodine oxyanions. The element is conveniently prepared by the oxidation of iodides in acid solution (using MnO_2). Industrial methods similarly use oxidation of iodides or reduction of iodates to iodides by sulfur(IV) oxide (sulfur dioxide) followed by oxidation, depending on the source of the raw materials. Iodine and its compounds are used in chemical synthesis, photography, pharmaceuticals, and dyestuffs manufacture. Trace amounts are essential in the diet to ensure proper production of thyroid hormones.

Iodine has the lowest electronegativity of the stable halogens and consequently is the least reactive. It combines only slowly with hydrogen to form *hydroiodic acid*, HI. Iodine also combines directly with many electropositive elements, but does so much more slowly than does bromine or chlorine. Because of the larger size of the iodine ion and the consequent low lattice energies, the iodides are generally more soluble than related bromides or chlorides. As with the other halides, iodides of Ag(I), Cu(I), Hg(I), and Pb(II) are insoluble unless complexing ions are present.

Iodine also forms a range of covalent iodides with the metalloids and nonmetallic elements, but these are generally less thermodynamically stable and are more readily hydrolyzed than chlorine or bromine analogs.

Four oxides are known, of which iodine(V) oxide, I_2O_5, is the most important. The other oxides, I_2O_4, I_4O_9, and I_2O_7 are much less stable and of uncertain structure. Like chlorine (but *not* bromine) iodine forms oxo-species based on IO^-, IO_2^-, IO_3^-, and IO_4^-. The chemistry of the other oxo-species in solution is complex. Elemental iodine reacts with alkalis in a similar way to bromine and chlorine.

The ionization potential of iodine is sufficiently low for it to form a number of compounds in which it is electropositive. It forms I^+ cations, for example, by reaction of solid silver nitrate with iodine solution. Iodine also exhibits the interesting property of forming solutions that are violet colored in nondonor type solvents, but dark brown solutions in donor solvents, such as ethanol. Even though iodine solutions were once common as antiseptic agents, the element is classified as toxic and care should be taken to avoid eye intrusions or excessive skin contact.

Symbol: I; m.p. 113.5°C; b.p. 184°C; r.d. 4.93 (20°C); p.n. 53; most common isotope ^{127}I; r.a.m. 126.90447.

iodine monochloride (ICl) A dark red liquid made by passing chlorine over iodine. It has properties similar to those of its constituent halogens. Iodine monochloride is used as a nonaqueous solvent and as an iodating agent in organic reactions.

iodine(V) oxide (iodine pentoxide; I_2O_5) A white crystalline solid made by heating iodic(V) acid to a temperature of 200°C. It is a very strong oxidizing agent.

Iodine oxide is the acid anhydride of iodic(V) acid, which is reformed when water is added to the oxide. Its main use is in titration work, measuring traces of carbon monoxide in the air.

iodine pentoxide *See* iodine(V) oxide.

iodine trichloride *See* diiodine hexachloride.

ion An atom or molecule that has a negative or positive charge as a result of losing or gaining one or more electrons. *See* electrolysis; ionization.

ion association The association of pairs of ions in electrolyte solutions occurring because of electrostatic attraction between the ions. Ion association means that complete dissociation of electrolytes does not occur. This has the effect that electrical conductivities that are found experimentally are lower than the expectations of the DEBYE–HÜCKEL THEORY. Ion association is a dynamic phenomenon in which there is rapid exchange between the ions in the electrolyte solution. A theory to take ion association into account, and hence improve on the Debye–Hückel theory, was proposed by Neils Bjerrum (1879–1958) in 1926. Bjerrum found that ion association becomes more significant as the dielectric constant of the solvent becomes lower.

ion exchange A process that takes place in certain insoluble materials, which contain ions capable of exchanging with ions having the same charge in the surrounding medium. ZEOLITES, the first ion exchange materials, were used for water softening. These have largely been replaced by synthetic resins made of an inert backbone material, such as polyphenylethene, to which ionic groups are weakly attached. If the ions exchanged are positive, the resin is a CATIONIC RESIN. An ANIONIC RESIN exchanges negative ions. When all available ions on the resin have been exchanged (e.g. sodium ions replaced by calcium ions from the medium) the material can usually be regenerated by passing a concentrated solution containing the original resin ion (e.g. sodium chloride) through it. In the example cited, the calcium ions will be replaced by sodium ions. Ion-exchange techniques are used for a vast range of purification and analytical purposes.

For example, solutions with ions that can be exchanged for OH^- and H^+ can be estimated by titrating the resulting solution with an acid or a base.

ionic bond *See* electrovalent bond.

ionic carbide *See* carbide.

ionic crystal A crystal composed of ions of two or more elements. The positive and negative ions are arranged in definite patterns and are held together by electrostatic attraction. Common examples are sodium chloride and cesium chloride.

ionic product The product of ionic concentrations in a solution. In the case of pure water the equilibrium constant is:

$$K_W = [H^+][OH^-]$$

as a result of a small amount of self-ionization.

ionic radius A measure of the effective radius of an ion in a compound. For an isolated ion, the concept is not very meaningful, since the ion is a nucleus surrounded by an 'electron cloud'. Values of ionic radii can be assigned, however, based on the distances between ions in crystals.

Different methods exist for determining ionic radii and often different values are quoted for the same ion. The two main methods are those of Goldschmidt and of Pauling. The Goldschmidt radii are determined by substituting data from one compound to another, to produce a set of ionic radii for different ions. The Pauling radii are assigned by a more theoretical treatment for apportioning the distance between an anion and a cation.

ionic strength For an ionic solution a quantity can be introduced that emphasizes the charges of the ions present:

$$I = \frac{1}{2}\Sigma_i m_i z^2_i$$

where m is the molality and z the ionic charge. The summation is continued over all the different ions in the solution, i.

ionization The process of producing ions. There are several ways in which ions may be formed from atoms or molecules. In certain chemical reactions ionization occurs by transfer of electrons; for example, sodium atoms and chlorine atoms react to form sodium chloride, which consists of sodium ions (Na^+) and chloride ions (Cl^-). Certain molecules can ionize in solution; acids, for example, form hydrogen ions as in the reaction

$$H_2SO_4 \rightarrow 2H^+ + SO_4^{2-}$$

The 'driving force' for ionization in a solution is solvation of the ions by molecules of the solvent. H^+, for example, is solvated as a hydroxonium (hydronium) ion, H_3O^+.

Ions can also be produced by ionizing radiation; i.e. by the impact of particles or photons with sufficient energy to break up molecules or detach electrons from atoms: $A \rightarrow A^+ + e^-$. Negative ions can be formed by capture of electrons by atoms or molecules: $A + e^- \rightarrow A^-$.

ionization energy *See* ionization potential.

ionization potential (IP; Symbol: I) The energy required to remove an electron from an atom (or molecule or group) in the gas phase, i.e. the energy required for the process:

$$M \rightarrow M^+ + e^-$$

It gives a measure of the ability of metals to form positive ions. The second ionization potential is the energy required to remove two electrons and form a doubly charged ion:

$$M \rightarrow M^{2+} + e^-$$

Ionization potentials stated in this way are positive; often they are given in electronvolts. *Ionization energy* is the energy required to ionize one mole of the substance, and is usually stated in kilojoules per mole (kJ mol^{-1}).

In chemistry, the terms 'second', 'third', etc., ionization potentials are usually used for the formation of doubly, triply, etc., charged ions. However, in spectroscopy and physics, they are often used with a different meaning. The second ionization potential is the energy to remove the second least strongly bound electron in forming a singly charge ion. For lithium ($ls^2 2s^1$) it would refer to removal of a 1s electron to produce an excited ion with the configuration $1s^1 2s^1$. Note also that ionization potentials are now stated as energies.

ion pair A positive ion and a negative ion in close proximity in solution, held by the attractive force between their charges. *See* electrolysis.

IP *See* ionization potential.

IR *See* infrared.

iridium A hard brittle white transition metal, the third element of group 9 (formerly part of subgroup VIIIB) of the periodic table. Iridium is found native and in association with platinum, to which it is chemically similar. Unusually high concentrations of iridium are also found worldwide in clay bands associated with the boundary marking the end of the Cretaceous Period 65 million years ago. This iridium is thought to have been deposited by the explosive impact of an ancient asteroid, which in turn may account for the subsequent mass extinctions of many organisms, including the dinosaurs.

Iridium is highly resistant to corrosion making it desirable for alloys for high-precision instruments and bearings. It is also used in electrical contacts, in pen points, in spark plugs, and in jewelry.

Symbol: Ir; m.p. 2410°C; b.p. 4130°C; r.d. 22.56 (17°C); p.n. 77; most common isotope ^{193}Ir; r.a.m. 192.217.

iron A malleable ductile medium-hard silvery-gray ferromagnetic transition element, the first element of group 8 (formerly part of subgroup VIIIB) of the periodic table. Iron occurs in many ores, especially hematite (Fe_2O_3), magnetite (Fe_3O_4), siderite ($FeCO_3$), and pyrite (FeS_2). It is extracted in blast furnaces using coke, limestone, and hot air. The coke and oxygen in the air form carbon monoxide, which then reduces the iron in the ore to the metal. The limestone removes acidic impurities and forms a layer of slag above the molten iron at the base of the furnace. Carbon STEEL is formed by reducing the carbon content to between 0.1 and 1.5%.

Iron is used to form structural supports for a wide variety of applications, including bridges, and as a catalyst in the Haber process for ammonia production. Iron is also an essential element in the diet, because it is required to make hemoglobin, a component of red blood cells which absorbs oxygen in the lungs and transports it to the tissues.

Iron corrodes in air and moisture to hydrated iron(III) oxide (rust).

The most stable oxidation state is +3, compounds of which are yellow or brown, but +2 (green) and +6 (easily reduced) states also exist. Solutions of iron(II) ions give a green precipitate with sodium hydroxide solution, whereas iron(III) ions give a brown precipitate. The concentration of iron(II) ions can be estimated in acid solution by titration with standard potassium manganate(VII) solution. Iron(III) ions must first be reduced to iron(II) ions with sulfur(IV) oxide (sulfur dioxide).

Symbol: Fe; m.p. 1535°C; b.p. 2750°C; r.d. 7.874 (20°C); p.n. 26; most common isotope ^{56}Fe; r.a.m. 55.845.

iron(II) chloride (ferrous chloride; $FeCl_2$) A compound prepared by passing dry hydrogen chloride gas over heated iron. White feathery anhydrous crystals are produced. Hydrated iron(II) chloride ($FeCl_2.6H_2O$) is prepared by reacting excess iron with dilute or concentrated hydrochloric acid. Green crystals of the hexahydrate are obtained on crystallization from solution. Iron(II) chloride is readily soluble in water, producing an acidic solution due to salt hydrolysis.

iron(III) chloride (ferric chloride; $FeCl_3$) A compound prepared in the anhydrous state as dark red crystals by passing dry chlorine over heated iron. The product sublimes and is collected in a cooled receiver as brownish-black iridescent scales. The hydrated salt ($FeCl_3.6H_2O$) is prepared by adding excess iron(III) oxide to concentrated hydrochloric acid. On crystallization yellow-brown crystals of the hexahydrate are formed. Iron(III) chloride is very soluble in water and undergoes salt hydrolysis. At temperatures below 400°C the anhydrous salt exists as a dimer, Fe_2Cl_6.

iron chromium oxide *See* chromite.

iron(II) oxide (ferrous oxide; FeO) A black powder formed by the careful reduction of iron(III) oxide using either carbon monoxide or hydrogen. It can also be prepared by heating iron(II) oxalate in the absence of air. It is only really stable at high temperatures and disproportionates slowly on cooling to give iron(III) oxide and iron. Iron(II) oxide can be reduced by heating in a stream of hydrogen. When exposed to air, it is oxidized to iron(III) oxide. It is a basic oxide, dissolving readily in dilute acids to form iron(II) salt solutions. If heated to a high temperature in an inert atmosphere, iron(II) oxide disproportionates to give iron and triiron tetroxide (Fe_3O_4).

iron(III) oxide (ferric oxide; Fe_2O_3) A rusty-brown solid prepared by the action of heat on iron(III) hydroxide or iron(II) sulfate. It occurs in nature as the mineral HEMATITE. Industrially it is obtained by roasting iron pyrites. Iron(III) oxide dissolves in dilute acids to produce solutions of iron(III) salts. It is stable at red heat, decomposes around 1300°C to give triiron tetroxide (Fe_3O_4), and can be reduced to iron by hydrogen at 1000°C. Iron(III) oxide is not ionic in character but has a structure similar to that of aluminum oxide.

iron(II) sulfate (ferrous sulfate; green vitriol; $FeSO_4.7H_2O$) A compound that occurs in nature as the mineral *melanterite* (or *copperas*). It is made industrially from iron pyrites. In the laboratory iron(II) sulfate is prepared by dissolving excess iron in dilute sulfuric acid. On crystallization, green crystals of the heptahydrate are obtained. Careful heating of the hydrated salt yields anhydrous iron(II) sulfate; on further heating the sulfate decomposes to give iron(III) oxide, sulfur(IV) oxide, and sulfur(VI) oxide. The hydrated crystals oxidize easily on exposure to air owing to the formation of basic iron(III) sulfate. A freshly prepared solution of iron(II) sulfate absorbs nitrogen monoxide (BROWN-RING TEST).

Iron(II) sulfate crystals are isomorphous with the sulfates of zinc, magnesium, nickel, and cobalt.

iron(III) sulfate (ferric sulfate; $Fe_2(SO_4)_3$) A compound prepared in the hydrated state by the oxidation of iron(II) sulfate dissolved in dilute sulfuric acid using an oxidizing agent, such as hydrogen peroxide or concentrated nitric acid. On crystallization, the solution deposits a white mass of the nonahydrate ($Fe_2(SO_4)_3.9H_2O$). The anhydrous salt can be prepared by gently heating the hydrated salt. Iron(III) sulfate decomposes on heating to give iron(III) oxide and sulfur(VI) oxide. It forms alums with the sulfates of the alkali metals.

irreversible change *See* reversible change.

irreversible reaction A reaction in which conversion to products is complete; i.e. there is little or no back reaction.

isobars **1.** Two or more nuclides that have the same NUCLEON NUMBERS but different proton numbers. Actinium-89 and thorium-90 are isobars.
2. Lines on a chart or graph joining points of equal pressure.

isoelectronic Describing compounds that have the same number of electrons. For example, carbon monoxide (CO) and nitrogen (N_2) are isoelectronic.

isomer *See* isomerism.

isomerism The existence of two or more chemical compounds with the same molecular formulae but different structural formulae or different spatial arrangements

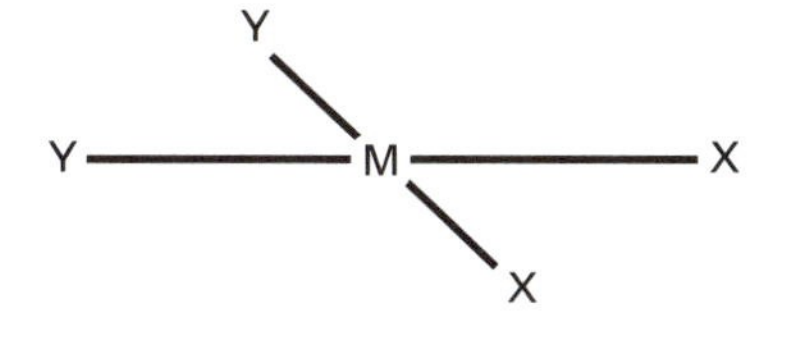

cis-isomer

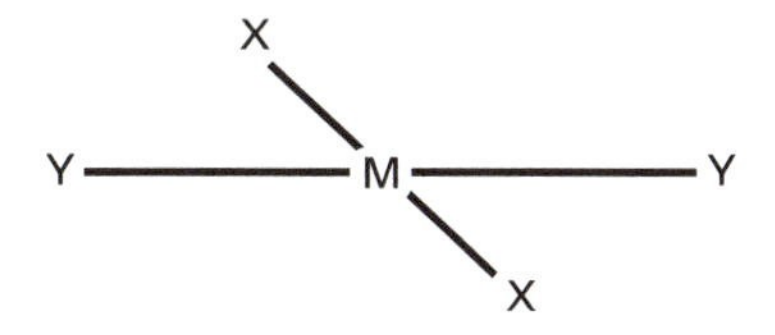

trans-isomer

Isomerism: isomers of a square-planar complex

of atoms. The different forms are known as *isomers.*

In *structural isomerism* the structural formulae of the compounds differ as to molecular structure or to the position of functional groups. For example, the compound C_4H_{10} may be butane (with a straight chain of carbon atoms) or 2-methyl propane ($CH_3CH(CH_3)CH_3$, with a branched chain). A particular case of this occurs in metal complexes when a ligand may coordinate to a metal ion in two ways. For example, NO_2 can coordinate through N (the *nitro ligand*) or through O (the *nitrido ligand*). Such ligands are said to be *ambidentate.* Complexes that differ only in the way in which the ligand coordinates are said to show *linkage isomerism.*

Stereoisomerism occurs when two compounds with the same molecular formulae and the same groups differ only in the arrangement of the groups in space. There are various types of stereoisomerism.

Cis-trans (or *syn-anti*) *isomerism* occurs when there is restricted rotation about a bond between two atoms (e.g. a double bond or a bond in a ring). Groups attached to each atom may be on the same side of the bond (the *cis-* or *syn-isomer*) or opposite sides (the *trans-* or *anti-isomer*). *Cis-trans* isomerism occurs in square-planar complexes of the type MX_2Y_2, where M is a metal ion and X and Y are different ligands. If the X ligands are adjacent the isomer is a cis-isomer; if they are opposite, it is a trans-isomer. This type of isomerism can also occur in octahedral complexes of the type MX_2Y_4. Octahedral complexes of the type MX_3Y_3 show a different type of stereoisomerism. If the three X ligands are

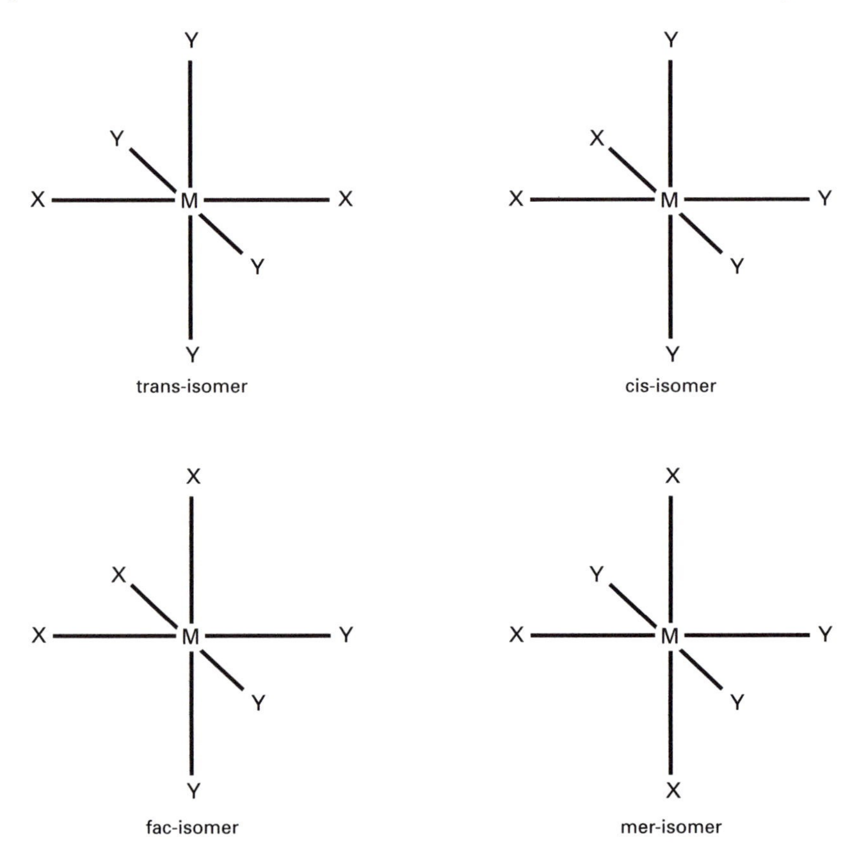

Isomerism: isomers of octahedral complexes

in a plane that includes the M ion (with the three Y ligands in a plane at right angles), then the structure is called the *mer-isomer* ('mer' stands for meridional). If the three X (and Y) ligands are all on a face of the octahedron, the structure is the *fac-isomer* ('fac' stands for facial). *Cis-trans* isomerism was formerly called *geometrical isomerism.*

Optical isomerism occurs when the compound has no plane of symmetry and can exist in left- and right-handed forms that are mirror images of each other. *See also* optical activity.

isomorphism The existence of compounds with the same crystal structure.

isomorphism, law of *See* Mitscherlich's law.

isotherm A line on a chart or graph joining points of equal temperature. *See also* isothermal change.

isothermal change A process that takes place at a constant temperature. Throughout an isothermal process, the system is in thermal equilibrium with its surroundings. For example, a cylinder of gas in contact with a constant-temperature bath may be compressed slowly by a piston. The work done appears as energy, which flows into the bath to keep the gas at the same temperature. Isothermal changes are contrasted with *adiabatic changes*, in which no energy enters or leaves the system, and the temperature changes. In practice no process is perfectly isothermal and none is perfectly adiabatic, although some can approximate in behavior to one of these ideals.

isotones Two or more nuclides that have the same neutron numbers but different proton numbers.

isotonic Describing solutions having the same osmotic pressure.

isotopes Two or more species of the same element differing in their mass numbers because of differing numbers of neutrons in their nuclei. The nuclei must have the same number of protons (an element is characterized by its proton number). Isotopes of the same element have very similar properties because they have the same electron configuration, but differ slightly in their physical properties. An unstable isotope is termed a *radioisotope* or *radioactive isotope*.

Isotopes of elements are useful in chemistry for studies of the mechanisms of chemical reactions. A standard technique is to *label* one of the atoms in a molecule by using an isotope of a component element, which is known as a tracer. It is then possible to follow the way in which this atom behaves throughout the course of the reaction. In labeling, radioisotopes are detected by counters; stable isotopes can also be used, and detected by a mass spectrum.

Isotopes are also used in kinetic studies. For example, if the bond between two atoms X–Y is broken in the rate-determining step, and Y is replaced by a heavier isotope of the element, Y*, the reaction rate will be slightly lower with the Y* present. This *kinetic isotope effect* is particularly noticeable with hydrogen and deuterium compounds, because of the large relative difference in mass.

isotope separation The separation of different isotopes of a chemical element, making use of the some differences in the physical properties of these isotopes. For small-scale isotope separation a MASS SPECTROMETER is frequently used. For large-scale isotope separation the methods used include diffusion in the gas phase (used for uranium from the gas uranium hexafluoride), distillation (which was used to produce heavy water), electrolysis, the use of centrifuges, and the use of lasers, with excitation of one isotope followed by its separation using an electromagnetic field.

isotopic mass (isotopic weight) The mass number of a given isotope of an element.

isotopic number The difference between the number of neutrons in an atom and the number of protons.

isotopic weight *See* isotopic mass.

J

Jahn–Teller effect A distortion of molecules that occurs so as to prevent the molecule having degenerate valence orbitals. The Jahn–Teller effect is observed in metal complexes. Some complexes that would be expected to be regular octahedra are distorted octahedra because of the effect. It was predicted theoretically by H. A. Jahn and Edward Teller in 1937.

jeweler's rouge Iron(III) oxide (Fe_2O_3), used as a mild abrasive.

joliotium Symbol: Jl A name formerly suggested for element-105, now known as dubnium (Db). *See* element.

joule Symbol: J The SI derived unit of all forms of energy and work, equal to the energy required or work done when the point of application of a force of one newton moves another object one meter in the direction of the force. 1 J = 1 N m. It is named for the British physicist James Joule (1818–89).

K

kainite A mineral form of hydrated crystalline magnesium sulfate ($MgSO_4$) containing also some potassium chloride. It is used as a fertilizer and source of potassium compounds.

kaolin *See* China clay.

kaolinite A hydrated aluminosilicate mineral, $Al_2(OH)_4Si_2O_5$, the major constituent of CHINA CLAY. It is formed by the weathering of FELDSPARS from granites.

katal Symbol: kat The SI derived unit of catalytic activity, equal to the amount of substance in moles that can be catalyzed per second. 1 kat = 1 mol s^{-1}.

kelvin Symbol: K The SI base unit of thermodynamic temperature. It is defined as the fraction 1/273.16 of the thermodynamic temperature of the triple point of water. Zero kelvin (0 K) is absolute zero. An interval of one kelvin is the same as an interval of one degree Celsius on the Celsius scale. It is named for British physicist Baron William Thompson Kelvin (1824–1907), who devised the ABSOLUTE TEMPERATURE scale that now bears his name in 1848.

kieselguhr *See* diatomite.

killed spirits Zinc(II) chloride solution when used as a flux for solder.

kilo- Symbol: k A prefix used with SI units denoting 10^3. For example, 1 kilometer (km) = 10^3 meters (m).

kilocalorie *See* calorie.

kilogram (kilogramme) Symbol: kg The SI base unit of mass, equal to the mass of the international prototype of the kilogram, a cylinder consisting of a platinum–iridium alloy kept in a vault at Sèvres in France. The kilogram is the only SI base unit that to date has yet to be defined in terms of physical constants.

kilogramme *See* kilogram.

kilowatt-hour Symbol: kWh An m.k.s. system unit of electrical energy often used to to measure electrical power consumption, equal to the energy transferred or work done by one kilowatt of power in one hour. It has a value of 3.6×10^6 joules in SI units.

kinetic energy *See* energy.

kinetic isotope effect *See* isotopes.

kinetics A branch of physical chemistry concerned with the study of rates of chemical reactions and the effect of physical conditions that influence the rate of reaction, e.g. temperature, light, concentration, etc. The measurement of these rates under different conditions gives information on the mechanism of the reaction, i.e. on the sequence of processes by which reactants are converted into products.

kinetic theory A theory explaining physical properties in terms of the motion of particles. The kinetic theory of gases assumes that the molecules or atoms of a gas are in continuous random motion and the pressure (p) exerted on the walls of a containing vessel arises from the bombardment by these fast moving particles. When the temperature is raised the speeds increase; so consequently does the pressure.

If more particles are introduced or the volume is reduced there are more particles to bombard unit area of the walls and the pressure also increases. When a particle collides with the wall it experiences a rate of change of momentum, which is proportional to the force exerted. For a large number of particles this provides a steady force per unit area (or pressure) on the wall.

Following certain additional assumptions, the kinetic theory leads to an expression for the pressure exerted by an ideal gas. The assumptions are:

1. The particles behave as if they are hard smooth perfectly elastic points.
2. They do not exert any appreciable force on each other except during collisions.
3. The volume occupied by the particles themselves is a negligible fraction of the volume of the gas.
4. The duration of each collision is negligible compared with the time between collisions.

By considering the change in momentum on impact with the walls it can be shown that

$$p = \rho c^{2/3}$$

where ρ is the density of the gas and c is the root-mean-square speed of the molecules. The mean-square speed of the molecules is proportional to the absolute temperature:

$$Nmc^2 = RT$$

See also degrees of freedom.

Kipp's apparatus An apparatus for the production of a gas from the reaction of a liquid on a solid. It consists of three globes, the upper globe being connected via a wide tube to the lower globe. The upper globe is the liquid reservoir. The middle globe contains the solid and also has a tap at which the gas may be drawn off. When the gas is drawn off the liquid rises from the lower globe to enter the middle globe and reacts with the solid, thereby releasing more gas. When turned off the gas released forces the liquid back down into the lower globe and up into the reservoir, thus stopping the reaction. It is named for Dutch chemist Petrus Kipp (1808–84).

Kroll process A method of obtaining certain metals by reducing the metal chloride with magnesium. Titanium can be obtained in this way:

$$TiCl_4 + 2Mg \rightarrow 2MgCl_2 + Ti$$

It is named for its inventor, Luxembourg metallurgist William J. Kroll (1889–1973), who developed it in 1932.

Kroto, Sir Harold Walter (1939–) British chemist. Along with his Sussex University colleague David Walton, Kroto had a long-standing interest in molecules containing carbon chains linked by alternate triple and single bonds. Such chains with five, seven, and nine carbon atoms had been identified by radioastronomers in space. In 1984 Kroto heard that the American chemist Richard SMALLEY had developed new techniques involving laser bombardment for the production of clusters of atoms. In 1985 he visited Smalley in Houston and persuaded him to direct his laser beam at a graphite target. Clusters of carbon atoms were indeed produced but, more interesting than the small chains he was looking for, Kroto found a mass-spectrum signal for a molecule of exactly 60 carbon atoms. The first suggestion was that it had a sandwichlike graphite structure. However, such a planar fragment would have reactive carbon atoms at the edges, whereas C_{60} appeared to be stable. Kroto called C_{60} BUCKMINSTERFULLERENE. He shared the 1996 Nobel Prize for chemistry with Smalley and with Robert CURL.

krypton A colorless odorless monatomic gas, the fourth member of the rare-gases; i.e. group 18 (formerly VIIIA or 0) of the periodic table. It occurs in minute quantities (0.001% by volume) in air, from which it is recovered by fractional distillation. Krypton is used in fluorescent lights, lasers, and pneumatic heart valves. It is known to form unstable compounds with fluorine.

Symbol: Kr; m.p. –156.55°C; b.p. –152.3°C; mass density 3.749 (0°C) kg m^{-3}; p.n. 36; most common isotope ^{84}Kr; r.a.m. 83.80.

kurchatovium Symbol: Ku A name formerly used for element-104 now known as rutherfordium (Rf). *See* element.

L

label *See* isotopes.

labile Describing a complex having ligands that can easily be replaced by more strongly bonded ligands.

lake A pigment made by combining an organic dyestuff with an inorganic compound (e.g. aluminum oxide).

lamellar compound A compound with a crystal structure composed of thin plates or layers. Silicates form many compounds with distinct layers. Typical examples are talc ($Mg_3(OH)_2Si_4O_{10}$) and pyrophyllite ($Al_2(OH)_2Si_4O_{10}$). *See also* intercallation compound.

lamp black A black pigment; a finely divided form of carbon formed by incomplete combustion of an organic compound.

Langmuir, Irving (1881–1957) American chemist. In 1919 Langmuir extended the model of electrons in atoms proposed by Gilbert LEWIS to try to account for the chemical valence of all atoms. In the 1920s he investigated the chemistry of surfaces very extensively. He won the 1932 Nobel Prize for chemistry for his work on chemical reactions at surfaces.

Langmuir isotherm An equation used to describe the adsorption of a gas onto a plane surface at a fixed temperature. It can be written in the form:

$$\theta = bp/(1 + bp),$$

where θ is the fraction of the surface covered by the gas, p is the gas pressure and b is a constant known as the *adsorption coefficient*, which is the equilibrium constant for the adsorption process. This equation was derived by the American chemist Irving Langmuir in 1916 using the kinetic theory of gases.

lanthanide contraction The decrease in atomic and ionic radii in the series of lanthanoid elements from lanthanum to lutetium as the atomic number increases. This contraction occurs because the 4f electrons in this series of elements are rather ineffective in shielding the outer electrons from the attraction of a positively charged nucleus. In the case of the tripositive ions the radius of the lanthanum ion is 1.17Å and the radius of the lutetium ion is 1.00Å. A similar effect, the ACTINOID CONTRACTION, occurs for 5f electrons.

lanthanides *See* lanthanoids.

lanthanoids (lanthanides; lanthanons; rare earths) A group of 15 silvery, reactive metals whose electronic configurations display back-filling of the 4f level. There is a maximum of 14 electrons in an f-orbital. The element lanthanum itself has no f-electrons (La $[Xe]5d^16s^2$) and is thus strictly not a lanthanoid, but it is included by convention, thus giving a closely related series of 15 elements. Excluding lanthanum, the elements all have $4f^x6s^2$ configurations, although gadolinium and lutecium have an additional $5d^1$ electron.

The characteristic oxidation state is M^{3+} and the great similarity in the size of the ions leads to a very close similarity of chemical properties and hence to great difficulties of separation using conventional methods. In addition, cerium can assume a Ce^{4+} state and ytterbium a Yb^{2+} state. Chromatographic and solvent-extraction methods have been specially developed for the lanthanoids.

lanthanons *See* lanthanoids.

lanthanum A soft ductile malleable silvery highly reactive metallic element of group 3 (formerly IIIB) of the periodic table. It is usually considered to be the first member of the LANTHANOID series. It is found associated with other lanthanoids in many minerals, including monazite and bastnaesite. Lanthanum is used in several alloys, especially those for lighter flints, because it sparks or ignites readily when scratched. It is also used as a catalyst, and in making optical glass to which it imparts enhanced refractive properties.

Symbol: La; m.p. 921°C; b.p. 3457°C; r.d. 6.145 (25°C); p.n. 57; most stable isotope ^{139}La; r.a.m. 138.9055.

lapis lazuli *See* lazurite.

laser An acronym for Light Amplification by Stimulated Emission of Radiation. A laser device produces high-intensity monochromatic coherent beams of light. In the laser process the molecules of a sample (such as ruby doped with Cr^{3+} ions) are promoted to an excited state. Because the sample is in a cavity between two reflective surfaces, when a molecule emits spontaneously, the photon so generated ricochets backward and forward. In this way other molecules are stimulated to emit photons of the same energy. If one of the reflective surfaces is partially transmitting this radiation can be tapped.

laser spectroscopy Spectroscopy that uses lasers. The special features of lasers, notably their ability to produce beams of coherent monochromatic radiation, have several important advantages over other spectroscopic techniques. For example, spectroscopy that makes use of the RAMAN EFFECT has greatly benefitted from the application of lasers because there are major experimental problems with other sources of light.

latent heat The heat evolved or absorbed when a substance changes its physical state, e.g. the latent heat of fusion is the heat absorbed when a substance changes from a solid to a liquid.

laterite A red fine-grained type of clay formed in tropical climates by the weathering of igneous rocks. Its color comes from the presence of iron(III) hydroxide.

lattice A regular three-dimensional arrangement of points. A lattice is used to describe the positions of the particles (atoms, ions, or molecules) in a crystalline solid. The lattice structure can be examined by x-ray diffraction techniques.

lattice energy The energy released when ions of opposite charge are brought together from infinity to form one mole of a given crystal. The lattice energy is a measure of the stability of a solid ionic substance, with respect to ions in the gas. *See also* Born–Haber cycle.

lattice vibrations The vibrations of the atoms or ions in a crystal lattice about their equilibrium positions. Such vibrations occur even at the absolute zero temperature because of zero point energy. At low temperatures the vibrations are well described by the simple harmonic motion of the atoms. As the temperature is increased the anharmonicity of the vibrations becomes more pronounced.

laughing gas *See* dinitrogen oxide.

Lavoisier, Antoine Laurent (1743–94) French chemist. Lavoisier is frequently referred to as the founder of modern chemistry. Perhaps his most significant contribution was to peform careful quantitative experiments that disproved the PHLOGISTON THEORY of combustion. This led him to establish that oxygen is one of the gases present in air. He also noticed the presence of an inert gas in air, which was subsequently named nitrogen. He summarized his work in the influential book *Elementary Treatise on Chemistry*, which stated the law of mass conservation in chemical reactions. Lavoisier, who had been a tax farmer, was executed in 1794, in the aftermath of the French Revolution.

law of conservation of energy *See* conservation of energy.

law of conservation of mass *See* conservation of mass.

law of constant composition *See* constant proportions.

law of constant heat summation *See* Hess's law.

law of constant proportions *See* constant proportions.

law of definite proportions *See* constant proportions.

law of equivalent proportions *See* equivalent proportions.

law of isomorphism *See* Mitscherlich's law.

law of mass action *See* mass action, law of.

law of octaves *See* Newlands' law.

law of reciprocal proportions *See* equivalent proportions.

lawrencium A radioactive synthetic transuranic element of the actinoid series of elements, first made in 1961 by bombarding californium-252 targets with boron nuclei. It can also be made by bombarding berkelium-249 targets with oxygen-18 nuclei. Several very short-lived isotopes have been synthesized.

Symbol: Lr; p.n. 103; m.p., b.p., and r.d. unknown; most stable isotope ^{262}Lr (half-life 261 minutes).

laws of chemical combination *See* chemical combination; laws of.

lazurite An uncommon typically azure blue mineral consisting of a silicate of sodium, calcium, and aluminum with some sulfur. It and its parent rock, *lapis lazuli*, are widely used for ornaments and as semiprecious gemstones. Lazurite was the original source of the pigment ultramarine.

LCAO (linear combination of atomic orbitals) *See* orbital.

leaching The washing out of a soluble material from an insoluble solid by use of a solvent. This process is often carried out in batch tanks or by dispersing the crushed solid in a liquid.

lead A dense dull gray soft metallic toxic element; the fifth member of group 14 (formerly IVA) of the periodic table. It is the end product of most radioactive decay series. It occurs in small quantities in a wide variety of minerals but only a few are economically important. The most important of these is galena (PbS), found in Australia, Mexico, the USA, and Canada. Other economically important minerals are anglesite ($PbSO_4$), litharge (PbO), and cerussite ($PbCO_3$).

Galena is often associated with zinc ores and the smelting operations of the lead and zinc industries are thus closely integrated. Lead ore from galena is first concentrated by froth flotation and the concentrate then roasted and reduced as follows:

$$2PbS + 3O_2 \rightarrow 2PbO + 2SO_2$$
$$PbS + 2O_2 \rightarrow PbSO_4$$
$$2PbO + 2C \rightarrow 2Pb + 2CO$$
$$2PbO + PbS \rightarrow 3Pb + SO_2$$
$$PbSO_4 + 2C \rightarrow Pb + 2CO + SO_2$$

Additionally, various silver sulfides are often found in small quantities with galena, meaning that silver is often recovered in economic quantities with crude lead.

The outer s^2p^2 configurations of both tin and lead give these metals similar properties. There is, however, a much greater predominance of the divalent state with lead. Both lead oxides are amphoteric, lead(II) oxide (PbO) leading to plumbites and lead(IV) oxide (PbO_2) to plumbates on dissolution in alkalis. Lead also forms mixed oxides Pb_2O_3 (a yellow solid, better written as $Pb(II)Pb(IV)O_3$) and Pb_3O_4 (a red powder, better written as $Pb_2(II)$-$Pb(IV)O_4$). Both lead and tin have low melting points and neither is attacked by dilute acids; they differ however in their reaction with concentrated nitric acid, with lead reacting as follows:

$$3Pb + 8HNO_3 \rightarrow 3Pb(NO_3)_2 + 2NO + 4H_2O$$

Tin, on the other hand, gives hydrated Sn(IV) oxide. Concentrated hydrochloric acid will not attack lead.

Lead, like tin, forms halides, PbX_2, with all the halogens. Lead(IV) chloride ($PbCl_4$) is the only known lead(IV) halide, whereas all halides of SnX_4 are known. The great stability of the Pb(II) state leads to Pb(IV) compounds being powerful oxidizing agents.

Lead is used in lead–acid storage batteries, alloys, radiation shielding, and water and sound proofing. It is also used in the petrochemical, munitions, paint, and glass industries, although to a far lesser extent than previously, due to the toxic nature of the metal and its compounds.

Symbol: Pb; m.p. 327.5°C; b.p. 1830°C; r.d. 11.35 (20°C); p.n. 82; most common isotope ^{208}Pb; r.a.m. 207.2.

lead(II) acetate *See* lead(II) ethanoate.

lead–acid battery A type of battery used in vehicles. It has two sets of plates: spongy lead plates connected in series to the negative terminal, and lead(IV) oxide plates connected to the positive terminal. The material of the electrodes is held in a hard lead-alloy grid. The plates are interleaved. The electrolyte is dilute sulfuric acid.

The e.m.f. of each cell when fully charged is about 2.2 volts (V). This value falls to a steady 2 V when current is drawn; a typical battery has six cells, giving an overall voltage of about 12 volts. As the battery begins to run down, the e.m.f. falls further. During discharge the electrolyte becomes more dilute and its relative density falls. To recharge the battery, current is passed through it in the opposite direction to the direction of current supply. This process reverses the cell reactions and increases the relative density of the electrolyte, which should be about 1.26–1.29 for a fully charged battery.

The electrolyte contains hydrogen ions (H^+) and sulfate ions (SO_4^{2-}). During discharge, H^+ ions react with the lead(IV) oxide to give lead(II) oxide and water

$$PbO_2 + 2H^+ + 2e^- \rightarrow PbO + H_2O$$

This reaction takes electrons from the lead(IV) oxide plates, causing the positive charge. A further reaction follows, which yields soft lead sulfate:

$$PbO + SO_4^{2-} + 2H^+ \rightarrow PbSO_4 + H_2O + 2e^-$$

Electrons are thus released to the negative electrode, producing the negative charge. During charging the reactions are reversed:

$$PbSO_4 + 2e^- \rightarrow Pb + SO_4^{2-}$$

$$PbSO_4 + 2H_2O \rightarrow PbO_2 + 4H^+ + SO_4^{2-} + 2e^-$$

lead(II) carbonate ($PbCO_3$) A white poisonous powder that occurs naturally as the mineral CERUSSITE. It forms rhombic crystals and can be precipitated by reacting a cold aqueous solution of a soluble lead(II) salt (e.g. lead(II) nitrate) with ammonium carbonate. It decomposes to lead(II) oxide and carbon dioxide above 315°C.

lead(II) carbonate hydroxide (white lead; $2PbCO_3.Pb(OH)_2$) The most important basic lead carbonate. It is manufactured electrolytically using high-purity lead anodes. It was once widely used as a pigment in white and colored paints and in ceramics. Because the compound is poisonous, its use for these purposes has been largely discontinued.

lead-chamber process A process formerly used to manufacture sulfuric acid by oxidizing sulfur(IV) oxide with nitrogen monoxide. The reaction, which was carried out in expensive large lead chambers, has now been replaced by the CONTACT PROCESS.

lead dioxide *See* lead(IV) oxide.

lead(II) ethanoate (lead(II) acetate; $Pb(CH_3CO_2)_2$) A poisonous compound usually occurring as the hydrate $Pb(CH_3CO_2)_2.3H_2O$, forming monoclinic crystals. At 100°C it loses ethanoic acid and water, forming a basic lead(II) ethanoate. It is extremely soluble in water – 50 g per 100 g of water at 25°C. Lead(II) ethanoate can be obtained as an anhydrous salt. Its chief asset is that it is one of the few common lead(II) salts that are soluble in water. It has been used to make varnishes and enamels.

lead(IV) hydride *See* plumbane.

lead monoxide *See* lead(II) oxide.

lead(II) oxide (lead monoxide; PbO) A yellow crystalline poisonous powder formed by roasting molten lead in air. *Litharge*, the most common form, is obtained when lead(II) oxide is heated above its melting point. If prepared below its melting point, another form, called *massicot*, is obtained. Litharge is used in the rubber industry, in the manufacture of paints and varnishes, and in the manufacture of lead glazes for pottery.

lead(IV) oxide (lead dioxide; PbO_2) A poisonous dark-brown solid forming hexagonal crystals. On heating to 310°C it decomposes into lead(II) oxide and oxygen. It can be prepared electrolytically or by reacting lead(II) oxide with potassium chlorate. Lead(IV) oxide is an oxidizing agent and has been used in the manufacture of matches.

lead(II) sulfate ($PbSO_4$) A poisonous white crystalline solid that occurs naturally as the mineral *anglesite*. It is almost insoluble in water and can be precipitated by reacting an aqueous solution containing sulfate ions with a solution of a soluble lead(II) salt (e.g. lead(II) ethanoate). It forms basic lead(II) sulfates when shaken together with lead(II) hydroxide and water. Due to their toxicity, basic lead(II) sulfates have been largely discontinued in their former use as pigments.

lead(II) sulfide (PbS) A black solid that occurs naturally as the mineral galena. It can be prepared as a precipitate by reacting a solution of a soluble lead(II) salt with hydrogen sulfide or with a solution of a soluble sulfide. Lead(II) sulfide has the capacity to rectify alternating current.

lead tetraethyl (tetraethyl lead; $Pb(CH_2CH_3)_4$) A poisonous liquid that is insoluble in water but soluble in organic solvents. It is manufactured by the reaction of an alloy of sodium and lead with chloroethane (CH_2ClCH_3). The product is obtained by steam distillation. Lead tetraethyl was once used as an additive in internal-combustion engine fuel to increase its octane number and thus prevent preignition (knocking). Lead-free compounds are now prescribed or encouraged for this purpose in many countries.

Leblanc process An obsolete process for the manufacture of sodium carbonate. Sodium chloride is converted to sodium sulfate by heating with sulfuric acid. This sulfate is then roasted in a rotary furnace where it is first reduced to the sulfide using carbon, and then immediately converted to the carbonate by the action of limestone. Sodium carbonate solution is obtained by leaching the product with water and this solution is subsequently dried and calcined to obtain the solid. The process was invented by French chemist Nicolas LeBlanc (1742–1806) in 1783. The Solvay process is now used instead.

Le Chatelier's principle If a system is at equilibrium and a change is made in the conditions, the equilibrium adjusts so as to oppose the change. The principle was first stated by French chemist Henri Le Chatelier (1850–1936) in 1888 and can be applied to the effect of temperature and pressure on chemical reactions. A good example is the Haber process for synthesis of ammonia:

$$N_2 + 3H_2 \rightleftharpoons 2NH_3$$

The 'forward' reaction

$$N_2 + 3H_2 \rightarrow 2NH_3$$

is exothermic. Thus, reducing the temperature displaces the equilibrium toward production of NH_3 (because this tends to increase temperature). Increasing the pressure also favors formation of NH_3, because this leads to a reduction in the total number of molecules (and hence pressure).

Leclanché cell A primary voltaic cell consisting, in its 'wet' form, of a carbon-rod anode and a zinc cathode, with a 10–20% solution of ammonium chloride as electrolyte. Manganese(IV) oxide mixed with crushed carbon in a porous bag or pot surrounding the anode acts as a depolarizing agent. The dry form (*dry cell*) is widely used for flashlight batteries, transistor radios, etc. It has a mixture of ammonium chloride, zinc chloride, flour, and gum forming an electrolyte paste. Sometimes the dry cell is arranged in layers to form a rectangular battery, which has a longer life than the cylinder type.

LEED (low-energy electron diffraction) A method of electron diffraction that is used to investigate surfaces. A surface is bombarded with a beam of low-energy electrons, with the diffracted electrons hitting a fluorescent screen. The electron beam has a low energy so that it interacts with the surface rather than the bulk of the material. The technique gives a great deal of useful information about surfaces and processes occurring at surfaces.

Lennard-Jones, Sir John Edward (1894–1954) British chemist. Lennard-Jones was a theoretical chemist who did a great deal of work on intermolecular forces. Starting in the late 1920s he helped develop and promote molecular orbital theory. In particular, in 1929 he showed that the theory readily explains the paramagnetism of molecular oxygen. He subsequently showed how molecular orbital theory can also explain the directionability of chemical bonds.

Lennard-Jones potential An expression used to represent the potential energy V of intermolecular interactions as a function of the intermolecular distance r. V can be written in the form $V = -A/r^6 + B/r^{12}$, where A and B are constants. The Lennard-Jones potential has been used extensively to study intermolecular interactions. It was proposed by the British theoretical chemist Sir John Lennard-Jones in 1924.

levo-form *See* optical activity.

levorotatory *See* optical activity.

Lewis, Gilbert Newton (1875–1946) American physical chemist. Lewis proposed the idea that valence and bonding of atoms is associated with electrons. He first put forward this idea in 1916, introducing the concept of a covalent bond in which the bonding between two atoms occurs because the atoms share a pair of electrons, with each atom supplying one electron. He also suggested that the outer electrons in atoms form octets. He developed his ideas more fully in the book *Valence and the Structure of Atoms and Molecules* (1923). Lewis also performed important work in chemical thermodynamics, with his book with Merle Randall entitled *Thermodynamics and the Free Energy of Chemical Substances* (1923) being very influential.

Lewis acid A substance that can accept an electron pair to form a coordinate bond; a *Lewis base* is an electron-pair donor. In this model, the neutralization reaction is seen as the acquisition of a stable octet by the acid, for example:

$$Cl_3B + :NH_3 \rightarrow Cl_3B:NH_3$$

Metal ions in coordination compounds are also electron-pair acceptors and therefore Lewis acids. The definition includes the 'traditional' Brønsted acids since H^+ is an electron acceptor, but in common usage the terms Lewis acid and Lewis base are reserved for systems without acidic hydrogen atoms.

Lewis base *See* Lewis acid.

Lewis formula *See* Lewis structure.

Lewis octet theory *See* octet.

Lewis structure (Lewis formula) A two-dimensional depiction, by means of chemical element and electron dot symbols, of the possible structure of a molecule or ion, showing each atom in relation to its neighbors, the bonds that hold the atoms together, and the lone pairs in each atom's outer shell.

L-form *See* optical activity.

Liebig condenser A simple type of laboratory condenser. It consists of a straight glass tube, in which the vapor is condensed, with a surrounding glass jacket through which cooling water flows. It is named for German chemist Justus von Liebig (1803–73).

ligand *See* complex; chelate.

ligand-field theory A theory that describes the properties of complexes. Ligand-field theory is an extension of CRYSTAL FIELD THEORY in which covalent bonding between the central atom or ion and the ligands is taken into account.

light (visible radiation) A form of electromagnetic radiation able to be detected by the human eye. Its wavelength range is between about 400 nanometers (nm) (far red) and about 700 nm (far violet). The boundaries are not precise, because individuals vary in their ability to detect extreme wavelengths; this ability also declines with age.

lignite (brown coal) The poorest grade of coal, containing up to 60% carbon and having a high moisture content.

lime *See* calcium hydroxide; calcium oxide.

limestone A natural form of CALCIUM CARBONATE. It is used in making calcium compounds, carbon dioxide, and cement.

limewater A solution of calcium hydroxide in water. If carbon dioxide is bubbled through lime water a milky precipitate of calcium carbonate forms. Prolonged bubbling of carbon dioxide turns the solution clear again as a result of the formation of soluble calcium hydrogencarbonate ($Ca(HCO_3)_2$).

limonite (bog iron ore) A geological field term for a group of yellowish-brown to dark-colored minerals consisting mainly of iron(III) hydroxide and hydrous oxides of iron. Limonite is a major iron ore. It is also used as a pigment.

Linde process A method of liquefying gases by compression followed by expansion through a nozzle. The process is used for producing liquid oxygen and nitrogen by liquefying air and then fractionating it.

line defect *See* defect.

line spectrum A SPECTRUM composed of a number of discrete lines corresponding to single wavelengths of emitted or absorbed radiation. Line spectra are produced by atoms or simple (monatomic) ions in gases. Each line corresponds to a change in electron orbit, with emission or absorption of radiation.

linkage isomerism *See* isomerism.

Lipscomb, William Nunn (1919–) American inorganic chemist. Lipscomb is noted for his work on boranes – hydrides of boron first investigated by Alfred STOCK in the early part of the 20th century. Using low-temperature x-ray diffraction analysis, Lipscomb tackled the problem of investigating the notoriously unstable boranes, producing evidence of some remarkable structures, totally original and completely unsuspected by earlier chemists. The basic concept of a multi-center bond was derived from a structure for diborane proposed by LONGUET-HIGGINS. Lipscomb was awarded the Nobel Prize for chemistry in 1976.

liquid The state of matter in which the particles of a substance are loosely bound by intermolecular forces. The weakness of these forces permits movement of the particles and consequently liquids can change their shape within a fixed volume. The liquid state lacks the order of the solid state. Thus, amorphous materials, such as glass, in which the particles are disordered and can move relative to each other, can be classed as liquids.

liquid air A pale blue liquid. It is a mixture of mainly liquid oxygen (boiling at –183°C) and liquid nitrogen (boiling at –196°C).

liquid crystal A substance that can flow

like a viscous liquid but has a considerable amount of molecular order. There are three types of liquid crystal. In a *smectic liquid crystal* there are layers of aligned molecules, with the long axes of the molecules being perpendicular to the layers. In a *cholesteric liquid crystal* there are also layers of aligned molecules, but with the axes of the molecules being parallel to the planes of the layers. In a *nematic liquid crystal* the molecules are all aligned in the same direction but not arranged in layers.

liter Symbol: L or l A metric unit of volume equal in SI units to one cubic decimeter (dm^3) or 10^{-3} meter^3, or in c.g.s. units to 1000 cm^3. One milliliter (mL or ml) is thus the same volume as one cubic centimeter (cm^3). However, the liter is not recommended for precise measurements since it was formerly defined as the volume of one kilogram of pure water at 4°C and standard pressure. By this older definition, one liter was the same as 1000.028 cm^3.

litharge *See* lead(II) oxide.

lithia *See* lithium oxide.

lithia water *See* lithium hydrogencarbonate.

lithium A light silvery moderately reactive alkali metal; the first element of group 1 (formerly IA) of the periodic table. It occurs in a number of complex silicates, such as spodumene, lepidolite, and petalite, and a mixed phosphate, tryphilite. It is a rare element, accounting for 0.0065% of the Earth's crust, and the lightest of the metals. Lithium ores are treated with concentrated sulfuric acid and lithium sulfate subsequently separated by crystallization. The element can be obtained by conversion to the chloride and electrolysis of the fused chloride.

Lithium has the electronic configuration $1s^22s^1$ and is entirely monovalent in its chemistry. The lithium ion is, however, much smaller than the ions of the other alkali metals; consequently it is polarizing and a certain degree of covalence occurs in its bonds. Lithium also has the highest ionization potential of the alkali metals.

The element reacts with hydrogen to form lithium hydride (LiH) a white high-melting solid that releases hydrogen at the anode during electrolysis (confirming the ionic nature Li^+H^-). Lithium reacts with oxygen to give Li_2O (sodium gives the peroxide) and with nitrogen to form Li_3N on fairly gentle warming. The metal itself reacts only slowly with water, giving the hydroxide (LiOH) but lithium oxide reacts much more vigorously to give again the hydroxide; the nitride is hydrolyzed to ammonia. With halogens the metal reacts to form halides (LiX).

Apart from the fluoride the halides are readily soluble both in water and in oxygen-containing organic solvents. In this property lithium partly resembles magnesium, which has a similar charge/size ratio. Compared to the other carbonates of group 1, lithium carbonate (Li_2CO_3) is thermally unstable, decomposing to Li_2O and CO_2. This is because the small Li^+ ion leads to particularly high lattice energies, favouring the formation of Li_2O. Lithium compounds impart a characteristic purple color to flames.

Lithium is used to make alloys for aircraft parts and other components where lightness is an important quality. It is also used to make lithium batteries and lithium grease, and to scavenge oxygen in metallurgy. Certain lithium salts are used to treat depression. Lithium is also used to certain glasses and ceramic glazes.

Symbol: Li; m.p. 180.54°C; b.p. 1347°C; r.d. 0.534 (20°C); p.n. 3; most common isotope ^7Li; r.a.m. 6.941.

lithium aluminum hydride *See* lithium tetrahydroaluminate(III).

lithium carbonate (Li_2CO_3) A white solid obtained by the addition of excess sodium carbonate solution to a solution of a lithium salt. It is soluble in the presence of excess carbon dioxide, forming lithium hydrogencarbonate. If heated (to 780°C) in a stream of hydrogen it undergoes thermal decomposition to give lithium oxide and carbon dioxide. Lithium carbonate forms monoclinic crystals and differs from the other group 1 carbonates in being only sparingly soluble in water. It has a close resemblance to the carbonates of magnesium and calcium – an example of a diagonal relationship in the periodic table. Lithium carbonate is used in the preparation of other lithium compounds and in the treatment of depression.

lithium chloride (LiCl) A white solid produced by dissolving lithium carbonate or oxide in dilute hydrochloric acid and crystallizing the product. Below 19°C the dihydrate ($LiCl.2H_2O$) is obtained; at 19°C the chloride loses one molecule of water. It becomes anhydrous at 93.5°C. The compound forms cubic crystals; the anhydrous salt is isomorphous with sodium chloride. Lithium chloride is probably the most deliquescent substance known and forms hydrates with 1, 2, and 3 molecules of water. It is used as a flux in welding aluminum.

lithium hydride (LiH) A white crystalline solid produced by direct combination of the elements at a temperature above 500°C. Lithium hydride has a cubic structure and is more stable than the other group 1 hydrides. Electrolysis of the fused salt yields hydrogen at the anode. Lithium hydride reacts violently with water to give lithium hydroxide and hydrogen; the reaction is highly exothermic. The compound is used as a reducing agent and in the preparation of other hydrides.

lithium hydrogencarbonate ($LiHCO_3$) A compound known only in solution, produced by the reaction between carbon dioxide, water, and lithium carbonate. When a solution of the hydrogencarbonate is heated, carbon dioxide and lithium carbonate are produced. Solutions of the hydrogencarbonate are sold and used in medicine under the name of *lithia water*.

lithium hydroxide (LiOH) A white solid made industrially as the monohydrate ($LiOH.H_2O$) by reacting calcium hydroxide with a lithium ore or with a salt made from the ore. Lithium hydroxide has a closer resemblance to the group 2 hydroxides than to the group 1 hydroxides. It is used in lubricants and batteries, and to absorb carbon dioxide.

lithium oxide (lithia; Li_2O) A white solid produced by burning metallic lithium in air above its melting point. It can also be produced by the thermal decomposition of lithium carbonate or hydroxide. Lithium oxide reacts slowly with water to form a solution of lithium hydroxide; the reaction is exothermic. The compound has a calcium-fluoride structure. Lithium oxide is used in lubricants, glass, ceramics, and welding and brazing fluxes.

lithium sulfate (Li_2SO_4) A white solid prepared by the addition of excess lithium oxide or carbonate to a solution of sulfuric acid. It is readily soluble in water from which it crystallizes as the monohydrate. In contrast to the other group 1 sulfates it does not form alums and it is not isomorphous with these other sulfates.

lithium tetrahydroaluminate(III) (lithium aluminum hydride; $LiAlH_4$) A white solid produced by the action of lithium hydride on aluminum chloride, the hydride being in excess. Lithium tetrahydroaluminate reacts violently with water, releasing hydrogen. It is a powerful reducing agent, much used in organic chemistry. In inorganic chemistry it is used in the preparation of hydrides.

litmus A natural pigment that changes color when in contact with acids and alkalis; above a pH of 8.3 it is blue and below a pH of 4.5 it is red. Thus it gives a rough indication of the acidity or basicity of a solution; because of the rather broad range over which it changes it is not used for precise work. Litmus is used both in solution and as litmus paper.

lixiviation The process of separating soluble components from a mixture by washing them out with water.

localized bond A molecular bond in which the electrons contributing to the bond remain between the two atoms concerned, i.e. the bonding orbital is localized. The majority of bonds are of this type. *Compare* delocalized bond.

lodestone A naturally occurring magnetic oxide of iron, a form of MAGNETITE (Fe_3O_4). A piece of lodestone is a natural magnet and lodestones were used as magnetic compasses in ancient times. *See* magnetite.

lone pair A pair of valence electrons having opposite spin that are located together on one atom, i.e. are not shared as in a covalent bond. Lone pairs occupy similar positions in space to bond pairs and account for the shapes of molecules. A molecule with a lone pair can donate this pair to an electron acceptor, such as H^+ or a metal ion, to form coordinate bonds. *See* complex.

long period *See* period.

Longuet-Higgins, Hugh Christopher (1923–) British theoretical chemist. One of his first contributions was to explain the properties of boron hydrides in terms of bridged structures. In the second half of the 1940s he performed calculations with Charles COULSON on conjugated molecules. Longuet-Higgins also applied statistical mechanics to chemical problems, particularly mixtures and polymer solutions.

Lotka–Volterra mechanism A possible mechanism for OSCILLATING REACTIONS. A reactant R is converted into a product P. The chemical reaction is in a steady state but is not in chemical equilibrium. There are three steps in this mechanism:

$$R + X \rightarrow 2X,$$
$$X + Y \rightarrow 2Y,$$
$$Y \rightarrow P.$$

Autocatalysis is involved in the first two steps of this process. It appears that oscillating chemical reactions have mechanisms that are different from the Lotka–Volterra mechanism. This type of mechanism does occur in certain types of complex system such as predator–prey relationships in biology. It was in the biological context that the mechanism was investigated by the Italian mathematician Vito Volterra (1860–1940).

lowering of vapor pressure A colligative property of solutions in which the vapor pressure of a solvent is lowered as a solute is introduced. When both solvent and solute are volatile the effect of increas-

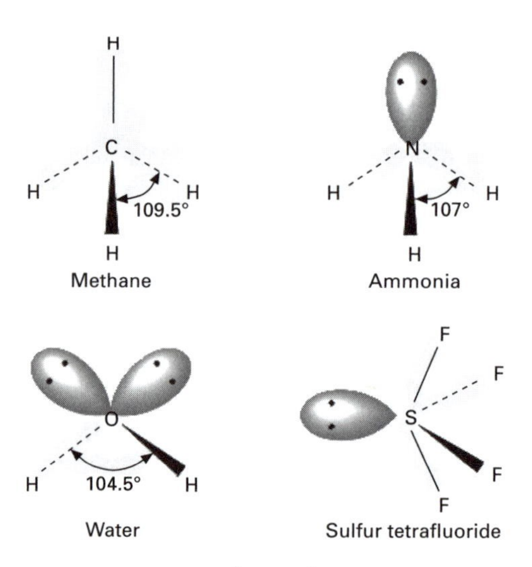

Lone pair

ing the solute concentration is to lower the partial vapor pressure of each component. When the solute is a solid of negligible vapor pressure the lowering of the vapor pressure of the solution is directly proportional to the number of species introduced rather than to their nature and the proportionality constant is regarded as a general solvent property. Thus the introduction of the same number of moles of any solute causes the same lowering of vapor pressure, if dissociation does not occur. If the solute dissociates into two species on dissolution the effect is doubled. The kinetic model for the lowering of vapor pressure treats the solute molecules as occupying part of the surface of the liquid phase and thereby restricting the escape of solvent molecules. The effect can be used in the measurement of relative molecular masses, particularly for large molecules, such as polymers. *See also* depression of freezing point; Raoult's law.

Lowry–Brønsted theory *See* acid; base.

lumen Symbol: lm The SI derived unit of luminous flux, equal to the luminous flux emitted in a solid angle of one steradian (Sr) by a point source of one candela (cd), radiating equally in all directions. 1 lm = 1 cd sr.

luminescence The emission of radiation from a substance in which the particles have absorbed energy and gone into excited states. They then return to lower energy states with the emission of electromagnetic radiation. If the luminescence persists after the source of excitation is removed it is called phosphorescence: if not, it is called fluorescence.

LUMO *See* frontier orbital theory.

lutetium A silvery element of the lanthanoid series of metals. It occurs in association with other lanthanoids, especially in the mineral monazite. Lutetium is the rarest of the naturally occurring elements. It is used as a catlyst.

Symbol: Lu; m.p. 1663°C; b.p. 3395°C; r.d. 9.84 (25°C); p.n. 71; most common isotope ^{175}Lu; r.a.m. 174.967.

lux Symbol: lx The SI derived unit of illumination, equal to the illumination produced by a luminous flux of one lumen falling on a surface of one square meter. 1 lx = 1 lm m^{-2}.

Lyman series *See* hydrogen atom spectrum.

lyophilic Solvent attracting. When the solvent is water, the word *hydrophilic* is often used. The terms are applied to:

1. Ions or groups on a molecule. In aqueous or other polar solutions ions or polar groups are lyophilic. For example, the –COO$^-$ group on a soap is the lyophilic (hydrophilic) part of the molecule.
2. The disperse phase in colloids. In lyophilic colloids the dispersed particles have an affinity for the solvent, and the colloids are generally stable. *Compare* lyophobic.

lyophobic Solvent repelling. When the solvent is water, the word *hydrophobic* is used. The terms are applied to:

1. Ions or groups on a molecule. In aqueous or other polar solvents, the lyophobic group is nonpolar. For example, the hydrocarbon group on a soap molecule is the lyophobic (hydrophobic) part.
2. The disperse phase in colloids. In lyophobic colloids the dispersed particles are not solvated and the colloid is easily solvated. Gold and sulfur sols are examples.

Compare lyophilic.

M

macromolecular crystal A crystal composed of atoms joined together by covalent bonds, which form giant three-dimensional or two-dimensional networks. Diamond and silicates are examples of macromolecular crystals.

macromolecule A large molecule; e.g. a natural or synthetic polymer.

Madelung constant A constant, denoted α, which appears in calculations of the cohesive energy of ionic crystals when the electrostatic interactions between ions in a lattice are summed. The Madelung constant is dimensionless and is a characteristic of the specific crystal structure. It is named for E. Madelung who introduced it in 1918.

magnesia *See* magnesium oxide.

magnesite *See* magnesium carbonate.

magnesium A light ductile malleable silver-white alkaline-earth metal; the second element of group 2 (formerly group IIA) of the periodic table. It has the electronic configuration of neon with two additional outer 3s electrons. The element accounts for 2.09% of the Earth's crust and is eighth in order of abundance. It occurs in a wide variety of minerals such as brucite ($Mg(OH)_2$), carnallite ($KCl.MgCl_2.6H_2O$), epsomite ($MgSO_4.7H_2O$), magnesite ($MgCO_3$), and dolomite ($CaCO_3.MgCO_3$); and also as magnesium chloride ($MgCl_2$) in sea water. The metal is obtained by several routes, depending on the mineral used, all eventually leading to the chloride which is then fused and subjected to electrolysis.

The element has a fairly low ionization potential and is electropositive. It reacts directly with oxygen, nitrogen, and sulfur on heating, to form the oxide, MgO, the nitride, Mg_3N_2, and the sulfide MgS, all of which are hydrolyzed by water to give the hydroxide. It burns in air with an intense white flame.

Magnesium also reacts directly with the halogens to form halides. These show a slight deviation in the general behavior of halides by dissolving normally in water and undergoing a significant amount of hydrolysis only when the solutions are evaporated:

$$MgX_2 + 2H_2O \rightarrow Mg(OH)_2 + 2HX$$

Magnesium hydroxide is considerably less soluble than the hydroxides of the heavier elements in group 2 and is a weaker base. Magnesium forms a soluble sulfate, chlorate, and nitrate. Like the heavier alkaline earths, it also forms an insoluble carbonate, but $MgCO_3$ is the least thermally stable of the group. Unlike calcium, magnesium does not form a carbide.

Magnesium metal is industrially important as a major component, along with aluminium and zinc, of lightweight alloys. The alloy surfaces develop an impervious oxide film which protects them from progressive deterioration. The metal is also used in flash bulbs, fireworks, and flares.

Because of its ionic nature magnesium forms very few coordination compounds. Donor–acceptor species in aqueous solutions are short-lived, but magnesium bromide and iodide have sufficient acceptor properties to dissolve in donor solvents such as alcohols. Magnesium in trace amounts is essential for living things, being required by animals for proper metabolism and by green plants as a component of chlorophyll.

The element crystallizes in the close-packed hexagonal structure.

Symbol: Mg; m.p. 649°C; b.p. 1090°C; r.d. 1.738 (20°C); p.n. 12; most common isotope ^{24}Mg; r.a.m. 24.3050.

magnesium bicarbonate *See* magnesium hydrogencarbonate.

magnesium carbonate ($MgCO_3$) A white solid that occurs naturally as the mineral *magnesite* and in association with calcium carbonate in dolomite. Magnesium carbonate is sparingly soluble in water but reacts with dilute acids to give the salt, carbon dioxide, and water. It thus finds use in medicine as a mild antacid. Magnesium carbonate also decomposes easily on heating to give magnesium oxide, which is an important refractory material.

magnesium chloride ($MgCl_2$) A solid that exists in a variety of hydrated forms, most commonly as the hexahydrate, $MgCl_2.6H_2O$. When heated, this hydrated form is hydrolyzed by its water of crystallization, with the evolution of hydrogen chloride and the formation of the oxide:

$$MgCl_2 + H_2O \rightarrow MgO + 2HCl$$

The anhydrous chloride must therefore be prepared by evaporating an aqueous solution in an atmosphere of hydrogen chloride. It is used in the preparation of cotton fabrics and artificial leather, as a laxative, and to fireproof wood. Magnesium metal is produced from the fused chloride by electrolysis.

magnesium hydrogencarbonate (magnesium bicarbonate; $Mg(HCO_3)_2$) The solid hydrogencarbonate is unknown at room temperature. The compound is produced in solution when water containing carbon dioxide dissolves magnesium carbonate:

$$MgCO_3 + H_2O + CO_2 \rightarrow Mg(HCO_3)_2$$

It is a cause of temporary hardness in water.

magnesium hydroxide ($Mg(OH)_2$) A white solid, occurring naturally as the mineral *brucite*, that dissolves sparingly in water to give an alkaline solution. It can be prepared by adding an alkali to a soluble magnesium compound. Magnesium hydroxide is used as an antacid, in sugar refining, and in uranium processing.

magnesium oxide (magnesia; MgO) A white solid that occurs naturally as the mineral periclase. It can be prepared by heating magnesium in oxygen or by thermal decomposition of magnesium hydroxide, carbonate, or nitrate. Magnesium oxide is weakly basic because of the attraction of the oxide ions for protons from water molecules:

$$O^{2-} + H_2O \rightarrow 2OH^-$$

Magnesium oxide melts at 2800°C, making it useful as a refractory lining for furnaces. It is also used as an antacid and to make reflective coatings for optics and windshields.

magnesium peroxide (MgO_2) A white insoluble solid prepared by reacting sodium or barium peroxide with a concentrated solution of a magnesium salt. Magnesium peroxide is used as a bleach for dyestuffs and silk.

magnesium sulfate ($MgSO_4$) A solid that occurs naturally in many salt deposits in combination with other minerals. One hydrated form, $MgSO_4.7H_2O$, is known as *Epsom salt* or *epsomite*. Epsom salt is made on a commercial scale by reacting magnesium carbonate with sulfuric acid. Magnesium sulfate is used as a purgative drug, as a treatment for indigestion, and as an antidote for barium and barbiturate poisoning. It is also used in the dyeing and sizing of textiles, in tanning, in synthetic fiber production, and as a source of magnesium in fertilizers.

magnetic constant *See* magnetic permeability.

magnetic moment The ratio of the torque exerted on a magnet, a loop that carries an electrical current, or a moving electrical charge in a magnetic field to the field strength of that magnetic field. An electron has an *orbital magnetic moment* because of the magnetic field generated by its orbital motion around its nucleus and a *spin magnetic moment* because of its quantum mechanical spin. Atomic nuclei can also have magnetic moments because of their spin. The existence of electron and nuclear magnetic moments is made use of in various branches of spectroscopy and in magnetochemistry.

magnetic permeability symbol μ. A quantity that characterizes the response of a medium to a magnetic field. The *permeability of free space*, which is denoted μ_0 and is also called the *magnetic constant*, is a constant that appears in Ampère's law relating an electric current to the magnetic field produced by the current. It has the value $4\pi \times 10^{-7}$ Hm^{-1} in S.I. units. The *magnetic permeability of a medium* μ relates the magnetic flux density **B** to the magnetic field strength **H** by the equation: $\mathbf{B} = \mu\mathbf{H}$. The *relative magnetic permeability*, μ_r, of a medium is defined by $\mu = \mu_r\mu_0$.

magnetic quantum number *See* atom.

magnetic separation A method of separating crushed mineral mixtures using the magnetic properties that a component may possess, such as ferromagnetism or diamagnetism.

magnetic susceptibility Symbol χ. A dimensionless quantity, related to the MAGNETIC PERMEABILITY of a medium, that characterizes the magnetic nature of that medium. The relation between the magnetic susceptibility and the relative magnetic permeability μ_r is given by $\chi = \mu_r^{-1}$. If the material is diamagnetic χ has a small negative value. If the material is paramagnetic χ has a small positive value. If the material is ferromagnetic χ has a large positive value.

magnetism The study of the nature and cause of magnetic force fields, and how different substances are affected by them. Magnetic fields are produced by moving charge – both on a large scale (as with current in a wire coil, forming an *electromagnet*), or on the small scale of the moving charges in atoms. It is generally assumed that the Earth's magnetism and that of other extraterrestrial bodies have the same cause.

Substances may be classified on the basis of how samples interact with fields. Different types of magnetic behavior result from the type of atom. *Diamagnetism*, which is common to all substances, is due to the orbital motion of electrons. *Paramagnetism* is due to electron spin, and is thus a property of materials containing unpaired electrons. It is particularly important in transition-metal chemistry, in which the complexes often contain unpaired electrons. Magnetic measurements can give information about the bonding in these complexes. *Ferromagnetism*, the strongest effect, also involves electron spin and the alignment of magnetic moments in domains.

magnetite A black mineral form of mixed iron(II)–iron(III) oxide, Fe_3O_4, a major type of iron ore. It is also used as a flux in making ceramics. *See also* lodestone; triiron tetroxide.

magnetochemistry The branch of chemistry concerned with the magnetic properties of molecules. Magnetochemistry is used extensively in the study of transition-metal complexes because measuring their MAGNETIC SUSCEPTIBILITY enables information about chemical bonding in the complex to be obtained using LIGAND FIELD THEORY.

magneton Symbol μ. A unit for measuring magnetic moments at the atomic and subatomic scales. The *Bohr magneton* μ_B is given by $\mu_B = eh/4\pi m_e$, where e and m are the charge and mass of an electron respectively, and h is the Planck constant. The Bohr magneton is the smallest unit the orbital magnetic moment of an electron in an atom can have. The *nuclear magneton* μ_N is given by replacing the mass of the electron by the mass of the proton and is therefore about 1840 times smaller than the Bohr magneton.

malachite A green mineral consisting of copper(II) carbonate and hydroxide

($CuCO_3.Cu(OH)_2$). It is used as an ore of copper, as a pigment, and in jewelry.

manganate(VI) A salt containing the ion MnO_4^{2-}.

manganate(VII) (permanganate) A salt containing the ion MnO_4^-. Manganate(VII) salts are purple and strong oxidizing agents. In basic solutions the dark green manganate(VI) ion is formed.

manganese A transition metal, the first element of group 7 (formerly VIIB) of the periodic table. It occurs naturally as oxides, e.g. pyrolusite (MnO_2) and *hausmannite* (trimanganese tetraoxide, Mn_3O_4). Nodules found at various locations on the ocean floor are about 25% manganese. The metal is recovered from its oxides after roasting by reduction with aluminum, carbon, or magnesium, followed by electrolysis. Managanese is moderately electropositive and combines with the nonmetals, excluding hydrogen, when heated. Its main use is in alloy steels made by adding pyrolusite to iron ore in an electric furnace. It decomposes cold water and dilute acids to give hydrogen. Manganese exhibits all possible positive oxidation states, with +7, +4, and (the most stable) +2 being the most common. Manganese(II) salts are pale pink. With alkali their solutions precipitate manganese(II) hydroxide, which rapidly oxidizes in air to brown manganese(III) oxide.

Symbol: Mn; m.p. 1244°C; b.p. 1962°C; r.d. 7.44 (20%C); p.n. 25; most common isotope ^{55}Mn; r.a.m. 54.93805.

manganese dioxide *See* manganese(IV) oxide.

manganese(II) oxide (manganous oxide; MnO) A green powder prepared by heating manganese(II) carbonate or oxalate in the absence of air. Alternatively it may be prepared by heating the higher manganese oxides in a stream of hydrogen. Manganese(II) oxide is a basic oxide and is almost insoluble in water. At high temperatures it is reduced by hydrogen to manganese. On exposure to air, manganese(II) oxide rapidly oxidizes. The compound has a crystal lattice similar to that of sodium chloride.

manganese(III) oxide (manganic oxide; manganese sesquioxide; Mn_2O_3) A black powder obtained by igniting manganese(IV) oxide or a manganese(II) salt in air at 800°C. The powder reacts slowly with cold dilute acids to form manganese(III) salts. Manganese(III) oxide occurs in nature as *braunite* ($3Mn_2O_3.MnSiO_3$) and as the monohydrate mineral *manganite* ($Mn_2O_3.H_2O$). The manganese(III) oxide crystal lattice contains Mn^{3+} and O^{2-} ions. With concentrated alkalis, it undergoes disproportionation to give manganese(II) and manganese(IV) ions.

manganese(IV) oxide (manganese dioxide; MnO_2) A black powder prepared by the action of heat on manganese(II) nitrate. A hydrated form occurs naturally as the mineral *pyrolusite*. Manganese(IV) oxide is insoluble in water. It is a powerful oxidizing agent: it reacts with hot concentrated hydrochloric acid to produce chlorine and with warm sulfuric acid to give oxygen. The dihydrate ($MnO_2.2H_2O$) is formed when potassium manganate(VII) is reduced in alkaline solution. It is used as a catalyst in the laboratory preparation of chlorine, as a depolarizer in electric dry cells, and in the glass industry. At 500–600°C, manganese(IV) oxide decomposes to give manganese(III) oxide (Mn_2O_3) and trimanganese tetroxide (Mn_3O_4). It has good electrical conductivity.

manganese sesquioxide *See* manganese(III) oxide.

manganic oxide *See* manganese(III) oxide.

manganin An alloy of copper, manganese, and nickel. It has a high electrical resistivity over a wide range of temperatures and is used to make resistors.

manganite *See* manganese(III) oxide.

manganous oxide *See* manganese(II) oxide.

manometer A device for measuring pressure. A simple type is a U-shaped glass tube containing mercury or other liquid. The pressure difference between the arms of the tube is indicated by the difference in heights of the liquid.

marble A dense metamorphic rock composed of recrystallized limestone or dolomite. It is used in construction and as a decorative stone.

Marcus, Rudolph Arthur (1923–) Canadian–American chemist. In the 1950s Marcus began to work on electron-transfer reactions. The addition, removal, and transfer of electrons is the driving force behind many basic chemical processes including photosynthesis, respiration, and the production of solar energy. Such reactions are, in principle, very simple, involving the movement of an electron from one ion to form another ion. The rates of such reactions can, however, vary widely. Marcus was able to explain electron-transfer reaction rates in terms of the way in which the solvent molecules, initially configured to solvate the reactant, reorganize to solvate the products. He was awarded the 1992 Nobel Prize for chemistry for his work in this field.

Marsh's test *See* arsine.

mass Symbol: m A measure of the quantity of matter in an object. Mass is determined in two ways: the *inertial mass* of an object determines its tendency to resist change in motion; the *gravitational mass* determines its gravitational attraction for other masses. The SI base unit of mass is the kilogram.

mass action, law of At constant temperature, the rate of a chemical reaction is directly proportional to the *active mass* of the reactants, the active mass being taken as the concentration. In general, the rate of reaction decreases steadily as the reaction proceeds; a measure of the concentration of any of the reactants will give a measure of the rate of reaction.

For the reaction A + B → products, the law of mass action states that

$$\text{rate} = k[\text{A}][\text{B}]$$

where [A] and [B] represent the concentration of the reactants in mol dm^{-3} and k is a constant dependent on the reaction. The active mass can equal the concentration in mol dm^{-3} only if there is no interaction or interference between the reacting molecules. Many systems exhibit such interactions and interference and consequently the concentration has to be multiplied by an activity coefficient in order to obtain the effective active mass. *See* activity coefficient.

mass concentration *See* concentration.

mass density *See* concentration.

mass-energy equation The equation $E = mc^2$, where E is the total energy (rest-mass energy + kinetic energy + potential energy) of a mass m, c being the speed of light. The equation is a consequence of Einstein's special theory of relativity; mass is a form of energy and energy also has mass. Conversion of rest-mass energy into kinetic energy (and thus heat) is the source of power in radioactive substances and the basis of nuclear-power generation.

massicot *See* lead(II) oxide.

mass number *See* nucleon number.

mass spectrograph *See* mass spectrometer.

mass spectrometer An instrument for producing ions in a gas and analyzing them according to their charge/mass ratio. The earliest experiments by J. J. Thomson (1856–1940) used a stream of positive ions from a discharge tube, which were deflected by parallel electric and magnetic fields at right angles to the beam. Each type of ion formed a parabolic trace on a photographic plate (a *mass spectrograph*).

In modern instruments, as originated by Francis Aston (1877–1945), the ions are produced by ionizing the gas with electrons. The positive ions are accelerated out of this ion source into a high-vacuum region. Here, the stream of ions is deflected and focused by a combination of electric and magnetic fields, which can be varied so that different types of ion fall on a detector. In this way, the ions can be analyzed according to their mass, giving a *mass spectrum* of the material. Mass spectrometers are used for accurate measurements of relative atomic mass and for analysis of isotope abundance. They can also be used to identify compounds and analyze mixtures. The relative proportions of different types of ions in a mixture is used to find the structure of new compounds. The characteristic spectrum can also identify compounds by comparison with standard spectra.

mass spectrum *See* mass spectrometer.

matrix **1.** A continuous solid phase in which particles of a different solid phase are embedded.
2. (plural **matrices**) A rectangular array of quantities which are frequently real or complex numbers arranged in rows or columns. A matrix is usually indicated by enclosing the array in square brackets. Matrices are used extensively in mathematics and its physical applications. If the sizes of the matrices are right it is possible to add and multiply matrices together. An important feature of the multiplication of two matrices A and B is that, in general, $AB \neq BA$; i.e. matrix multiplication is noncommutative.

matrix mechanics A formulation of quantum mechanics in which the fundamental equations are stated in terms of matrices. This formulation was stated by the German physicist Werner Heisenberg in 1925 as a set of mathematical rules for physical quantities of interest. The German physicist Max Born recognized that the rules for quantum mechanics corresponded to matrix algebra. Born, Heisenberg, and another German physicist Pascual Jordan developed matrix mechanics further. Matrix mechanics is historically important because it was the first formulation of quantum mechanics. In 1926 the Austrian physicist Erwin Schrödinger and others showed that matric mechanics and WAVE MECHANICS are equivalent, and so give the same answers to physical and chemical problems. However, matrix mechanics is used less frequently to solve problems than is wave mechanics.

matte A mixture of iron and copper sulfides obtained at an intermediate stage in smelting copper ores.

maxwell Symbol: Mx A unit of magnetic flux used in the c.g.s. system. It is equal to 10^{-8} Wb in SI units.

Maxwell–Boltzmann distribution The statistical distribution of the speeds of molecules in a gas as a function of both the absolute temperature and the mass of the molecules. It was derived by the British physicist James Clerk Maxwell (1831–79) and the Austrian physicist Ludwig Boltzmann (1844–1906) in the second half of the 19th century using statistical methods.

mean bond energy *See* bond energy.

mechanism A step-by-step description of the events taking place in a chemical reaction. It is a theoretical framework accounting for the fate of bonding electrons and illustrates which bonds are broken and which are formed. For example, in the chlorination of methane to give chloromethane:

step 1

$$Cl{:}Cl \rightarrow 2Cl\bullet$$

step 2

$$Cl\bullet + CH_4 \rightarrow HCl + CH_3\bullet$$

step 3

$$CH_3\bullet + Cl{:}Cl \rightarrow CH_3Cl + Cl\bullet$$

mega- Symbol: M A prefix used with SI units, denoting 10^6. For example, 1 megahertz (MHz) = 10^6 hertz (Hz).

meitnerium A radioactive synthetic transactinide metallic element created by bombarding bismuth-209 nuclei with iron-58 nuclei in a particle accelerator. Only a few short-lived atoms of the element have ever been created.

Symbol: Mt; m.p., b.p., and r.d. unknown; p.n. 109; only known isotope ^{266}Mt (half-life about 3.4×10^{-3}s).

melting (fusion) The process by which a solid is converted into a liquid by heat or pressure.

melting point The temperature at which a solid is in equilibrium with its liquid phase at standard pressure and above which the solid melts. This temperature is always the same for a particular pure solid. Ionically bonded solids generally have much higher melting points than those in which the forces are covalent or intermolecular.

membrane A thin pliable sheet of tissue or other material acting as a boundary. The membrane may be either natural (as in cells, skin, etc.) or synthetic modifications of natural materials (cellulose derivatives or rubbers). In many physicochemical studies membranes are supported on porous materials, such as porcelain, to provide mechanical strength. Membranes are generally permeable to some degree.

Membranes can be prepared to permit the passage of other molecules and micromolecular material. Because of permeability effects, concentration diffferences at a membrane give rise to a whole range of membrane-equilibrium studies, of which osmosis, dialysis, and ultrafiltration are examples. *See also* semipermeable membrane.

Mendeleev, Dmitri Ivanovich (1834–1907) Russian chemist. Mendeleev is remembered for developing the periodic table of chemical elements in a classic paper published in 1869 entitled *On the Relation of the Properties to the Atomic Weights of Elements*. Other scientists such as Julius Lothar Meyer and John Newlands had similar ideas at about the same time but Mendeleev developed his ideas much more fully, including the predictions of the existence and properties of hitherto unknown elements such as gallium, scandium, and germanium, which were discovered subsequently and found to have the properties Mendeleev had predicted.

mendelevium A synthetic silver-white radioactive transuranic element of the actinoid series, first created by bombarding einsteinium-253 isotope with alpha particles. Several short-lived isotopes have been synthesized.

Symbol: Md; m.p. 1021°C; b.p. 3074°C; r.d. unknown, p.n. 101; most stable isotope ^{258}Md (half-life 57 minutes).

mercuric chloride *See* mercury(II) chloride.

mercuric oxide *See* mercury(II) oxide.

mercuric sulfide *See* mercury(II) sulfide.

mercurous chloride *See* mercury(I) chloride.

mercurous oxide *See* mercury(I) oxide.

mercurous sulfide *See* mercury(I) sulfide.

mercury A heavy tin-silver white transition metal, the third element of group 12 (fromerly IIB) of the periodic table. It is the only metal that is a liquid at room temperature (20–22°C). It occurs naturally as the mineral cinnabar (HgS, mercury(II) sulfide); small drops of metallic mercury also occur native in cinnabar and in some volcanic rocks. Elemental mercury vapor is very poisonous as are many mercury compounds. Mercury is used in thermometers and barometers, special AMALGAMS, scientific apparatus, electrical switches, mercury-vapor lamps, and in mercury cells. Mercury compounds are used as fungicides,timber preservatives, and detonators.

Symbol: Hg; m.p. –38.87°C; b.p. 356.58°C; r.d. 13.546 (20°C); p.n. 80; most common isotope ^{202}Hg; r.a.m. 200.59. *See also* zinc group.

mercury cell A voltaic or electrolytic cell in which one or both of the electrodes consists of mercury or an amalgam. Amalgam electrodes are used in the DANIELL CELL and the WESTON CADMIUM CELL. Flowing mercury electrodes are also used in certain electrolytic cells. In the production of chlorine, sodium chloride solution is electrolyzed in a cell with carbon anodes and a flowing mercury cathode. At the cathode, sodium metal is formed, which forms an amalgam with the mercury. *See also* polarography.

mercury(I) chloride (mercurous chloride; calomel; Hg_2Cl_2) A white precipitate prepared by adding dilute hydrochloric acid to a mercury(I) salt solution or by subliming mercury(II) chloride with mercury. Mercury(I) chloride is sparingly soluble in water and is blackened by both ammonia gas, which it absorbs, and by alkalis. It is used in the calomel electrode and was formerly used as a purgative.

mercury(II) chloride (mercuric chloride; $HgCl_2$) A colorless crystalline compound prepared by direct combination of mercury with cold dry chlorine. Mercury(II) chloride is soluble in water; it also dissolves in concentrated hydrochloric acid because of the formation of complex ions, $HgCl_4^{2-}$ and $HgCl_3^-$. On heating, mercury(II) chloride sublimes forming a white translucent mass. It is extremely poisonous, but in dilute solution (1:1000) it is used as an antiseptic. It is also used as a fungicide.

mercury(I) oxide (mercurous oxide; Hg_2O) The black precipitate formed on addition of sodium hydroxide solution to a solution of mercury(I) nitrate is thought by some to be mercury(I) oxide; others, who doubt its existence, think that the blackness of the precipitate is due to some free mercury. X-ray examination of this black compound has shown it to be an intimate mixture of mercury(II) oxide and mercury.

mercury(II) oxide (mercuric oxide; HgO) A poisonous compound formed as a yellow powder by the addition of sodium(I) hydroxide to a solution of mercury(II) nitrate or as a red solid by heating mercury at 350°C for a long time. The difference in color is simply due to particle size. If the oxide is strongly heated it decomposes to give mercury and oxygen.

mercury(I) sulfide (mercurous sulfide; Hg_2S) The brownish-black precipitate originally formed when mercury is treated with cold concentrated sulfuric acid over an extended period is thought to be mercury(I) sulfide. Alternatively it may be prepared by the action of hydrogen sulfide or an alkaline sulfide on a mercury(I) salt solution. As soon as the mercury(I) sulfide is formed it disproportionates to give mercury(II) sulfide and mercury.

mercury(II) sulfide (mercuric sulfide; HgS) A compound that occurs in nature as the minerals CINNABAR (a red solid) and *metacinnabarite* (a black solid). It is prepared as a black precipitate by the action of hydrogen sulfide on a soluble mercury(II) salt solution. On heating, mercury(II) sulfide sublimes and becomes red. It is insoluble in dilute hydrochloric and nitric acids but will dissolve in concentrated nitric acid and aqua regia. Mercury(II) sulfide is used as the pigment *vermilion.*

mer-isomer *See* isomerism.

***meso*-form** *See* optical activity.

mesomerism *See* resonance.

meta- Certain acids regarded as formed from an anhydride and water are named meta acids to distinguish them from the more hydrated ortho acids. For example, H_2SiO_3 ($SiO_2 + H_2O$) is metasilicic acid; H_4SiO_4 ($SiO_2 + 2H_2O$) is orthosilicic acid.
See also para-.

metacinnabarite *See* mercury(II) sulfide.

metaiodic(VII) acid *See* iodic(VII) acid.

metal *See* metals.

metal carbonyl A coordination compound formed between a metal and CAR-

BONYL GROUPS. Transition metals form many such compounds.

metallic bond A bond formed between atoms of a metallic element in its zero oxidation state and in an array of similar atoms. The outer electrons of each atom are regarded as contributing to a free electron 'gas', which occupies the whole crystal of the metal. It is the attraction of the positive atomic cores for the negative electron gas that provides the strength of the metallic bond.

Quantum mechanical treatment of the electron gas restricts its energy to a series of 'bands'. It is the behavior of electrons in these bands that gives rise to semiconductors. *See also* delocalization.

metallic crystal A crystal formed by metal atoms in the solid state. Each atom contributes its valence (outer) electrons to a free electron 'gas'. These electrons are free to migrate through the solid while the remaining ions are arranged in a lattice. The ability of electrons to move through the lattice accounts for the electrical and thermal conductivity of metals.

metallocene A sandwich compound in which a metal atom or ion is coordinated to two cyclopentadienyl ions. Ferrocene ($Fe(C_5H_5)_2$) is the commonest example.

metalloid Any of a class of chemical elements that are intermediate in properties between METALS and NONMETALS. Examples are germanium, arsenic, and tellurium, all of which are semiconductors.

There is, in fact, no clear-cut distinction between metals and nonmetals. In the periodic table, there is a change from metallic to nonmetallic properties across the table, and an increase in metallic properties down a group. Consequently there are elements (boron, silicon, germaium, arsenic, antimony, and telluvium) that form a diagonal running to the right down the table near its right edge and which exhibit these intermediate properties. Other elements are also sometimes considered to be metalloids, according to their chemical properties. Tin, for instance, forms salts with acids but also forms stannates with alkalis. Its oxide is amphoteric. Tin also has metallic (white tin) and nonmetallic (gray tin) allotropes.

metallurgy The study of metals, especially methods of extracting metals from their ores and the formation and properties of alloys.

metals Any of a class of chemical elements with certain characteristic properties. In everyday usage, metals are elements (and alloys) such as iron, aluminum, and copper, which are typically lustrous malleable solids and good conductors of heat and electricity. Such properties do not always apply because some metals, e.g. bismuth, are poor conductors, mercury is a liquid at room temperature, etc.

In chemistry, metals are distinguished by their chemical properties into two main groups. Reactive metals, such as the alkali metals and alkaline-earth metals, are electropositive elements. They are high in the electromotive series and tend to form compounds by losing electrons to give positive ions. They have basic oxides and hydroxides. This typical metallic behavior decreases across the periodic table and increases down a group in the table.

The other group of metals is the transition elements, which are less reactive, have variable valences, and tend to form complexes. In the solid and liquid states metals have METALLIC BONDS, formed by positive ions with free electrons. *See also* metalloid; nonmetal.

metaphosphoric acid *See* phosphoric(V) acid.

metastable species An excited state of an atom, ion, or molecule, that has a relatively long lifetime before reverting to the ground state. Metastable species are intermediates in some reactions.

metastable state A condition of a system or object in which it appears to be in stable equilibrium but, if disturbed, can settle into a lower energy state. For example, supercooled water is liquid below 0°C (at standard pressure). When a small crystal of ice or dust (for example) is introduced, rapid freezing occurs.

metathesis *See* double decomposition.

meter (metre) Symbol: m The SI base unit of length, defined as the distance traveled by light in vacuum in 1/299 792 458 second. This definition was adopted in 1983 to replace the 1960 definition of a length equal to 1 650 763.73 wavelengths in vacuum corresponding to the transition between the levels $2p^{10}$ and $5d^5$ of the krypton-86 atom.

methane The simplest hydrocarbon, CH_4. It is the main component of natural gas.

methanide *See* carbide.

methyl orange An acid–base indicator that is red in solutions below a pH of 3.1 and yellow above a pH of 4.4. Because the transition range is clearly on the acid side, methyl orange is suitable for the titration of an acid with a moderately weak base, such as sodium carbonate.

methyl red An acid–base indicator that is red in solutions below a pH of 4.4 and yellow above a pH of 6.0. It is often used for the same types of titration as methyl orange but the transition range of methyl red is nearer neutral (pH 7) than that of methyl orange. The two molecules are structurally similar.

metre *See* meter.

metric system A decimal system of units based on the meter or centimeter as the base unit of length, the gram or kilogram as the base unit of mass, and the second as the base unit of time.

metric ton *See* tonne.

mho *See* siemens.

mica A member of an important group of aluminosilicate minerals that have a characteristic three-layered structure. The three main types are biotite, lepidolite, and muscovite, which differ in their content of other elements (such as potassium, magnesium, and iron). Mica flakes are used as electrical insulators, dielectrics, and small heat-proof windows.

micro- Symbol: μ A prefix used with SI units, denoting 10^{-6}. For example, 1 micrometer (μm) = 10^{-6} meter (m).

micron Symbol: μ A m.k.s.a. unit of length equal to 10^{-6} meter in SI units.

microwaves A form of electromagnetic radiation, ranging in wavelength from about 1 mm (where it merges with infrared) to about 120 mm (bordering on radio waves). Microwaves are produced by various electronic devices including the klyston; they are often carried over short distances in tubes of rectangular section called *waveguides*.

Spectra in the microwave region can give information on the rotational energy levels of certain molecules. *See also* electromagnetic radiation.

milk of lime *See* calcium hydroxide.

milli- Symbol: m A prefix used with SI units, denoting 10^{-3}. For example, 1 millimeter (mm) = 10^{-3} meter (m).

millimeter of mercury *See* mmHg.

mineral A naturally occurring inorganic compound, having a characteristic crystalline structure and definite chemical composition; a mineral's physical properties are also more or less constant. In contrast, rocks are composed of mixtures of individual minerals. *See* rock.

mineral acid An inorganic acid, especially an acid used commercially in large quantities. Examples are hydrochloric, nitric, and sulfuric acids.

mirror image A shape that is identical to another except that its structure is reversed as if viewed in a mirror. If an object is not symmetrical it cannot be superimposed on its mirror image. For example, the left hand is the mirror image of the right hand.

misch metal A pyrophoric alloy of cerium and other lanthanoids, used in lighter flints.

miscible Denoting combinations of substances that, when mixed, give rise to only one phase; i.e. substances that dissolve in each other. *See* solid solution; solution.

Mitscherlich's law (law of isomorphism) The law stating that substances that crystallize in isomorphous forms (i.e. have identical crystalline forms and form mixed crystals) have similar chemical compositions. The law can be used to indicate the formulae of compounds. For instance, the fact that chromium(III) oxide is isomorphous with Fe_2O_3 and Al_2O_3 implies that its formula is Cr_2O_3.

mixed indicator A mixture of two or more indicators so as to decrease the pH range, heighten the color, etc. over or at which the change occurs.

mixed oxides *See* oxygen.

mixture Two or more substances forming a system in which there is no chemical bonding between the two. In homogeneous mixtures (e.g. solutions or mixtures of gases) the molecules of the substances are mixed, and there is only one PHASE. In heterogeneous mixtures (e.g. certain alloys) different phases can be distinguished. Mixtures differ from chemical COMPOUNDS in that:

1. The chemical properties of the components of a mixture are the same as those of the pure substances.
2. The mixture can be separated by physical means (e.g. distillation or crystallization) or mechanically.
3. The proportions of the components can vary. Some mixtures (e.g. certain solutions) can vary in proportions only between definite limits.

m.k.s.a. system A coherent metric system of units for mechanics and electromagnetics based on the meter, the kilogram, the second, and the ampere. It superseded the m.k.s. system and was in turn superseded by the SI in 1960.

m.k.s. system A metric system of units for mechanics based on the meter, the kilogram, and the second. It superseded the c.g.s. system and was in turn superseded by the m.k.s.a. system in 1946.

mmHg (millimeter of mercury) A c.g.s. unit of pressure defined as the pressure that will support a column of mercury one millimeter high under specified conditions. It is equal to 133.322 4 Pa in SI units. It continues to be used in health care.

mobile phase *See* chromatography.

molal concentration *See* concentration.

molality *See* concentration.

molar 1. Denoting a physical quantity divided by the amount of substance. In almost all cases the amount of substance will be in moles. For example, volume (V) divided by the number of moles (n) is molar volume $V_m = V/n$.
2. A *molar solution* contains one mole of solute per cubic decimeter of solvent.

molar heat capacity *See* Dulong and Petit's law.

molarity *See* concentration.

mole Symbol: mol The SI base unit of amount of substance, defined as the amount of substance that contains as many elementary entities as there are atoms in 0.012 kilogram of ^{12}C. The elementary entities may be atoms, molecules, ions, electrons, photons, etc., and they must be specified. The amount of substance is proportional to the number of entities, the constant of proportionality being the AVOGADRO CONSTANT. One mole contains $6.022\,192 \times 10^{23}$ entities. One mole of an element with relative atomic mass *A* has a mass of *A* grams (this mass was formerly called one *gram-atom* of the element).

molecular beam A beam of molecules (or atoms or ions). Molecular beams are used in spectroscopy and in the study of intermolecular forces, chemical reactions, and surfaces.

molecular crystal A crystal in which molecules, as opposed to atoms, occupy lattice points. Examples include iodine and solid carbon dioxide (dry ice). Because the forces holding the molecules together are weak, molecular crystals have low melting points. When the molecules are small, the crystal structure approximates to a close-packed arrangement. *See* close packing.

molecular dipole moment *See* dipole moment.

molecular formula The formula of a compound showing the number and types of the atoms present in a molecule, but not the arrangement of the atoms. For example, H_2O_2 represents the molecular formula of hydrogen peroxide but not its empirical formula (HO), or the arrangement of its atoms (HOOH). The molecular formula can be determined only if the molecular mass is known. *Compare* empirical formula; general formula; structural formula.

molecularity The total number of reacting molecules in the individual steps of a chemical reaction. Thus, a unimolecular step has molecularity 1, a bimolecular step 2, etc. Molecularity is always an integer, whereas the order of a reaction need not necessarily be so. The molecularity of a reaction gives no information about the mechanism by which it takes place.

molecular orbital *See* orbital.

molecular sieve A substance through which molecules of a limited range of sizes can pass, enabling volatile mixtures to be separated. Zeolites and other metal aluminum silicates can be manufactured with pores of constant dimensions in their molecular structure. When a sample is passed through a column packed with granules of this material, some of the molecules enter these pores and become trapped. The remainder of the mixture passes through the interstices in the column. The trapped molecules can be recovered by heating. Molecular-sieve chromatography is widely used in chemistry and biochemistry laboratories. A modified form of molecular sieve is used in *gel filtration*. The sieve is a continuous gel made from a polysaccharide. In this case, molecules larger than the largest pore size are totally excluded from the column.

molecular spectrum The absorption or emission spectrum that is characteristic of a molecule. Molecular spectra are usually band spectra.

molecular symmetry The set of symmetry operations (rotations, reflections, etc.) that can be applied to a molecule. This set forms the *point group* of the molecule. Molecular symmetry is analyzed systematically using GROUP THEORY.

molecular weight *See* relative molecular mass.

molecule A particle formed by the combination of atoms in a whole-number ratio. A molecule of an element (combining atoms are the same, e.g. O_2) or of a compound (different combining atoms, e.g. HCl) retains the properties of that element or compound. Thus, any quantity of a compound is a collection of many identical molecules. Molecular sizes are characteristically 10^{-10} to 10^{-9} m.

Many molecules of natural products are so large that they are regarded as giant molecules (MACROMOLECULES); they may contain thousands of atoms and have complex structural formulae that require very advanced techniques to identify. *See also* formula; relative molecular mass.

mole fraction The number of moles of a given component in a mixture divided by the total number of moles present of all the components. The mole fraction of component A is

$$n_A/(n_A + n_B + n_C + \ldots)$$

where n_A is the number of moles of A, etc.

molybdenum A hard brittle silver-white transition metal, the second element of group 6 (formerly VIB) of the periodic table. It occurs naturally in the minerals molybdenite (MoS_2) and wulfenite ($PbMoO_4$). Molybdenite, the main ore, is converted to the oxide by roasting, which is then reduced to the metal with hydrogen. Molybdenum is used in alloy steels, incandescent bulbs, and catalysts. In trace amounts it is also an essential dietary element for animals. The compound ammonium molybdate, dissolved in nitric acid, is used as a test for phosphates(V). Purified molybdenum(IV) sulfide (MoS_2) is used in lubricants to enhance viscosity.

Symbol: Mo; m.p. 2620°C; b.p. 4610°C; r.d. 10.22 (20°C); p.n. 42; most common isotope ^{98}Mo; r.a.m. 95.94.

monatomic Denoting a molecule, radical, or ion consisting of only one atom. For example, helium is a monatomic gas and H• is a monatomic radical.

Mond process *See* nickel.

monobasic acid An acid that has only one active proton, such as hydrochloric acid. *Compare* dibasic acid.

monoclinic crystal *See* crystal system.

monohydrate A salt that has a single molecule of water of crystallization, such as sodium carbonate monohydrate, $Na_2CO_3.H_2O$. *See* water of crystallization.

monomer The molecule, group, or compound from which a dimer, trimer, or polymer is formed.

monotropy The existence of a single allotrope of an element that is always more stable than the other allotrope(s) regardless of temperature; phosphorus, for example, exhibits monotropy. The phase diagram for a monotropic element shows that one allotrope always has a lower vapor pressure than the other(s) at all temperatures; this is the stable allotrope. *See also* allotropy; enantiotropy.

monovalent (univalent) Having a valence of one.

mordant An inorganic compound used to fix dye in cloth. The mordant (e.g. aluminum hydroxide or chromium salts) is precipitated in the fibers of the cloth, and the dye then absorbs in the particles.

Moseley's law Lines in the x-ray spectra of elements have frequencies that depend on the proton number of the element. For a set of elements, a graph of the square root of the frequency of x-ray emission against proton number is a straight line (for spectral lines corresponding to the same transition).

Mössbauer effect The emission of gamma rays with a very narrow spread of wavelengths from a solid. Usually, when gamma rays are emitted by a radioactive nucleus they have a broad spread of wavelengths due to the recoil of the emitting nuclei. However, if the emitting nuclei are embedded in a crystal the whole lattice recoils, meaning that there is very little effect on the wavelength of the emitted gamma rays.

In *Mössbauer spectroscopy*, a source of gamma rays is mounted on a moving support close to a sample. Detectors monitor gamma rays scattered by the sample. The wavelength of the gamma rays from the source can be changed by varying the speed at which the source is moved toward the sample (the Doppler effect). The spectrum is a plot of detector response against speed of source. A sharp decrease in response at a particular speed indicates resonance absorption of gamma ray by a nuclide in the sample. The wavelength (energy) at which this occurs depends on the energy levels in the sample nucleus, but these are affected to some extent by the surrounding electrons (i.e. there are *chemical shifts*).

mother liquor The solution remaining after the formation of crystals.

Mulliken, Robert Sanderson (1896–1986) American chemist and physicist. Mulliken devoted most of his career to understanding the electronic structure of molecules and molecular spectra in terms of quantum mechanics. He was awarded the 1966 Nobel Prize for chemistry for his 'fundamental work concerning chemical bonds and the electronic structure of molecules by the molecular-orbital method'.

multicenter bond A two-electron bond formed by the overlap of orbitals from more than two atoms (usually 3). The bridging in diborane, B_2H_6, is believed to take place by overlap of an sp^3 hybrid orbital from each boron atom with the 1s orbital on the hydrogen atom. This multicenter bond is called a three-center two-electron bond. The molecule is electron-deficient. *See also* boron hydride; electron-deficient compounds.

multidecker sandwich compound *See* sandwich compound.

multidentate ligand A ligand that possesses at least two sites at which it can coordinate.

multiple bond A bond between two atoms involving more than one pair of electrons, e.g. a double bond or a triple bond. This additional bonding arises from overlap of atomic ORBITALS that are perpendicular to the internuclear axis and gives rise to an increase in electron density above and below the internuclear axis. Such bonds are called pi bonds. If the sigma bond axis is taken as the z-axis (i.e. the internuclear axis) then overlap of orbitals along the x-axis gives rise to a pi bond in the xz plane. Similarly orbitals on the y-axis form a pi bond in the yz plane.

multiple proportions, law of Proposed by John Dalton in 1804, the law states that when two elements A and B combine to form more than one compound, the masses of B that combine with a fixed mass of A are in small, whole-number ratios. For example, in dinitrogen oxide, N_2O, nitrogen monoxide, NO, and dinitrogen tetroxide, N_2O_4, the amounts of nitrogen combined with a fixed mass of oxygen are in the ratio 4:2:1.

multiple-range indicator *See* universal indicator.

Mumetal (*Trademark*) A magnetic alloy containing about 75% nickel, the remainder being iron, copper, and chromium and sometimes molybdenum. It is easily magnetized and demagnetized, has a high permeability, and is used in transformer cores and electromechanical equipment.

muriate An obsolete name for a chloride; e.g. muriate of potash (KCl), muriatric acid (HCl).

muriatic acid *See* muriate.

N

nano- Symbol: n A prefix used with SI units, denoting 10^{-9}. For example, 1 nanometer (nm) = 10^{-9} meter (m).

nanotechnology The technology of devices at the nanometer scale. In such small devices the quantum mechanical nature of electrons has to be taken into account fully. Instruments such as the ATOMIC FORCE MICROSCOPE which can identify and manipulate individual atoms are very useful in nanotechnology.

nanotubes Tubular structures with diameters of a few nanometers. Examples of nanotubes are the carbon structures known as bucky tubes, which have a structure similar to that of BUCKMINSTERFULLERENE. Interest has been shown in nanotubes as possible microscopic probes in experiments, as semiconductor materials, and as a component of composite materials.

nascent hydrogen A particularly reactive form of hydrogen, which is believed to exist briefly between its generation (e.g. by the action of dilute acid on magnesium) and its appearance as bubbles of normal hydrogen gas. It is thought that part of the free energy of the production reaction remains with the hydrogen molecules for a short time. Nascent hydrogen may be used to produce the hydrides of phosphorus, arsenic, and antimony, which are not readily formed from ordinary hydrogen.

natron A naturally occurring mineral form of hydrated SODIUM CARBONATE ($Na_2CO_3.10H_2O$), found on the beds of dried-out soda lakes.

natural abundance *See* abundance.

negative catalyst *See* catalyst.

neodymium A soft toxic silvery yellow element belonging to the lanthanoid series of metals. It occurs in association with other lanthanoids in such minerals as allanite, bastnaesite, cerite, and monazite. Neodymium is used in various alloys, as a catalyst, in compound form in carbon-arc searchlights, etc., and in the glass industry.

Symbol: Nd; m.p. 1021°C; b.p. 3068°C; r.d. 7.0 (20°C); p.n. 60; most common isotope ^{140}Nd;r.a.m. 144.24.

neon An inert colorless odorless monatomic gas the second member of the rare gases; i.e. group 18 (formerly VIIIA or 0) of the periodic table. It has a very high ionization potential and forms no compounds. It occurs in minute quantities (0.0018% by volume) in air and is obtained from liquid air by fractional distillation. It is used in neon signs and lamps (emitting a characteristic orange-red light), electrical equipment, Geiger counters, and gas lasers.

Symbol: Ne; m.p. –248.67°C; b.p. –246.05°C; mass density 0.9 kg m^{-3} (0°C); p.n. 10; most common isotope ^{20}Ne; r.a.m. 20.18.

neptunium A toxic radioactive silvery element of the actinoid series of metals that was the first transuranic element to be synthesized (1940). Found on Earth only in minute quantities in uranium ores, it is also obtained as a by-product from uranium fuel elements bombarded by neutrons in nuclear reactors. It can be converted to plutonium-238, which can be used as a nuclear power source.

Symbol: Np; m.p. 640°C; b.p. 3902°C; r.d. 20.25 (20°C); p.n. 93; most stable isotope ^{237}Np (half-life 2.14×10^6 years).

Nernst, Hermann Walther (1864–1941) German physical chemist. Nernst is best remembered for his contributions to electrochemistry and for discovering the third law of thermodynamics. His work on electrochemistry included the concept of the solubility product and the use of buffer solutions. In 1906 he stated a theorem concerning the entropy of crystals at absolute zero which, in slightly different form, became known as the third law of thermodynamics. He also studied photochemistry and wrote an influential book entitled *Theoretical Chemistry* (1893). He was awarded the 1920 Nobel Prize for chemistry.

neutralization The stoichiometric reaction of an acid and a base in volumetric analysis. The neutralization point or end point is detected with indicators.

neutral oxide An oxide that forms neither an acid nor a base on reaction with water. Examples include carbon monoxide (CO), dinitrogen oxide (N_2O), and water (H_2O). *Compare* acidic oxide; amphoteric; basic oxide.

neutron diffraction A method of structure determination used for solids, liquids, and gases that makes use of the quantum mechanical wave nature of neutrons. Thermal neutrons with average kinetic energies of about 0.025 eV have a wavelength of about 0.1 nanometer, making them suitable for investigating the structure of matter at the atomic level. Neutron diffraction is particularly useful for identifying the positions of hydrogen atoms. These are difficult to establish using x-rays since x-rays interact with electrons, and hence are scattered weakly by hydrogen atoms, which have only one electron. Protons scatter neutrons strongly, meaning that the positions of protons can readily be determined using neutron scattering. Neutron scattering has also been used extensively in investigations of the magnetic ordering in solids because there is an interaction between the magnetic moments of the neutrons and the magnetic moments of atoms in the sample.

neutron number Symbol: N The number of neutrons in the nucleus of an atom; i.e. the nucleon number (A) minus the proton number (Z).

Newlands' law (law of octaves) The observation that, when the elements are arranged in order of increasing relative atomic mass, there is a similarity between members that are eight elements apart (inclusive). For example, lithium has similarities to sodium, beryllium to magnesium, and so on. Newlands discovered this relationship in 1863 before the announcement of Mendeléev's famous periodic law (1869). It is now recognized that Newlands' law arises from there being eight members in each short period. It is named for the British chemist John Newlands (1837–98).

newton Symbol: N The SI derived unit of force, equal to the force needed to accelerate one kilogram by one meter second^{-2}. $1\ N = 1\ kg\ m\ s^{-2}$.

Nichrome (*Trademark*) Any of a group of nickel–chromium–iron alloys, containing 60–80% nickel and about 16% chromium; small amounts of other elements, such as carbon or silicon, may be added. They can withstand very high temperatures and their high electrical resistivity makes them suitable for use in heating elements.

nickel A ductile malleable relaatively hard corrosion-resistant transition metal, the first member of group 10 (formerly part of subgroup VIIIB) of the periodic table. Nickel occurs in nature, most importantly, as the sulfides millerite (NiS) and pentlandite $(Ni,Fe)_9S_8$. It is extracted by the *Mond process*, which involves roasting the ore to obtain the oxide. The oxide is then reduced using carbon monoxide, followed by the formation and subsequent decomposition of volatile NICKEL CARBONYL $Ni(CO)_4$. Nickel is used as a catalyst, in coinage alloys, and in stainless steel. Its compounds in its main oxidation state (+2) are usually green.

Symbol: Ni; m.p. 1453°C; b.p. 2732°C; r.d. 8.902 (25°C); p.n. 28; most common isotope ^{58}Ni; r.a.m. 58.6934.

nickel carbonyl (tetracarbonyl nickel(0); $Ni(CO)_4$) A colorless liquid with a highly toxic vapor. It is a typical example of a transition metal complex, prepared by passing carbon monoxide over finely divided nickel at a temperature of approximately 60°C. The product is purified by fractional distillation. Nickel carbonyl decomposes on heating to give pure nickel. It is used as a catalyst and in the preparation of nickel by the Mond process.

nickelic oxide *See* nickel(III) oxide.

nickel–iron accumulator (Edison cell; Nife or NIFE cell) A type of secondary cell in which the electrodes are formed by steel grids. The positive electrode is impregnated with a nickel–nickel hydride mixture and the negative electrode is impregnated with iron oxide. Potassium hydroxide solution forms the electrolyte. A nickel–iron cell is lighter and more durable than a lead accumulator, and it can work with higher currents. Its e.m.f. is approximately 1.3 volts.

nickelous oxide *See* nickel(II) oxide.

nickel(II) oxide (nickelous oxide; NiO) A pale green powder formed by heating nickel(II) hydroxide, nitrate, or carbonate in the absence of air. It is a basic oxide, dissolving in dilute acids to give green solutions of nickel(II) salts. Nickel(II) oxide may be reduced to the metal by carbon, carbon monoxide, or hydrogen.

nickel(III) oxide (nickelic oxide; Ni_2O_3) A black powder formed by the action of heat on nickel(II) oxide in air. It can also be prepared by igniting nickel(II) nitrate or carbonate. It exists also as the dihydrate, $Ni_2O_3.2H_2O$.

nickel-silver (German silver) Any of a group of alloys containing nickel, copper, and zinc (but no silver) in various proportions. They are white or silvery in color, can be highly polished, and have good corrosion-resistance. They are used, for example, as base metals in silver plating, chromium plating, and enameling.

nielsbohrium Symbol: Ns A name formerly suggested for element-107, now known as bohrium (Bh). *See* element.

NIFE cell (Nife cell) *See* nickel–iron accumulator.

niobium A soft silvery malleable ductile transition metal, the second member of group 5 (formerly VB) of the periodic table. It occurs naturally chiefly in the minerals columbite ($(Fe,Mn)(Nb,Ta)_2O_6$) and tantalite ($(Fe,Mn)(Ta,Nb)_2O_6$). It is used in welding, special steels for high temperature environments, and in superconductor alloys. It was formerly known as columbium.

Symbol: Nb; m.p. 2468°C; b.p. 4742°C; r.d. 8.570 (20°C); p.n. 41; most common isotope ^{93}Nb; r.a.m. 92.90638.

niter *See* potassium nitrate.

nitrate A salt of nitric acid.

nitration A reaction introducing the nitro ($-NO_2$) group into a compound. Nitration is usually carried out using a mixture of concentrated nitric and sulfuric acids, although the precise conditions differ from compound to compound. The attacking species is NO_2^+ (the *nitryl ion*).

nitric acid (HNO_3) A colorless fuming corrosive liquid that is a strong acid. Nitric acid can be made in a laboratory by the distillation of a mixture of an alkali metal nitrate and concentrated sulfuric acid. Commercially it is prepared by the catalytic oxidation of ammonia and is supplied as concentrated nitric acid, which contains 68% of the acid and is often colored yellow by dissolved oxides of nitrogen.

Nitric acid is a strong oxidizing agent. Most metals are converted to their nitrates with the evolution of oxides of nitrogen (the composition of the mixture of the oxides depends on the temperature and on the concentration of the nitric acid used). Some nonmetals (e.g. sulfur and phosphorus) react to produce oxyacids.

nitric oxide *See* nitrogen monoxide.

nitride A compound of nitrogen with a more electropositive element such as magnesium or calcium.

nitriding *See* case hardening.

nitrification The action of nitrifying bacteria in converting ammonia and nitrites to nitrates. It is an important stage in the nitrogen cycle. *See also* nitrogen cycle; nitrogen fixation.

nitrite A salt of nitrous acid.

nitrogen A colorless odorless gaseous element, the first member of group 15 (formerly VA) of the periodic table. It is a very electronegative element existing in the uncombined state as gaseous diatomic N_2 molecules. The nitrogen atom has the electronic configuration $[He]2s^22p^3$. It is typically nonmetallic and its bonding is primarily by polarized covalent bonds. With electropositive elements the nitride ion N^{3-} may be formed.

Nitrogen accounts for about 78% of the atmosphere (by volume) and it also occurs as sodium nitrate in various mineral deposits. It is separated for industrial use by the fractional distillation of liquid air. Pure nitrogen is prepared by thermal decomposition of azides:

$$2NaN_3 \rightarrow 2Na + 3N_2$$

The 'active nitrogen' gas obtained by passing an electric discharge through nitrogen is a mixture of ordinary nitrogen molecules and excited single nitrogen atoms.

As a diatomic molecule nitrogen is effectively triple-bonded and has a high dissociation energy (940 kJ mol^{-1}). It is therefore relatively inert at room temperature and reacts readily only with lithium and other highly electropositive elements. Consequently it is used to provide an inert atmosphere for many metallurgical processes and electrical devices. The direct combination of nitrogen and hydrogen occurs at elevated temperatures and pressures (400–600°C, 100 megapascals) and is the basis of the industrially important Haber process for the manufacture of ammonia. Liquid nitrogen is used in cryogenics.

Nitrogen forms a number of oxides:

1. Dinitrogen oxide (nitrous oxide, N_2O) – a neutral oxide.
2. Nitrogen monoxide (nitric oxide, NO) – a neutral oxide.
3. *Nitrogen(III) oxide* (nitrogen sesquioxide, N_2O_3) – an unstable oxide that is the anhydride of nitrous acid.
4. Nitrogen dioxide (NO_2), which gives a mixture of nitrous and nitric acids in water.
5. Dinitrogen tetroxide (N_2O_4), which is in equilibrium with nitrogen dioxide.
6. *Dinitrogen pentoxide* (N_2O_5) – the anhydride of nitric acid.
7. *Nitrogen trioxide* (NO_3) – an unstable white solid formed by the reaction of dinitrogen pentoxide and ozone (O_3).

Nitrogen also forms a number of binary halides such as NF_3 and NCl_3, and halogen azides such as ClN_3. Except for NF_3 these are all very explosive, and even NF_3, which is reputedly stable, has been known to explode.

Transition metals form nitrides that are different from the ionic or covalent nitrides, the stoichiometries are, for example, ZrN, W_2N, Mn_4N. They are often called 'interstitial' nitrides. They are very hard and their formation is the basis of nitriding – surface hardening by exposing hot metal to ammonia gas (*see* case hardening).

Nitrogen has two isotopes; ^{14}N, the common isotope, and ^{15}N (natural abundance 0.366%). The latter is used as a marker in mass spectrometric studies.

Symbol: N; m.p. –209.86°C; b.p. –195.8°C; mass density 1.2506 kg m^{-3} (0°C); p.n. 7; r.a.m. 14.

nitrogen cycle The circulation of nitrogen and its compounds in the environment, one of the major natural cycles of an element. Nitrogen occurs mainly as a diatomic gas in the atmosphere and as nitrates in the soil. *Denitrification* of these nitrates converts them to nitrogen. Plants also take up nitrates and are eaten by animals; the plants and animals convert them to proteins, which after the death of the organisms are broken down to yield ammonia, which NITRIFICATION converts back to nitrates. NITROGEN FIXATION by bacteria or lightning also gives rise to nitrates.

nitrogen dioxide (NO_2) A brown gas produced by the dissociation of dinitrogen tetroxide N_2O_4, with which it is in equilibrium, the dissociation being complete at 140°C. Further heating causes dissociation to colorless nitrogen monoxide and oxygen:

$$2NO_2(g) = 2NO(g) + O_2(g)$$

Nitrogen dioxide can also be made by the action of heat on metal nitrates, although not the nitrates of the alkali metals or some of the alkaline-earth metals.

nitrogen fixation (fixation of nitrogen) A reaction in which atmospheric nitrogen is converted into nitrogen compounds. Nitrogen fixation occurs naturally as part of the life cycle processes of certain bacteria present in the soil and in the roots of leguminous plants. Small amounts of nitrogen(II) oxide (NO) are also formed in thunderstorms by direct combination of the elements during lightning discharges. Fixation of nitrogen is important because nitrogen is an essential element for life, and because vast amounts of nitrogenous fertilizers are used for agriculture. Former processes for fixing nitrogen were the Birkeland-Eyde process (reacting N_2 and O_2 in an electric arc – the equivalent of a lightning discharge in nature), and the cyanamide process (heating calcium dicarbide with nitrogen to give $CaCN_2$). The major method employed today is the Haber process for making ammonia.

nitrogen monoxide (nitric oxide; NO) A colorless gas that is insoluble in water but dissolves in a solution containing iron(II) ions owing to the formation of the complex ion $(FeNO)^{2+}$: the nitrogen monoxide can be released by heating. Nitrogen monoxide can be prepared in the lab by the action of nitric acid on copper turnings and purified by using a solution of iron(II) ions to absorb the product. Commercially, nitrogen monoxide is prepared by the catalytic oxidation of ammonia or by the direct union of nitrogen and oxygen in an electric arc. Nitrogen monoxide is the most heat-stable of the oxides of nitrogen, only decomposing above 1000°C. At ordinary temperatures it combines immediately with oxygen to give nitrogen dioxide:

$$2NO(g) + O_2(g) \rightarrow 2NO_2(g)$$

nitrogen oxides (NO_x) Any of the various compounds formed between nitrogen and oxygen. *See* dinitrogen oxide; nitrogen; nitrogen dioxide; nitrogen monoxide.

nitrous acid (HNO_2) A weak acid known only in solution or as a gas, obtained by acidifying a solution of a nitrite. It readily decomposes on warming or shaking to nitrogen monoxide and nitric acid. It is used extensively in the dyestuffs industry. Nitrous acid and the nitrites are normally reducing agents but in certain circumstances they can behave as oxidizing agents, e.g. with sulfur dioxide and hydrogen sulfide.

nitrous oxide *See* dinitrogen oxide.

nitryl ion (nitronium ion) The positive ion NO_2^+. *See* nitration.

NMR *See* nuclear magnetic resonance.

nobelium A synthetic radioactive transuranic element of the actinoid series, created by bombarding a curium isotope with the nuclei of a carbon isotope. Several very short-lived isotopes have been synthesized.

Symbol: No; m.p. 863°C; b.p., r.d. unknown; p.n. 102; most stable isotope ^{259}No (half-life 58 minutes).

noble gases *See* rare gases.

nonionic detergents *See* detergents.

nonlocalized bond *See* delocalized bond.

nonmetal Any of a class of chemical elements. Nonmetals lie right of the diagonal formed by the METALLOIDS in the periodic table. They are electronegative elements with a tendency to form covalent compounds or negative ions. They have acidic oxides and hydroxides. In the solid state, nonmetals are either covalent volatile crystals or macromolecular (giant-covalent) crystals. *See also* metals.

nonpolar compound A compound that has molecules with no permanent dipole moment. Examples of nonpolar compounds are hydrogen and carbon dioxide.

nonpolar solvent *See* solvent.

nonstoichiometric compound A chemical compound whose molecules contain fractional numbers of atoms. For example, titanium(IV) oxide in the form of the mineral rutile has the chemical formula $TiO_{1.8}$.

normality The number of gram equivalents per cubic decimeter of a given solution.

normal modes of vibration The basic vibrations of a polyatomic molecule. All vibrational motion of a polyatomic molecule is a superposition of the normal modes of vibration of the molecule. If N is the number of atoms in a molecule the number of modes of vibration is $3N–5$ for a linear molecule and $3N–6$ for a nonlinear molecule. Each of these vibrational modes has a characteristic frequency, although it is possible for some of them to be degenerate. For example, in a linear triatomic molecule there are four normal modes of vibration since $3N–5 = 4$ for $N = 3$. These vibrational modes are: (a) symmetric stretching (breathing) vibrations; (b) antisymmetric stretching vibrations; and (c) two bending vibrations, which are degenerate.

normal solution A solution that contains one gram equivalent weight per liter of solution. Values are designated by the symbol N, e.g. 0.2N, N/10, etc. Because there is not a clear definition of equivalent weight suitable for all reactions, a solution may have one value of normality for one reaction and another value in a different reaction. Because of this many workers prefer the molar solution notation.

Norrish, Ronald George Wreyford (1897–1978) British physical chemist. Norrish made important contributions to the study of fast chemical reactions, particularly those initiated by light. Between 1949 and 1965 he developed the techniques of flash photolysis and kinetic spectroscopy with George PORTER to study fast chemical reactions. Norrish and Porter shared the 1967 Nobel Prize for chemistry with Manfred EIGEN for this work. In his later years Norrish studied chain reactions and the kinetics of polymerization.

NO_x *See* nitrogen oxides.

NTP *See* STP.

nuclear magnetic resonance (NMR) A method of investigating nuclear spin. In an external magnetic field the nucleus of an atom can have certain quantized energy states, corresponding to certain orientations of the spin magnetic moment. Hydrogen nuclei, for instance, can have two energy states, and transitions between the two occur by absorption of radiofrequency radiation. In chemistry, this is the basis of a spectroscopic technique for investigating the structure of molecules. Radiofrequency radiation is fed to a sample and the magnetic field is changed slowly. Absorption of the radiation is detected when the difference between the nuclear levels corresponds to absorption of a quantum of radiation. This difference depends slightly on the electrons around the nucleus – i.e. the position of the atom in the molecule. Thus a different absorption frequency is seen for each type of hydrogen atom. In ethanol, for example, there are three frequencies, corresponding to hydrogen atoms on the CH_3, CH_2, and OH groups. The intensity of absorption also depends on the number of hydrogen atoms (3:2:1).

NMR spectroscopy is a powerful method of finding the structures of complex compounds.

nucleon number (mass number) Symbol: *A* The number of nucleons (protons plus neutrons) in an atomic nucleus.

nucleus The compact positively charged center of an atom, made up of one or more nucleons (protons and neutrons) around which is a cloud of one or more electrons. The density of nuclei is about 10^{15} kg m^{-3}. The number of protons in the nucleus defines the element, being its proton number (also known as its atomic number). The nucleon number, or atomic mass number, is the sum of the protons and neutrons. The simplest nucleus is that of a hydrogen atom, ^{1}H, being simply one proton (mass 1.67×10^{-27} kg). The most massive nucleus that occurs in any quantity on Earth is ^{238}U of 92 protons and 146 protons (mass 4×10^{-25} kg, radius 9.54×10^{-15} m). Only certain combinations of protons and neutrons form stable nuclei. Others undergo spontaneous decay.

A nucleus is depicted by a symbol indicating nucleon number (mass number), proton number (atomic number), and element name. For example, ${}_{1}^{2}{}_{1}^{3}Na$ represents a nucleus of sodium having 11 protons and mass 23, hence there are (23 – 11) = 12 neutrons.

nuclide A nuclear species with a given number of protons and neutrons; for example, ^{23}Na, ^{24}Na, and ^{24}Mg are all different nuclides. Thus:

^{23}Na has 11 protons and 12 neutrons
^{24}Na has 11 protons and 13 neutrons
^{24}Mg has 12 protons and 12 neutrons

The term is applied to the nucleus and often also to the atom in which it is found.

O

occlusion 1. The process in which small amounts of one substance are trapped in the crystals of another; for example, pockets of liquid occluded during crystallization from a solution.
2. Absorption of a gas by a solid; for example, the occlusion of hydrogen by palladium.

ocher A clay or rock containing significant amounts of iron(III) oxide (Fe_2O_3), used as a yellow, orange, or brown pigment.

octahedral complex *See* complex.

octahydrate A crystalline compound containing eight molecules of water of crystallization per molecule of compound.

octavalent Having a valence of eight.

octaves, law of *See* Newlands' law.

octet A stable shell of eight electrons in an atom. The completion of the octet gives rise to particular stability, which is the basis of the *Lewis octet theory*:
1. The rare gases have complete octets and are essentially chemically inert.
2. The bonding in small covalent molecules is frequently achieved by the central atom completing its octet by sharing electrons with surrounding atoms, e.g. CH_4, H_2O.
3. The ions formed by electropositive and electronegative elements are generally those with a complete octet, e.g. Na^+, Ca^{2+}, O^{2-}, Cl^-.

oersted Symbol: Oe A c.g.s. unit of magnetic field strength. It is equal in SI units to $10^3/4\pi$ A m^{-1}.

ohm Symbol: Ω The SI derived unit of electrical resistance, equal to a resistance that passes a current of one ampere when there is an electric potential difference of one volt across it. 1 Ω = 1 V A^{-1}.

oil of vitriol Sulfuric(VI) acid.

oleum (disulfuric(VI) acid; fuming sulfuric acid; pyrosulfuric acid; $H_2S_2O_7$) A colorless fuming liquid formed by dissolving sulfur(VI) oxide in concentrated sulfuric acid. It gives sulfuric acid on dilution with water. *See* contact process.

olivine A greenish rock-forming silicate mineral ($(Mg,Fe)_2SiO_4$) containing iron and magnesium. Clear, green, gemstone quality olivine is called *peridot*.

onium ion An ion formed by addition of a proton (H^+) to a previously neutral molecule. The hydronium ion (H_3O^+) and ammonium ion (NH_4^+) are examples.

Onsager, Lars (1903–76) Norwegian-born American chemist. Onsager made several important contributions to theoretical chemistry and physics. In 1926 he improved on the Debye-Hückel theory of electrolytes by taking the Brownian motion of ions into account. He subsequently investigated the dielectric constants of matter that contains polar molecules. In 1931 he published fundamental work on nonequilibrium thermodynamics. He was awarded the 1968 Nobel Prize for chemistry for this work.

(R)-lactic acid

(S)-lactic acid

Optical activity: two forms of lactic acid

oolite Any sedimentary rock consisting of an aggregate of small spherical masses, most typically composed of calcium carbonate ($CaCO_3$), in a fine matrix.

opal A naturally occurring hydrated noncrystalline form of silica ($SIO_2 \cdot_n H_2O$), valued as gemstones. Various layers within precious opal can reflect and refract light to give a play of spectral colors. Varieties include black, fire, and milky white opal.

open-hearth process An obsolete method of making steel by heating pig iron in an open-hearth furnace by burning a mixture of producer gas and air. Oxygen is injected through pipes to oxidize impurities and burn off carbon, and lime is used to form a slag. The basic oxygen process has replaced it.

optical activity The ability of certain compounds to rotate the plane of polarization of plane-polarized light when the light is passed through them. Optical activity can be observed in crystals, gases, liquids, and solutions. The amount of rotation depends on the concentration of the active compound.

Optical activity is caused by the interaction of the varying electric field of the light with the electrons in the molecule. It occurs when the molecules are asymmetric – i.e. they have no plane of symmetry. Such molecules have a mirror image that cannot be superimposed on the original molecule. In organic compounds this usually means that they contain a carbon atom attached to four different groups, forming a chiral center. The two mirror-image forms of an asymmetric molecule are *optical isomers*. One isomer will rotate the polarized light in one sense and the other by the same amount in the opposite sense. Such isomers are described as *dextrorotatory* (symbol *d* or (+)) or *levorotatory* (symbol *l* or (–)), according to whether they rotate the plane to the 'right' or 'left' respectively (rotation to

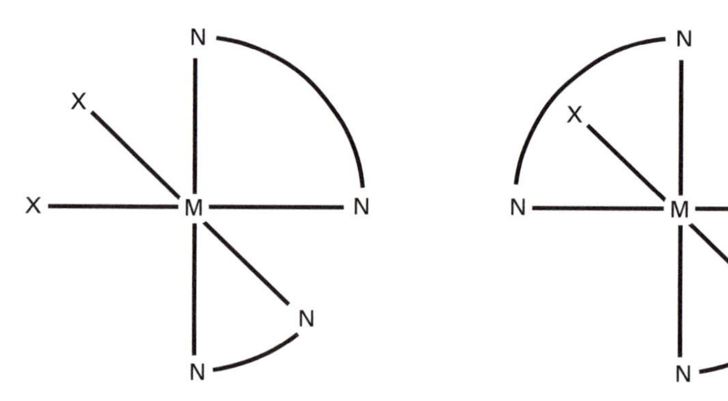

Optical activity: two forms of an octahedral complex. The bidentate ligands are represented by a curved line

the left is clockwise to an observer viewing the light coming toward the observer). A mixture of the two isomers in equal amounts does not show optical activity. Such a mixture is sometimes called the (±) or *dl-form*, a *racemate*, or a *racemic mixture*

Optical isomers have identical physical properties (apart from optical activity) and cannot be separated by fractional crystallization or distillation. Their general chemical behavior is also the same, although they do differ in reactions involving other optical isomers. Many naturally occurring substances are optically active (only one optical isomer exists naturally) and biochemical reactions occur only with the natural isomer. For instance, the natural form of glucose is *d*-glucose and living organisms cannot metabolize the *l*-form.

The terms 'dextrorotatory' and 'levorotatory' refer to the effect on polarized light. A more common method of distinguishing two optical isomers is by their *D-form* (*dextro-form*) or *L-form* (*levo-form*). This convention refers to the absolute structure of the isomer according to specific rules. Sugars are related to a particular configuration of glyceraldehyde (2,3-dihydroxypropanal). For alpha amino acids the 'corn rule' is used: the structure of the acid $RC(NH_2)(COOH)H$ is drawn with H at the top; viewed from the top the groups spell CORN in a clockwise direction for all D-amino acids (i.e. the clockwise order is $-COOH, R, NH_2$). The opposite is true for L-amino acids. Note that this convention does not refer to optical activity: D-alanine is dextrorotatory but D-cystine is levorotatory.

An alternative is the R/S system for showing configuration. There is an order of priority of attached groups based on the proton number of the attached atom:
I, Br, Cl, SO_3H, $OCOCH_3$, OCH_3, OH, NO_2, NH_2, $COOCH_3$, $CONH_2$, $COCH_3$, CHO, CH_2OH, C_6H_5, C_2H_5, CH_3, H

Hydrogen has the lowest priority. The chiral carbon is viewed such that the group of lowest priority is hidden behind it. If the other three groups are in descending priority in a clockwise direction, the compound is R-. If descending priority is anticlockwise it is S-.

The existence of a carbon atom bound to four different groups is not the strict condition for optical activity. The essential point is that the molecule should be asymmetric. Inorganic octahedral complexes, for example, can show optical isomerism (*see illustration*). It is also possible for a molecule to contain *asymmetric carbon atoms* and still have a plane of symmetry. One structure of tartaric acid has two parts of the molecule that are mirror images, thus having a plane of symmetry. This (called the *meso-form*) is not optically active. *See also* resolution.

optical isomers *See* isomerism; optical activity.

optical rotary dispersion (ORD) The phenomenon in which the amount of rotation of plane-polarized light by an optically active substance depends on the wavelength of the light. Plots of rotation against wavelength can be used to give information about the molecular structure of optically active compounds.

optical rotation Rotation of the plane of polarization of plane-polarized light by an optically active substance.

orbital A region around an atomic nucleus in which there is a high probability of finding an electron. The modern picture of the atom, according to quantum mechanics, does not have electrons moving in fixed elliptical orbits. Instead, there is a finite probability that the electron will be found in any small region at all possible distances from the nucleus. In the hydrogen atom the probability is low near the nucleus, increases to a maximum, and falls off to infinity. It is useful to think of a region in space around the nucleus – in the case of hydrogen the region within a sphere – within which there is a high chance of finding the electron. Each of these regions, called an *atomic orbital*, corresponds to a subshell and can 'contain' either a single electron or two electrons with opposite spins. Another way of visualizing an or-

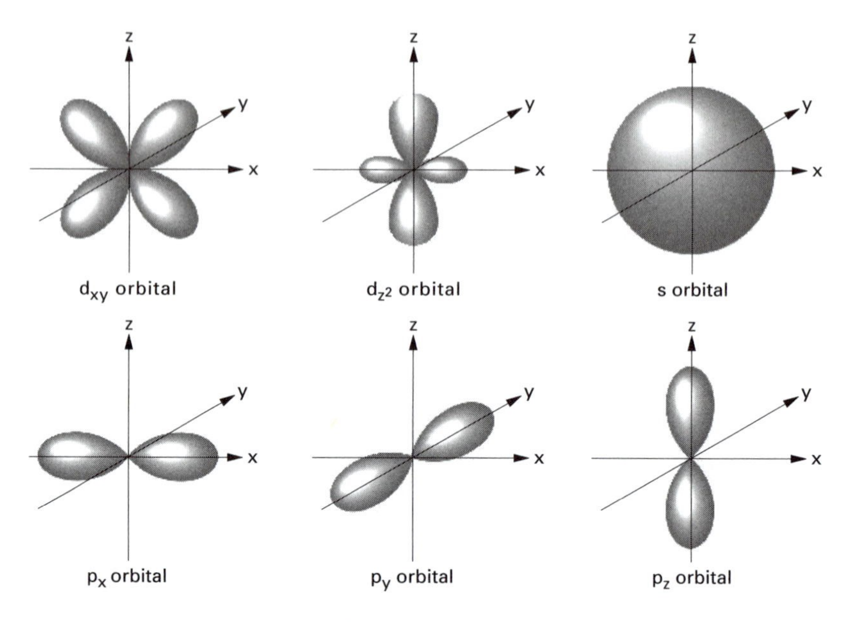

Orbital: atomic orbitals

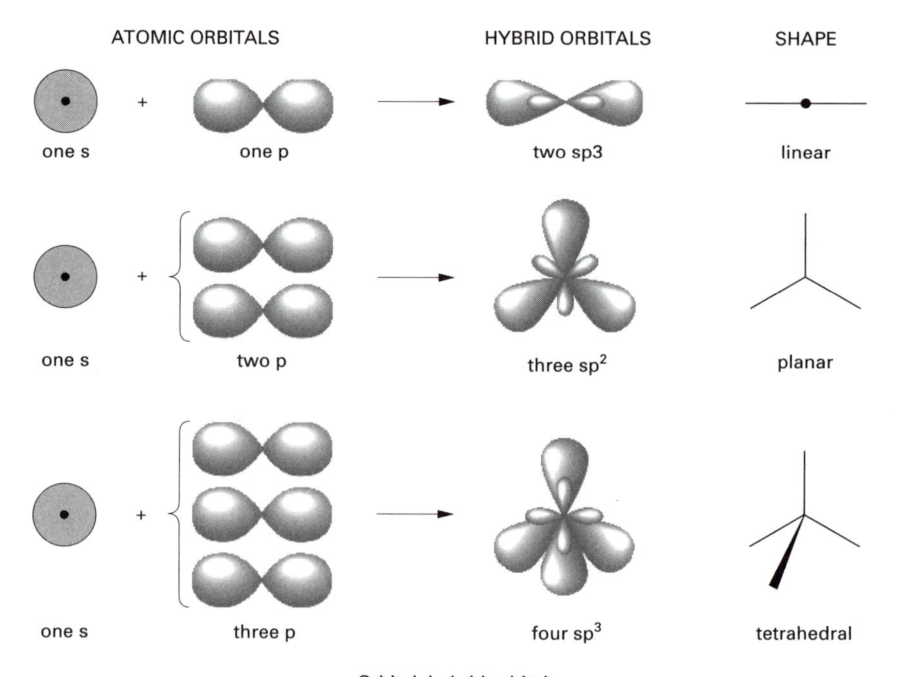

Orbital: hybrid orbitals

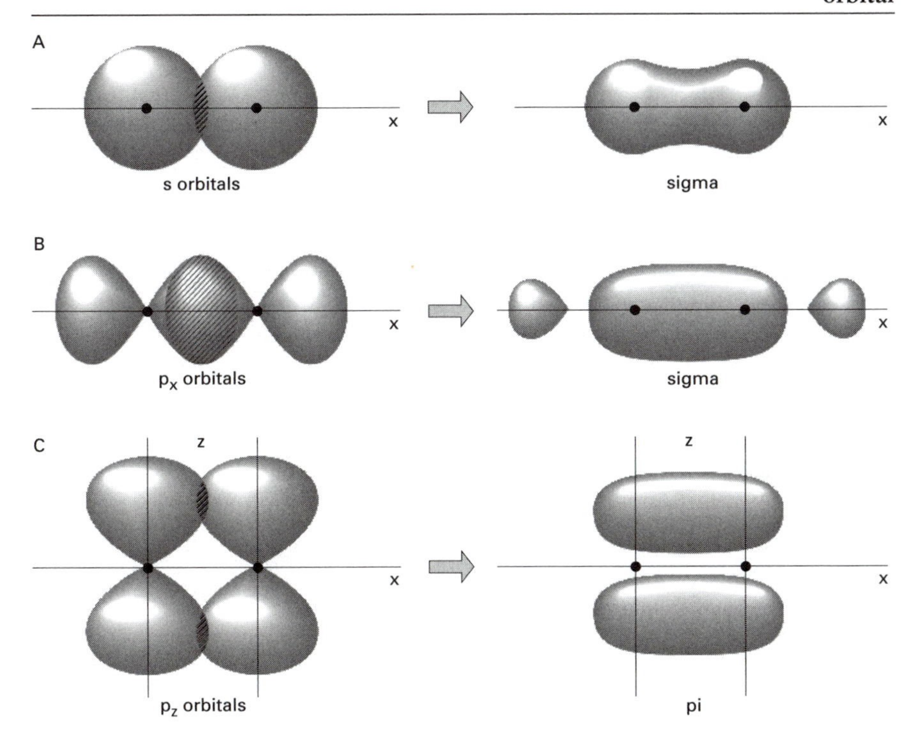

Orbital: molecular orbitals

bital is as a cloud of electron charge (the cloud being the average distribution of the charge with time).

Similarly, in molecules the electrons move in the combined field of the nuclei and can be assigned to *molecular orbitals*. In considering bonding between atoms it is useful to treat molecular orbitals as formed by overlap of atomic orbitals.

It is possible to calculate the shapes and energies of atomic and molecular orbitals by QUANTUM THEORY. The shapes of atomic orbitals depend on their orbital angular momentum. For each shell there is, at a maximum, one s orbital, three p orbitals, five d orbitals, and seven f orbitals. The s orbitals are spherical, the p orbitals each have two lobes, the d orbitals have more complex shapes, typically with four lobes, and the f orbitals are more complex yet, often with eight lobes.

Molecular orbitals are formed by overlap of atomic orbitals, and again there are different types. If the orbital is completely symmetrical about an axis between the nuclei, it is a *sigma orbital*. This can occur, for instance, by overlap of two s orbitals, as in the hydrogen atom, or two p orbitals with their lobes along the axis. However, two p orbitals overlapping at right angles to the axis form a different type of molecular orbital – a *pi orbital* – with regions above and below the axis. Pi orbitals are also formed by overlap of d orbitals. Each molecular orbital can contain a pair of electrons, forming a sigma bond or pi bond. A double bond, for example the bond in ethene, is a combination of a sigma bond and a pi bond.

Hybrid orbitals are atomic orbitals formed by combinations of s, p, and d atomic orbitals, and are useful in describing the bonding in compounds. There are various types. In carbon, for instance, the electron configuration is $1s^2 2s^2 2p^2$. Carbon, in its outer (valence) shell, has one filled s orbital, two filled p orbitals, and one 'empty' p orbital. These four orbitals may hybridize (sp^3 hybridization) to act as four equal orbitals arranged tetrahedrally, each with one electron. In methane, each hybrid orbital overlaps with a hydrogen s orbital to form a sigma bond. Alternatively, the s and two of the p orbitals may hybridize (sp^2 hybridization) and act as three orbitals in a plane at 120°. The remaining p orbital is at right angles to the plane, and can form pi bonds. Finally, sp hybridization may occur, giving two orbitals in a line. More complex types of hybridization, involving d orbitals, explain the geometries of inorganic complexes. For example, square-planar complexes can be formed by sp^2d hybridization; octahedral complexes can involve sp^3d^2 hybridization.

The combination of two atomic orbitals in fact produces two molecular orbitals. One – the *bonding orbital* – has a concentration of electron density between the nuclei, and thus tends to hold the atoms together. The other – the *antibonding orbital* – has slightly higher energy and tends to repel the atoms. If both atomic orbitals are filled, the two molecular orbitals are also filled and cancel each other out – there is no net bonding effect. If each atomic orbital has one electron, the pair occupies the lower energy bonding orbital, producing a net attraction.

ORD *See* optical rotary dispersion.

order The sum of the indices of the concentration terms in the expression that determines the rate of a chemical reaction. For example, in the expression

$$\text{rate} = k[A]^x[B]^y$$

x is called the order with respect to A, y the order with respect to B, and $(x + y)$ the order overall. The values of x and y are not necessarily equal to the coefficients of A and B in the molecular equation. Order is an experimentally determined quantity derived without reference to any equation or mechanism. Fractional orders do occur. For example, in the reaction

$$CH_3CHO \rightarrow CH_4 + CO$$

the rate is proportional to $[CH_3CHO]^{1.5}$ i.e. it is of order 1.5.

ore A mineral source of a chemical element.

ore dressing *See* beneficiation.

organic chemistry The chemistry of compounds of carbon. Originally the term *organic chemical* referred to chemical compounds present in living matter, but now it covers any carbon compound with the exception of certain simple ones, such as the carbon oxides, carbonates, cyanides, and cyanates. These are generally studied in inorganic chemistry. The vast numbers of synthetic and natural organic compounds exist because of the ability of carbon to form chains of atoms (catenation). Other elements are involved in organic compounds; principally hydrogen and oxygen but also nitrogen, halogens, sulfur, and phosphorus.

organometallic compound An organic compound containing a carbon–metal bond. Tetraethyl lead, $(C_2H_5)_4Pb$, is a well-known example.

ortho- **1.** Designating the form of a diatomic molecule in which both nuclei have the same spin direction; e.g. orthohydrogen, orthodeuterium.
2. Certain acids, regarded as formed from an anhydride and water, were named ortho acids to distinguish them from the less hydrated meta acids. For example, H_4SiO_4 (from $SiO_2 + 2H_2O$) is orthosilicic acid; H_2SiO_3 ($SiO_2 + H_2O$) is metasilicic acid. *See also* meta-; para-.

orthoboric acid *See* boric acid.

orthohydrogen *See* hydrogen.

orthophosphates *See* phosphoric(V) acid.

orthophosphoric acid *See* phosphoric(V) acid.

orthophosphorous acid *See* phosphonic acid.

orthorhombic crystal *See* crystal system.

osmiridium A very hard naturally occurring alloy consisting chiefly of osmium and iridium, used to make pen nibs.

osmium A hard bluish white transition element, the third element of group 8 (formerly part of subgroup VIIIB) of the periodic table. It is found associated with platinum. Osmium is the densest of all metals. It has a characteristic chlorinelike smell resulting from the production of osmium(VIII) oxide (osmium tetroxide, OsO_4) when heated in air. The metal is used in catalysts and in alloys for pen nibs, pivots, and electrical contacts.

Symbol: Os; m.p. 3045°C; b.p. 5027°C; r.d. 22.59 (20°C); p.n. 76; most common isotope $_{192}Os$; r.a.m. 190.23.

osmium(IV) oxide (osmium tetroxide; OsO_4) A volatile yellow crystalline solid with chlorinelike penetrating poisonous odor, used as an oxidizing agent and, in aqueous solution, as a catalyst for organic reactions.

osmosis Systems in which a solvent is separated from a solution by a semipermeable membrane approach equilibrium by solvent molecules on the solvent side of the membrane migrating through it to the solution side; this process is called osmosis and always leads to dilution of the solution. The phenomenon is quantified by measurement of the osmotic pressure. The process of osmosis is of fundamental importance in transport and control mechanisms in biological systems; for example, plant growth and general cell function.

Osmosis is a colligative property and its theoretical treatment is similar to that for the lowering of vapor pressure. The membrane can be regarded as equivalent to the liquid-vapor interface, i.e. one that permits free movement of solvent molecules but restricts the movement of solute molecules. The solute molecules occupy a certain area at the interface and therefore inhibit solvent egress from the solution. Just as the development of a vapor pressure in a closed system is necessary for liquid–vapor equilibrium, the development of an OSMOTIC PRESSURE on the solution side is necessary for equilibrium at the membrane.

osmotic pressure Symbol: π The pressure that must be exerted on a solution to prevent the passage of solvent molecules into it when the solvent and solution are separated by a semipermeable membrane. The osmotic pressure is therefore the pressure required to maintain equilibrium between the passage of solvent molecules through the membrane in either direction and thus prevent the process of osmosis from proceeding. The osmotic pressure can be measured by placing the solution, contained in a small perforated thimble covered by a semipermeable membrane and fitted with a length of glass tubing, in a beaker of the pure solvent. Solvent molecules pass through the membrane, diluting the solution and thereby increasing the volume on the solution side and forcing the solution to rise up the glass tubing. The process continues until the pressure exerted by the solvent molecules on the membrane is balanced by the hydrostatic pressure of the solution in the tubing. A sample of the solution is then removed and its concentration determined. Osmosis is a colligative property; therefore the method can be applied to the determination of relative molecular masses, particularly for large molecules, such as proteins, but it is restricted by the difficulty of preparing good semipermeable membranes.

Because the osmotic pressure is a colligative property it is directly proportional to the molar concentration of the solute if the temperature remains constant; thus π is proportional to the concentration n/V, where n is the number of moles of solute and V the solvent volume. The osmotic pressure is also proportional to the absolute temperature. Combining these two proportionalities gives $\pi V = nCT$, which has the same form as the gas equation, $PV = nRT$, and experimental values of C are similar to those for R, the universal gas constant. This gives considerable support to the kinetic theory of colligative properties.

Ostwald, Friedrich Wilhelm (1854–1932) Russian-born German physical chemist. Ostwald was one of the main figures involved in establishing physical chemistry as a distinct subject. In 1888 he put forward the OSTWALD DILUTION LAW for the degree of dissociation of an electrolyte. His own research was mainly concerned with catalysts. This led to the discovery of the OSTWALD PROCESS for converting ammonia to nitric acid by passing air and ammonia over a platinum catalyst. Ostwald won the 1909 Nobel Prize for chemistry for his work on catalysis. He also wrote an influential two-volume textbook entitled *Textbook of General Chemistry* (1885, 1887).

Ostwald's dilution law *See* dissociation.

ox *See* oxalic acid.

oxalato *See* oxalic acid.

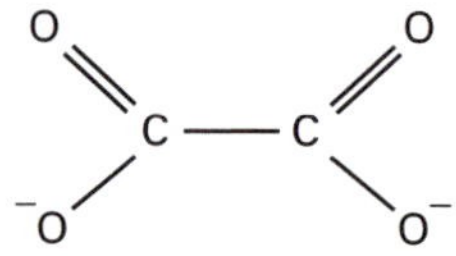

Oxalic acid: the oxalato ligand

oxalic acid A dibasic organic carboxylic acid, $CH_2(COOH)_2$. It is of interest in inorganic chemistry because the carboxylate ion functions as a bidentate ligand (the *oxalato* ligand) in forming certain complexes. In the names of complexes it is given the symbol *ox*.

oxidant *See* oxidation.

oxidation An atom, an ion, or a molecule is said to undergo oxidation or to be oxidized when it loses electrons. The process may be effected chemically, i.e. by reaction with an *oxidizing agent* (*oxidant*), or electrolytically, in which case oxidation occurs at the anode. For example,

$$2Na + Cl_2 \rightarrow 2Na^+ + 2Cl^-$$

where chlorine is the oxidizing agent and sodium is oxidized, and

$$4CN^- + 2Cu^{2+} \rightarrow C_2N_2 + 2CuCN$$

where Cu^{2+} is the oxidizing agent and CN^- is oxidized.

The *oxidation state* of an atom is indicated by the number of electrons lost or effectively lost by the neutral atom, i.e. its *oxidation number*. The oxidation number of a negative ion is negative. The process of oxidation is the converse of reduction. *See also* redox.

oxidation number *See* oxidation.

oxidation–reduction *See* redox.

oxidation state *See* oxidation.

oxide A compound of OXYGEN and another element. Oxides can be made by direct combination of the elements or by heating a hydroxide, carbonate, or other salt. *See also* acidic oxide; basic oxide; neutral oxide; peroxide.

oxidizing acid An acid that acts as an oxidizing agent, e.g. sulfuric acid or nitric acid. Because it is oxidizing, nitric acid can dissolve metals below hydrogen in the electrochemical series:

$$2HNO_3 + M \rightarrow MO + 2NO_2 + H_2O$$
$$MO + 2HNO_3 \rightarrow H_2O + M(NO_3)_2$$

oxidizing agent *See* oxidation.

oxyacid An ACID in which the replaceable hydrogen atom is part of a hydroxyl group, including inorganic acids such as phosphoric(V) acid and sulfuric acid.

oxygen A colorless odorless diatomic gas, the first member of group 16 (formerly VI) of the periodic table. It has the electronic configuration $[He]2s^22p^4$ and its chemistry involves the acquisition of electrons to form either the di-negative ion O^{2-} or two covalent bonds. In each case the oxygen atom attains the configuration of the rare gas neon.

Oxygen is the most plentiful element in the Earth's crust, accounting for over 49% by mass. It accounts for 23% of the volume of the atmosphere and is a constituent of the majority of minerals and rocks (e.g. silicates, SiO_2, carbonates, $CaCO_3$, aluminosilicates, Al_2SiO_5) as well as being the major constituent of the sea. Oxygen is an essential element for almost all living things. It is also used in solid propellants for rockets, in steel making and welding, in medical environments and breathing apparatus, and in the manufacture of various chemicals. Elemental oxygen has two allotropes: the diatomic molecule O_2 and the less stable molecule OZONE (trioxygen O_3), which is formed by passing an electric discharge through oxygen gas. There are several convenient laboratory routes by which to produce oxygen:

1. Heat on unstable oxides.
$$2HgO \rightarrow 2Hg + O_2$$
2. Decomposition of peroxides.
$$2H_2O_2 \rightarrow 2H_2O + O_2$$
3. Heat on '-ate' compounds, such as chlorates or manganate(VII).
$$2KClO_3 \rightarrow KCl + 3O_2$$
$$2KMnO_4 \rightarrow K_2MnO_4 + MnO_2 + O_2$$

Industrially oxygen is obtained by the fractional distillation of liquid air.

Oxygen reacts with elements from all groups of the periodic table, including xenon in group 18 (e.g. XeO_3, XeO_4, $XeOF_2$, XeO_2^{2-}).

The nature of the oxides has long been a convenient guide to metallic and nonmetallic character. Thus, metals typically combine with oxygen to form basic oxides, which if soluble react with water to form bases via the OH^- ion. The more electropositive the metal, the more vigorous the reaction. Most of the group 1 and group 2 metals react vigorously when forming oxides, O^{2-}, (e.g., Na_2O, CaO); peroxides, O_2^{2-}, (e.g., Na_2O_2, BaO_2); and superoxides, O_2^-, (e.g., RbO_2, CsO_2). Oxides of metallic elements are solids and x-ray studies show that discrete O^{2-} ions do exist in the solid state even though they do not exist in solution:
$$O^{2-} + H_2O \rightarrow 2OH^-$$

The nonmetals typically form acidic oxides such as SO_2, ClO_2, and NO_2. When soluble in water they generally dissolve to form acids. The less electronegative elements among the nonmetals generally form weaker acids, e.g. $B(OH)_3$ and H_2CO_3. The more electronegative elements form stronger acids. The higher the oxidation state of the central atom the stronger the acid (e.g. HOCl is weaker than $HClO_2$, which is weaker than $HClO_4$). Nonmetal oxides may be gaseous (SO_2, CO_2), liquid (N_2O_4), or solid (P_2O_5). The weakly nonmetallic elements such as silicon may require bases for dissolution and in some cases the oxides have extended polymeric structures, e.g., SiO_2. In between the acidic and basic oxides are those that are amphoteric, i.e. they behave acidically to strong bases and basically toward strong acids. This behavior is associated with elements that are physically metallic but are chemically only weakly cationic, e.g., ZnO gives Zn^{2-} with acids and zincates, $Zn(OH)_4^{2-}$, with bases. There is also a small class of *mixed oxides* that can be regarded as salts between a 'basic' and an 'acidic' oxide of the same metal: for example,
$$FeO + Fe_2O_3 \rightarrow Fe(FeO_2)_2 = Fe_3O_4$$
$$2PbO + PbO_2 \rightarrow Pb_3O_4$$

The neutral oxides are essentially insoluble nonmetal oxides such as N_2O, NO, CO, and F_2O, although extreme conditions can lead to some degree of acidic behavior in the case of CO, e.g.,
$$CO + OH^- \rightarrow HCOO^-$$

Oxygen is very limited in its catenation but an interesting range of O–O species is formed by the following: peroxide, O_2^{2-}; superoxide, O_2^-; oxygen gas, O_2; oxygenyl cation, O_2^+. The O–O bond gets both progressively shorter and stronger in the series. The peroxides are formed with Na, Ca, Sr, and Ba and are formally related to hydrogen peroxide (prepared by the action of acids on ionic peroxides). The superoxides of K, Rb, and Cs are prepared by burning the metal in air (the less electropositive metals Na, Mg, and Zn form less stable superoxides). Oxygenyl cation, O_2^+, is formed by the reaction of PtF_6 (which has a high electron affinity) with oxygen at room temperature. Other oxygenyl cation species can be made with group 15 fluorides, e.g. $O_2^+AsF_6^-$.

Oxygen occurs in three natural isotopic forms, ^{16}O (99.76%), ^{17}O (0.0374%), and ^{18}O (0.2039%); the rarer isotopes are used in detailed studies of the behavior of oxygen-containing groups during reactions (tracer studies).

Symbol: O; m.p. –218.4°C; b.p. –182.962°C; d. 1.429 kg m^{-3} (0°C); p.n. 8; r.a.m. 15.9994.

ozone (trioxygen; O_3) A poisonous blue diamagnetic allotrope of oxygen made naturally by high-energy ultraviolet radiation in the stratosphere and to a lesser extent by photochemical reactions involving pollutants such as NO_x. Ozone can also be made artificially by passing oxygen through a silent electric discharge. Ozone is unstable and decomposes to oxygen on warming. In the stratosphere it forms a layer of ozonized air that acts to screen the Earth from harmful short-wave ultraviolet radiation. In recent decades there has been considerable evidence to suggest that this layer is being depleted by fluorocarbons and other compounds produced by industry. Consequently most industrialized nations have agreed to severely restrict their production and use of these chemicals.

Ozone is a strong oxidizing agent. Its concentration can be determined by reacting it with iodide ions and titrating the iodine formed with standard sodium thiosulfate(VI).

P

paint A suspension of powdered pigment in a liquid vehicle, used for decorative and protective surface coatings. In oil-based paints, the vehicle contains solvents and a drying oil or synthetic resin. In water-based or emulsion paints, the vehicle is mostly water with an emulsified resin.

palladium A soft silvery white ductile transition metal, the second element of group 10 (formerly part of subgroup VIIIB) of the periodic table. It occurs native in association with platinum and also in nickel, copper, gold, and silver ores, from which it is obtained as a by-product. Although it will dissolve in acid, it will react with oxygen only at high temperatures, making it resistant to corrosion in most environments. Consequently it finds use in dentistry, surgical instruments, electrical relays, and jewelry (*white gold* is an alloy of gold and palladium). It is also used as a catalyst in hydrogenation processes. Hydrogen is purified by diffusing it through a hot palladium barrier.

Symbol: Pd; m.p. 1552°C; b.p. 3140°C; r.d. 12.02 (20°C); p.n. 46; most common isotope ^{106}Pd; r.a.m. 106.42.

paper chromatography A technique widely used for the analysis of mixtures. Paper CHROMATOGRAPHY usually employs a specially produced paper as the stationary phase. A base line is marked in pencil near the bottom of the paper and a small sample of the mixture is spotted onto it using a capillary tube. The paper is then placed vertically in a suitable solvent, which rises up to the base line and beyond by capillary action. The components within the sample mixture dissolve in this mobile phase and are carried up the paper. However, the paper holds a quantity of moisture and some components will have a greater tendency than others to dissolve in this moisture than in the mobile phase. In addition, some components will cling to the surface of the paper. Therefore, as the solvent moves through the paper, certain components will be left behind and separated from each other.

When the solvent has almost reached the top of the paper, the paper is removed and quickly dried. The paper is developed to locate the positions of colorless fractions by spraying with a suitable chemical, e.g. ninhydrin, or by exposure to ultraviolet radiation. The components are identified by comparing the distance they have traveled up the paper with standard solutions that have been run simultaneously, or by computing an R_F value. A simplified version of paper chromatography uses a piece of filter paper. The sample is spotted at the center of the paper and solvent passed through it. Separation of the components of the mixture again takes place as the mobile phase spreads out on the paper.

Paper chromatography is an application of the partition law.

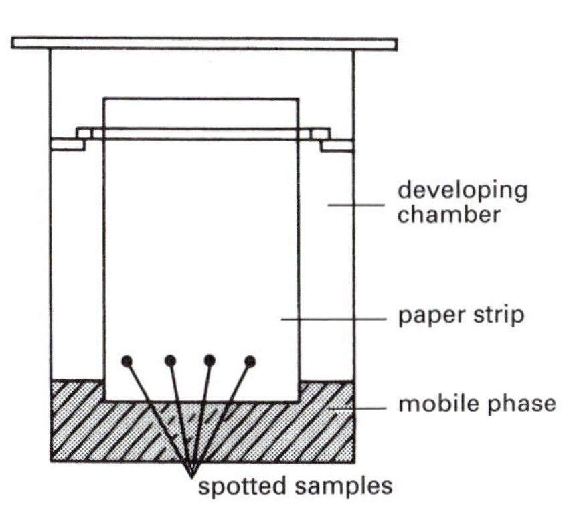

Paper chromatography

para- Designating the form of a diatomic molecule in which both nuclei have opposite spin directions; e.g. parahydrogen, paradeuterium.

See also meta-; ortho-.

parahydrogen *See* hydrogen.

paramagnetism *See* magnetism.

partial ionic character The electrons of a covalent bond between atoms or groups with different electronegativities will be polarized toward the more electronegative constituent; the magnitude of this effect can be measured by the ionic character of the bond. When the effect is small the bond is referred to simply as a polar bond and is adequately treated using dipole moments; as the effect grows stronger the theoretical treatment requires other contributions to ionic character. Examples of partial ionic character (from nuclear quadrupole resonance) are H–I, 21%; H–Cl, 40%; Li–Br, 90%.

partial pressure *See* Dalton's law.

partition coefficient If a solute dissolves in two nonmiscible liquids, the partition coefficient is the ratio of the concentration in one liquid to the concentration in the other liquid.

pascal Symbol: Pa The SI derived unit of pressure, equal to a pressure of one newton per square meter ($1\ \text{Pa} = 1\ \text{N m}^{-2}$).

Paschen–Back effect *See* Zeeman effect.

Paschen series A series of lines in the infrared spectrum emitted by excited hydrogen atoms. The lines correspond to the atomic electrons falling into the third lowest energy level and emitting energy as radiation. The wavelength (λ) of the radiation in the Paschen series is given by $1/\lambda = R(1/3^2 - 1/n^2)$ where n is an integer and R is the Rydberg constant. The series is named for German physicist Louis Paschen (1865–1947), who discovered it in 1908. *See also* spectral series.

passive Describing a metal that is unreactive because of the formation of a thin protective layer. Iron, for example, does not dissolve in concentrated nitric(V) acid because of the formation of a thin oxide layer.

Pauli, Wolfgang (1900–58) Austrian-born Swiss theoretical physicist. Pauli is best known for his enunciation of the *Pauli exclusion principle* in 1925. This enabled the electronic structure of atoms to be understood, particularly how the shell structure of atoms, and hence the periodic table of the elements, comes about. Pauli won the 1945 Nobel Prize for physics for this work. Pauli made many other important contributions to physics including his prediction of the neutrino in beta decay, the incorporation of spin into quantum mechanics, and the explanation of paramagnetism in metals. He also wrote several classic books and reviews on quantum mechanics.

Pauli exclusion principle *See* atom.

Pauling, Linus Carl (1901–94) American chemist. Pauling was one of the greatest scientists of the 20th century. His early work was on determining the structure of complex minerals by x-ray diffraction. In 1928–29 this led to *Pauling's rules* governing the structure of complex minerals. Pauling was also a major pioneer in the application of quantum mechanics to chemical bonding. In 1931 he published a classic paper entitled *The Nature of the Chemical Bond* that explained how a chemical bond is formed from a pair of electrons. Pauling also introduced the concept of hybridization to explain the chemical bonding of the carbon atom. He also considered partially ionic bonds and put together his ideas about chemical bonding in his great book *The Nature of the Chemical Bond*, the first edition of which was published in 1939. In the mid-1930s Pauling turned his attention to molecules of biological interest and was one of the founders of molecular biology. Together with Robert Corey, he showed that many proteins have helical shapes. He also worked on sickle-cell anemia. With E.

Bright Wilson he co-authored the book *Introduction to Quantum Mechanics* (1935) and wrote the influential books *General Chemistry* (1948) and *College Chemistry* (1950). Pauling was awarded the 1954 Nobel Prize for chemistry. In the 1950s he became concerned with nuclear weapons, particularly their testing in the atmosphere. This led to him being awarded the 1962 Nobel Prize for peace.

p-block elements The elements of the main groups 13 (B to Tl), 14 (C to Pb), 15 (N to Bi), 16 (O to Po), 17 (F to At), and 18 (He to Rn). They are so called because they have outer electronic configurations that have occupied p levels.

pearl A lustrous spherical accretion of nacre, comprised chiefly of calcium carbonate ($CaCO_3$), that builds up on a particle of foreign matter inside the shells of oysters and some other mollusks.

pearl spar *See* dolomite.

peat A brown or black material formed by the partial decomposition of plant remains in marshy ground, the first stage in the formation of coal. It is used to make charcoal and compost. When dried, it can also be burned as a fuel.

PEC *See* photoelectrochemical cell.

pentahydrate A crystalline solid containing five molecules of water of crystallization per molecule of compound.

pentavalent (quinquevalent) Having a valence of five.

percentage composition A way of expressing the composition of a chemical compound in terms of the percentage (by mass) of each of the elements that make it up. It is calculated by dividing the mass of each element (taking into account the number of atoms present) by the relative molecular mass of the whole molecule. For example, methane (CH_4) has a relative molecular mass of 16 and its percentage composition is 12/16 = 75% carbon and (4 × 1)/16 = 25% hydrogen.

perchloric acid *See* chloric(VII) acid.

perfect gas (ideal gas) *See* gas laws; kinetic theory.

peridot *See* olivine.

period One of the horizontal rows in the conventional periodic table. Each period represents the elements arising from progressive filling of the outer shell (i.e. the addition of one extra electron for each new element), the elements being arranged in order of ascending proton number. In a strict sense hydrogen and helium represent one period but convention refers to the elements lithium to neon (8 elements) as the first *short period* (n = 2), and the elements sodium to argon (8 elements) as the second *short period* (n = 3). With entry to the n = 4 level there is filling of the 4s orbital, then back-filling of the 3d orbitals, before the 4p are filled. Thus this set is based on the filling of a total of 18 electrons (potassium to krypton) and is called a *long period*. The next set, rubidium to xenon, is similarly a long period. *See also* Aufbau principle.

periodic acid *See* iodic(VII) acid.

periodic law The law upon which the modern periodic table is based. Enunciated in 1869 by Mendeléev, this law stated that the properties of the elements are a periodic function of their atomic masses: if arranged in order of increasing atomic mass then elements having similar properties occur at fixed intervals. Certain exceptions or gaps in the table lead to the view that the nuclear charge is a more characteristic function, thus the modern statement of the periodic law is that the physical and chemical properties of elements are a periodic function of their proton number.

periodic table A table of the elements arranged in order of increasing proton number to show similarities in chemical behavior between elements. Horizontal rows of elements are called *periods*. Across a period there is a general trend from metallic to nonmetallic behavior. Vertical columns of related elements are called *groups*. Down a group there is an increase in atomic size and in electropositive (metallic) behavior.

Originally the periodic table was arranged in eight groups with the alkali metals as group I, the halogens as group VII, and the rare gases as group 0. The transition elements were placed in a block in the middle of the table. Groups were split into two subgroups. For example, group I contained the main-group elements, Li, Na, K, Rb, Cs, in subgroup IA and the subgroup IB elements, Cu, Ag, Au. The system was confusing because there was a difference in usage for subgroups and the current form of the table has 18 groups (*see* Appendix).

See also group; period.

permanent gas A gas that cannot be liquefied by pressure alone – that is, a gas above its critical temperature. *See* critical temperature.

permanent hardness A type of water hardness caused by the presence of dissolved calcium, iron, and magnesium sulfates or chlorides. This form of hardness cannot be removed by boiling. *Compare* temporary hardness. *See also* water softening.

permanganate *See* manganate(VII).

Permutit (*Trademark*) A substance used to remove unwanted chemicals that have dissolved in water. It is a zeolite consisting of a complex chemical compound, sodium aluminum silicate. When hard water is passed over this material, calcium and magnesium ions exchange with sodium ions in the Permutit. This is a good example of ION EXCHANGE. Once all the available sodium ions have been used up, the Permutit can be regenerated by washing it with a saturated solution of sodium chloride. The excess sodium ions then exchange with the calcium and magnesium ions in the Permutit.

peroxide 1. An oxide containing the $^{-}O{-}O^{-}$ ion.

2. A compound containing the –O–O– group.

peroxosulfuric(VI) acid (Caro's acid; H_2SO_5) A white crystalline solid formed by reacting hydrogen peroxide with sulfuric acid. It is a powerful oxidizing agent.

perturbation theory An approximation technique in which a system is divided into two parts, with one part that can be solved exactly and a second part that makes the system too complicated to be solved exactly. For perturbation theory to be applicable it is necessary that the second part is a small correction to the first part. The effects of the small corrections are expressed as an infinite series, called a *perturbation series*, which modifies the results that could be calculated exactly. A classic example of perturbation theory is the modification of the orbits of planets around the Sun due to their gravitational interactions with other planets. In quantum mechanics an application of perturbation theory is the calculation of the effects of small electric or magnetic fields on atomic energy levels.

peta- Symbol: P A prefix used with SI units, denoting 10^{15}.

petrochemicals Organic chemicals obtained from crude oil or natural gas.

petrol *See* petroleum.

petroleum (crude oil) A mixture of hydrocarbons formed originally in shallow basins from decayed, chiefly marine animals and plants, found today beneath the ground trapped between layers of sedimentary rock. It is obtained by drilling. Different oilfields produce petroleum with differing compositions. The mixture is separated into fractions by fractional distillation in a vertical column. The main fractions are:

Diesel oil (*gas oil*) in the boiling range 220–350°C, consisting mainly of C_{13}–C_{25} hydrocarbons. It is used in diesel engines.

Kerosene (*paraffin*) in the range 160–250°C, consisting mainly of C_{11} and C_{12} hydrocarbons. It is a fuel both for domestic heating and jet engines.
Gasoline (*petrol*) in the range 40–180°C, consisting mainly of C_5–C_{10} hydrocarbons. It is used as motor fuel for gasoline engines and as a raw material for making other chemicals.
Refinery gas, consisting of C_1–C_4 gaseous hydrocarbons such as propane.

In addition, lubricating oils and paraffin wax are obtained from the residue. The black material left is composed of bitumen and asphalt.

pewter A tarnish-resistant alloy of tin and lead, today usually containing 80–90% tin. Modern pewter is hardened by antimony and copper, which in the best grades completely replaces the lead content. It is less dense than old pewter, and having a bright or a soft satiny finish, is often used for decorative and ceremonial purposes.

pH The logarithm to base 10 of the reciprocal of the hydrogen-ion concentration of a solution. In pure water at 25°C the concentration of hydrogen ions is 1.00×10^{-7} mol l^{-1}, thus the pH equals 7 at neutrality. An increase in acidity increases the value of $[H^+]$, decreasing the value of the pH below 7. An increase in the concentration of hydroxide ion $[OH^-]$ proportionately decreases $[H^+]$, therefore increasing the value of the pH above 7 in basic solutions. Potential of hydrogen or pH values can be obtained approximately by using indicators. More precise measurements use electrode systems.

phase One of the physically separable parts of a chemical system. For example, a mixture of ice (solid phase) and water (liquid phase) consists of two phases. A system consisting of only one phase is said to be homogeneous. A system consisting of more than one phase is said to be heterogeneous.

phase diagram A graphical representation of the state in which a substance will occur at a given pressure and temperature. The lines show the conditions under which more than one phase can coexist at equilibrium. For one-component systems (e.g. water) the point at which all three phases can coexist at equilibrium is called the triple point and is the point on the graph at which the pressure–temperature curves intersect.

phase equilibrium A state in which the proportions of the various phases in a chemical system are fixed. When two or more phases are present at fixed temperature and pressure a dynamic condition will be established in which individual particles will leave one phase and enter another; at equilibrium this will be balanced by an equal number of particles making the reverse change, leaving the total composition unchanged.

phase rule In a system at equilibrium the number of phases (P), number of components (C), and number of degrees of freedom (F) are related by the formula:

$$P + F = C + 2$$

phenolphthalein An acid–base indicator that is colorless in acid solutions and becomes red if the pH rises above the transition range of 8–9.6. It is used as the indicator in titrations for which the end point lies clearly on the basic side (pH > 7), e.g. oxalic acid or potassium hydrogentartrate against caustic soda.

philosopher's stone *See* alchemy.

phlogiston theory A former theory of combustion (disproved in the 18th century by Lavoisier). Flammable compounds were supposed to contain a substance (phlogiston), which was presumed to be released when they burned, leaving ash.

phosgene *See* carbonyl chloride.

phosphates *See* phosphoric(V) acid.

phosphide A compound of phosphorus with a more electropositive element.

phosphine (phosphorus(III) hydride;

PH_3) A very poisonous colorless gas that is slightly soluble in water. It has a characteristic fishy smell. It can be made by reacting water and calcium phosphide or by the action of yellow phosphorus on a concentrated alkali. Phosphine usually ignites spontaneously in air because of contamination with diphosphane. It decomposes into its elements if heated to 450°C in the absence of oxygen and it burns in oxygen or air to yield phosphorus oxides. It reacts with solutions of metal salts to precipitate phosphides. Like its nitrogen analog ammonia it forms salts, called *phosphonium salts*. It also forms complex addition compounds with metal ions.

phosphinic acid (hypophosphorous acid; H_3PO_2) A white crystalline solid. It is a monobasic acid forming the anion $H_2PO_2^-$ in water. The sodium salt, and hence the acid, can be prepared by heating white (or yellow) phosphorus with sodium hydroxide solution. The free acid and its salts are powerful reducing agents.

phosphonic acid (phosphorous acid; orthophosphorous acid; H_3PO_3) A colorless deliquescent solid that can be prepared by the action of water on phosphorus(III) oxide or phosphorus(III) chloride. It is a dibasic acid producing the anions $H_2PO_3^-$ and HPO_3^{2-} in water. The acid and its salts are weak reducing agents. On warming, phosphonic acid decomposes to phosphine and phosphoric(V) acid.

phosphonium ion The ion PH_4^+, derived from phosphine.

phosphonium salts *See* phosphine.

phosphor A substance that shows luminescence or phosphorescence.

phosphorescence **1.** The absorption of energy by atoms followed by emission of electromagnetic radiation. Phosphorescence is a type of luminescence, and is distinguished from fluorescence by the fact that the emitted radiation continues for some time after the source of excitation has been removed. In phosphorescence the excited atoms have relatively long lifetimes before they make transitions to lower energy states. However, there is no defined time distinguishing phosphorescence from fluorescence.

2. In general usage the term is applied to the emission of 'cold light' – light produced without a high temperature. The name comes from the fact that white phosphorus glows in the dark as a result of a chemical reaction with oxygen. The light comes from excited atoms produced directly in the reaction – not from the heat produced. It is thus an example of *chemiluminescence*. There are also a number of biochemical examples termed *bioluminescence*; for example, phosphorescence is sometimes seen in the sea emanating from marine organisms, or on rotting wood emanating from certain fungi (known as 'fox fire').

phosphoric(V) acid (orthophosphoric acid; H_3PO_4) A white solid that can be made by reacting phosphorus(V) oxide with water or by heating yellow phosphorus with nitric acid. The naturally occurring *phosphates* (*orthophosphates*, M_3PO_4) are salts of phosphoric(V) acid. Aqueous solutions of phosphoric(V) acid have a sharp taste and have been used in the manufacture of soft drinks. At about 220°C phosphoric(V) acid is converted to *pyrophosphoric acid* ($H_4P_2O_7$). Pure pyrophosphoric acid in the form of colorless crystals is made by warming solid phosphoric(V) acid with phosphorus(III) chloride oxide. *Metaphosphoric acid* is obtained as a polymer, $(HPO_3)_n$, by heating phosphoric(V) acid to 320°C. Both meta- and pyrophosphoric acids change to phosphoric(V) acid in aqueous solution; the change takes place faster at higher temperatures. Phosphoric(V) acid is tribasic, forming three series of salts with the anions $H_2PO_4^-$, HPO_4^{2-}, and PO_4^{3-}. These series of salts are acidic, neutral, and alkaline respectively. With the exception of lithium phophate the alkali metal phosphates are soluble in water, but other metal phosphates are insoluble. Phosphates are useful in water softening and as fertilizers.

phosphorous acid *See* phosphonic acid.

phosphorus A reactive solid nonmetallic element, the second element in group 15 (formerly VA) of the periodic table. It has the electronic configuration $[Ne]3s^2 3p^3$ and is therefore formally similar to nitrogen. It is, however, very much more reactive than nitrogen and is never found in nature in the uncombined state. Phosphorus is widespread throughout the world; economic sources are rock phosphate (consisting largely of calcium phosphate, $Ca_3(PO_4)_2$) and the apatites, variously occurring as both fluoroapatite ($3Ca_3(PO_4)_2.CaF_2$) and as chloroapatite ($3Ca_3(PO_4)_2CaCl_2$). Guano formed from the skeletal phosphate of fish in sea-bird droppings is also an important source of phosphorus. The largest amounts of phosphorus compounds produced are used in fertilizers and detergents, with smaller amounts being employed in safety matches, pesticides, special alloys and glasses, food and drinks, and fireworks. Phosphorus is an essential constituent of living tissue, bones, and nucleic acids, and plays a very important part in metabolic processes and muscle action. Phosphorus is also required for the development of vigorous root systems in green plants.

The element is obtained industrially by the reduction of rock phosphate using sand (SiO_2) and coke (C) in an electric furnace. The phosphorus is distilled off as P_4 molecules and collected under water as white phosphorus (defined below):

$$2Ca_3(PO_4)_2 + 6SiO_2 + 10C \rightarrow P_4 + 6CaSiO_3 + 10CO$$

There are three common allotropes of phosphorus and several other modifications of these, some of which have indefinite structures. The common forms and their methods of preparation are:

1. *white* (or *yellow*) *phosphorus* recovered by distillation of phosphorus as shown above; soluble in some organic solvents
2. *red phosphorus* obtained by the action of heat on white phosphorus; insoluble in organic solvents
3. *black phosphorus* obtained by the action of heat on white phosphorus under high pressures using an Hg catalyst.

Phosphorus has a sufficiently high ionization potential for the chemistry of its compounds to be almost entirely covalent. Phosphorus compounds are formally P(III) and P(V) compounds. The element forms a hydride, PH_3 (phosphine), which is analogous to ammonia, but it is not formed by direct combination. Phosphorus(V) hydrides are not known.

When phosphorus is burned in air the product is the highly hygroscopic white solid P_4O_{10} (commonly called *phosphorus pentoxide*, P_2O_5); this is the P(V) oxide. In a limited supply of air the P(III) oxide, P_4O_6 (called phosphorus(III) oxide) is obtained, contaminated with unreacted phosphorus. The two oxides both react with water to form acids:

$$P_4O_6 + 6H_2O \rightarrow 4H_3PO_3 \text{ phosphonic acid}$$

$$P_4O_{10} + 2H_2O \rightarrow 4HPO_3 \text{ metaphosphoric acid}$$

$$P_4O_{10} + 6H_2O \rightarrow 4H_3PO_4 \text{ phosphoric acid.}$$

There are also many polyphosphoric acids and salts with different ring sizes and different chain lengths. Pyrophosphoric acid is obtained by melting phosphoric acid; further dehydration above 200°C leads to metaphosphoric acids:

$$2H_3PO_4 \rightarrow H_4P_2O_7 + H_2O$$

$$H_4P_2O_7 \rightarrow 2HPO_3 + H_2O$$

Polyphosphates with molecular masses from 1000 to 10 000 are industrially important in detergent formulations, while polyphosphates with lower mass numbers are used to complex metal ions (sequestration).

Phosphorus forms the P(III) halides, PX_3, with all the halogens, and P(V) halides, PX_5, with X = F, Cl, Br. The P(III) halides are formed by direct combination, but because of hazards in the reaction of phosphorus and fluorine, PF_3 is prepared from PCl_3.

The P(III) halide molecules are all pyramidal (as is PH_3) and are readily hydrolyzed. The P(V) halides are also readily hydrolyzed by water in two stages giving first the oxyhalide, then phosphoric acid:

$$PX_5 + H_2O \rightarrow POX_3 + 2HX$$

$$POX_3 + 3H_2O \rightarrow H_3PO_4 + 3HX$$

In the gas phase the trihalides are pyramidal and the pentahalides are trigonal bipyramids. However, in tetrachloromethane solution phosphorus pentachloride is a dimer with two chlorine bridges, exhibiting the ready expansion of the coordination number to six. In the solid state, (PCl_5) is $[PCl_4]^+[PCl_6]^-$ (the first tetrahedral, the second octahedral) obtained by transfer of a chloride ion.

Phosphorus interacts directly with several metals and metalloids to give:

1. Ionic phosphides such as Na_3P, Ca_3P_2, Sr_3P_2.
2. Molecular phosphides such as P_4S_3 and P_4S_5.
3. Metallic phosphides such as Fe_2P.

Symbol: P; m.p. 44.1°C (white) 410°C (red under pressure); b.p. 280.5°C; r.d. 1.82 (white) 2.2 (red) 2.69 (black) (all at 20°C); p.n. 15; most common isotope ^{31}P; r.a.m. 30.973762.

phosphorus(III) bromide (phosphorus tribromide; PBr_3) A colorless liquid made by reacting phosphorus with bromine. It is readily hydrolyzed by water to phosphonic acid and hydrogen bromide. Phosphorus(III) bromide is important in organic chemistry, being used to replace a hydroxyl group with a bromine atom.

phosphorus(V) bromide (phosphorus pentabromide; PBr_5) A yellow crystalline solid that sublimes easily. It can be made by the reaction of bromine and phosphorus(III) bromide. Phosphorus(V) bromide is readily hydrolyzed by water to phosphoric(V) acid and hydrogen bromide. Its main use is in organic chemistry to replace a hydroxyl group with a bromine atom.

phosphorus(III) chloride (phosphorus trichloride; PCl_3) A colorless liquid formed from the reaction of phosphorus with chlorine. It is rapidly hydrolyzed by water to phosphonic acid and hydrogen chloride. Phosphorus(III) chloride is used in organic chemistry to replace a hydroxyl group with a chlorine atom.

phosphorus(V) chloride (phosphorus pentachloride; PCl_5) A white easily sublimed solid formed by the action of chlorine on phosphorus(III) chloride. It is hydrolyzed by water to phosphoric(V) acid and hydrogen chloride. Its main use is as a chlorinating agent in organic chemistry to replace a hydroxyl group with a chlorine atom.

phosphorus(III) chloride oxide (phosphorus trichloride oxide; phosphorus oxychloride; phosphoryl chloride, $POCl_3$) A colorless liquid that can be obtained by reacting phosphorus(III) chloride with oxygen or by distilling phosphorus(III) chloride with potassium chlorate. The reactions of phosphorus(III) chloride oxide are similar to those of phosphorus(III) chloride. Water hydrolysis yields phosphoric(V) acid. Phosphorus(III) chloride oxide forms complexes with many metal ions.

phosphorus(III) hydride *See* phosphine.

phosphorus(III) oxide (phosphorus trioxide; P_2O_3) A white waxy solid with a characteristic smell of garlic; it usually exists as tetrahedral P_4O_6 molecules in which the phosphorus atoms are linked to each other through oxygen bridges. It is soluble in organic solvents, and is made by burning phosphorus in a limited supply of air. Phosphorus(III) oxide oxidizes in air to phosphorus(V) oxide at room temperature and inflames above 70°C. It dissolves slowly in cold water or dilute alkalis to give phosphonic acid or one of its salts. Phosphorus(III) oxide reacts violently with hot water to form phosphine and phosphoric(V) acid.

phosphorus(V) oxide (phosphorus pentoxide; P_2O_5) A highly deliquescent white powder that is soluble in organic solvents. It usually exists as P_4O_{10} molecules. Phosphorus(V) oxide can be prepared by burning phosphorus in an abundant supply of oxygen. It readily combines with water to form phosphoric(V) acid and is therefore used as a drying agent for gases. It is a useful dehydrating agent because it is able to remove the elements of water from certain oxyacids and other compounds containing oxygen and hydrogen; for example, it reacts with nitric acid (HNO_3) to give nitrogen pentoxide (N_2O_5) on heating.

phosphorus oxychloride *See* phosphorus(III) chloride oxide.

phosphorus pentabromide *See* phosphorus(V) bromide.

phosphorus pentachloride *See* phosphorus(V) chloride.

phosphorus pentoxide *See* phosphorus(V) oxide.

phosphorus tribromide *See* phosphorus(III) bromide.

phosphorus trichloride *See* phosphorus(III) chloride.

phosphorus trichloride oxide *See* phosphorus(III) chloride oxide.

phosphorus trioxide *See* phosphorus(III) oxide.

phosphoryl chloride *See* phosphorus(III) chloride oxide.

phot A unit of illumination in the c.g.s. system defined as an illumination of one lumen per square centimeter. In SI units it is equal to 10^4 lux.

photochemical reaction A reaction brought about by light; examples include the bleaching of colored material, the reduction of silver halides, and the photosynthesis of carbohydrates. Chemical changes occur only when the reacting atoms or molecules absorb photons of the appropriate energy. The amount of substance that reacts is proportional to the quantity of energy absorbed. For example, in the reaction between hydrogen and chlorine, it is not the concentrations of hydrogen or chlorine that dictate the rate of reaction but the intensity of the radiation.

photochemistry The branch of chemistry dealing with reactions induced by light or ultraviolet radiation.

photoelectric effect The emission of electrons from a solid (or liquid) surface when it is irradiated with electromagnetic radiation. For most materials the photoelectric effect occurs with ultraviolet radiation or radiation of shorter wavelength; some show the effect with visible radiation.

In the photoelectric effect, the number of electrons emitted depends on the intensity of the radiation and not on its frequency. The kinetic energy of the electrons that are ejected depends on the frequency of the radiation. This was explained, by Einstein, by the idea that electromagnetic radiation consists of streams of photons. The *photon energy* is $h\nu$, where h is the Planck constant and ν the frequency of the radiation. To remove an electron from the solid a certain minimum energy must be supplied, known as the *work function*, ϕ. Thus, there is a certain minimum threshold frequency ν_0 for radiation to eject electrons: $h\nu_0 = \theta$. If the frequency is higher than this threshold the electrons are ejected. The maximum kinetic energy (W) of the electrons is given by *Einstein's equation*:

$$W = h\nu - \phi$$

The photoelectric effect also occurs with gases. *See* photoionization.

photoelectrochemical cell (PEC) A type of voltaic cell in which one of the electrodes is a light-sensitive semiconductor (such as gallium arsenide). A current is generated by light falling on the semiconductor, causing electrons to pass between the electrode and the electrolyte and produce chemical reactions in the cell.

photoelectron An electron ejected from a solid, liquid, or gas by the PHOTOELECTRIC EFFECT or by PHOTOIONIZATION.

photoelectron spectroscopy *See* photoionization.

photoemission The emission of photoelectrons by the photoelectric effect or by photoionization.

photoionization The ionization of atoms or molecules by electromagnetic radiation. Photons absorbed by an atom may have sufficient photon energy to free an electron from its attraction by the nucleus. The process is

$$M + h\nu \rightarrow M^+ + e^-$$

where h is the Planck constant and ν is the frequency of the radiation.

As in the photoelectric effect, the radiation must have a certain minimum threshold frequency. The energy of the photoelectrons ejected is given by $W = h\nu - I$, where W is the kinetic energy and I is the ionization potential of the atom or molecule. Analysis of the energies of the emitted electrons gives information on the ionization potentials of the substance – a technique known as *photoelectron spectroscopy*.

photolysis A chemical reaction that is produced by electromagnetic radiation (light or ultraviolet radiation). Many photolytic reactions involve the formation of free radicals.

photon energy *See* photoelectric effect.

physical change A change to a substance that does not alter its chemical properties. Physical changes (e.g. melting, boiling, and dissolving) are comparatively easy to reverse.

physical chemistry The branch of chemistry concerned with the physical properties of compounds and how these depend on the chemical bonds involved.

physisorption *See* adsorption.

piano-stool compound *See* sandwich compound.

pi bond *See* orbital.

pico- Symbol: p A prefix used with SI units, denoting 10^{-12}. For example, 1 picofarad (pF) = 10^{-12} farad (F).

pi complex A complex in which the metal ion is bound to a pi-electron system, as in FERROCENE.

pig iron The crude iron produced from a blast furnace, containing carbon, silicon, and other impurities. The molten iron is run out of the furnace into channels called 'sows', which branch out into a number of offshoots called 'pigs', in which the metal is allowed to cool.

pigment An insoluble, particulate coloring material as used in PAINT.

pi orbital *See* orbital.

pipette A device used to transfer a known volume of solution from one container to another; in general, several samples of equal volume are transferred for individual analysis from one stock solution. Pipettes are of two types, bulb pipettes, which transfer a known and fixed volume, and graduated pipettes, which can transfer variable volumes.

pitchblende (uraninite) A highly radioactive black mineral consisting mostly of uranium(VI) oxide, UO_3 along with uranium radioactive decay series elements such as thorium and radium. It is the chief ore of uranium and radium.

pK The logarithm to the base 10 of the reciprocal of an acid's dissociation constant (K_a):

$$\log_{10}(1/K_a)$$

planar chromatography *See* chromatography; thin-layer chromatography.

Planck constant Symbol: h A fundamental constant; the ratio of the energy (W) carried by a photon to its frequency (ν). A basic relationship in the quantum theory of radiation is $W = h\nu$. The value of h is $6.626\ 196 \times 10^{-34}$ J s. The Planck constant appears in many relationships in which some observable measurement is quantized (i.e. can take only specific discrete values rather than any of a range of values). It is named for German physicist

Max Karl Planck (1858–1947), who estimated its value in 1900.

plane polarization A type of polarization of electromagnetic radiation in which the vibrations take place entirely in one plane.

plane-polarized *See* polarization.

plasma A very hot mixture of ions and electrons, as in an electrical discharge. Sometimes, a plasma is described as a fourth state of matter.

plaster of Paris *See* calcium sulfate.

platinum A silvery-white malleable ductile transition metal the third element of group 10 (formerly part of subgroup VIIIB) of the periodic table. It occurs native or in association with other platinum metals in Australia, Canada, Russia, and South Africa. It is also found in certain copper and nickel ores, from which it is recovered during refining. It is resistant to oxidation and is not attacked by acids (except aqua regia) or alkalis. Platinum is used as a catalyst for ammonia oxidation (to make nitric acid) and in catalytic converters. It is also used in thermocouples and surgical instruments, and to make jewelry.

Symbol: Pt; m.p. 1772°C; b.p. 3830 ± 100°C; r.d. 21.45 (20°C); p.n. 78; most common isotope ^{195}Pt; r.a.m. 195.08.

platinum black A finely divided black form of platinum produced, as a coating, by evaporating platinum onto a surface in an inert atmosphere. Platinum-black coatings are used as pure absorbent electrode coatings in experiments on electric cells, and as catalysts. They are also used, like carbon-black coatings, to improve the ability of a surface to absorb radiation.

platinum(II) chloride ($PtCl_2$) A gray-brown powder prepared by the partial decomposition of platinum(IV) chloride. It may also be prepared by passing chlorine over heated platinum. Platinum(II) chloride is insoluble in water but dissolves in concentrated hydrochloric acid to form a complex acid.

platinum(IV) chloride ($PtCl_4$) A reddish-brown hygroscopic solid prepared by the action of heat on chloroplatinic acid. Crystals of the pentahydrate, $PtCl_4.5H_2O$, are formed; anhydrous platinum(IV) chloride is prepared by treating the hydrated crystals with concentrated sulfuric acid. The chloride dissolves in water to produce a strongly acidic solution. Platinum(IV) chloride forms a series of hydrates having 1, 2, 4, and 5 molecules of water; the tetrahydrate is the most stable. The strongly acidic solution produced on dissolving it in water probably contains $H_2[PtCl_4(OH)_2]$.

platinum–iridium An alloy of platinum containing up to 30% iridium. Hardness and resistance to chemical attack increase as the iridium content is increased. It is used in jewelry, electrical contacts, and hypodermic needles. The international prototype of the kilogram is a platinum–iridium alloy.

platinum metals The second and third transition series metals ruthenium (Ru), osmium (Os), rhodium (Rh), iridium (Ir), palladium (Pd), and platinum (Pt), sometimes classed together due to their related properties and similar compounds. All the metals, for example, are hard and corrosion-resistant.

plumbane (lead(IV) hydride; PbH_4) A colorless unstable gas that can be obtained by the action of acids on a mixture of magnesium and lead pellets.

plumbic Designating a lead(IV) compound.

plumbous Designating a lead(II) compound.

plutonium A radioactive silvery element of the actinoid series of metals. It is a transuranic element found on Earth only in minute quantities in uranium ores, but is readily obtained, as ^{239}Pu, by neutron bombardment of natural uranium. The readily fissionable ^{239}Pu is a major nuclear fuel and nuclear explosive. Plutonium emits high levels of alpha particles and is thus highly toxic; the particles accumulate in bone where they can cause leukemia and radiation poisoning.

Symbol: Pu; m.p. 641°C; b.p. 3232°C; r.d. 19.84 (25°C); p.n. 94; most stable isotope ^{244}Pu (half-life 8.2×10^7 years).

point defect *See* defect.

point group *See* molecular symmetry.

poison **1.** A substance that destroys catalyst activity.
2. Any substance that endangers biological activity, whether by physical or chemical means.

polar Describing a compound with molecules that have a permanent dipole moment. Hydrogen chloride and water are examples of polar compounds.

polar bond A covalent bond in which the bonding electrons are not shared equally between the two atoms. A bond between two atoms of different electronegativity is said to be polarized in the direction of the more electronegative atom, i.e. the electrons are drawn preferentially toward the atom. This leads to a small separation of charge and the development of a bond DIPOLE MOMENT as in, for example, hydrogen fluoride, represented as H→F or as $H^{\delta+}$–$F^{\delta-}$ (F is more electronegative).

The charge separation is much smaller than in ionic compounds; molecules in which bonds are strongly polar are said to display partial ionic character. The effect of the electronegative element can be transmitted beyond adjacent atoms. *See also* intermolecular forces.

polarimeter (polariscope) An instrument for measuring optical activity. *See* optical activity.

polariscope *See* polarimeter.

polarizability The ease with which an electron cloud is deformed, i.e. polarized. In ions, an increase in size or negative charge leads to an increase in polarizability. The concept is of particular use in the treatment of covalent contributions to predominantly ionic bonds embodied in Fajans' Rules. Thus ions, such as I^-, Se^{2-}, Te^{2-}, are especially prone to covalent character, particularly in combination with ions of small size and high positive charge (i.e. high ionic potential).

polarization **1.** The restriction of the vibrations in a transverse wave so that the vibration occurs in a single plane. Electromagnetic radiation, for instance, is a transverse wave motion. It can be thought of as an oscillating electric field and an oscillating magnetic field, both at right angles to the direction of propagation and at right angles to each other. Usually, the electric vector is considered since it is the electric field that interacts with charged particles of matter and causes the effects. In 'normal' unpolarized radiation, the electric field oscillates in all possible directions perpendicular to the wave direction. On reflection or on transmission through certain substances (e.g. Polaroid) the field is confined to a single plane. The radiation is then said to be *plane-polarized*. If the tip of the electric vector describes a circular helix as the wave propagates, the light is said to be *circularly polarized*.
2. *See* polarizability.
3. The reduction of current in a voltaic cell, caused by the build-up of products of the chemical reaction. Commonly, the cause is the build-up of a layer of bubbles (e.g. of hydrogen). This reduces the effective area of the electrode, causing an increase in the cell's internal resistance. It can also produce a back e.m.f. Often a substance such as manganese(IV) oxide (a *depolarizer*) is added to prevent hydrogen build-up.

polar molecule A molecule in which the individual polar bonds are not perfectly symmetrically arranged and are therefore not 'in balance'. Thus the charge separation in the bonds gives rise to an overall charge separation in the molecule as, for example, in water. Such molecules possess a dipole moment.

polarogram *See* polarography.

polarography An analytical method in which current is measured as a function of potential. A special type of cell is used in which there is a small, easily polarizable cathode (the *dropping-mercury electrode*, consisting of a thin tube through which mercury is slowly dripped into the solution) and a large nonpolarizable anode (reference cell). The analytical reaction takes place at the cathode and is essentially a reduction of the cations, which are discharged according to the order of their electrode potential values. The data is expressed in the form of a *polarogram*, which is a plot of current against applied voltage. As the applied potential is increased a point is reached at which the ion is discharged. There is a step-wise increase in current, which levels off because of polarization effects. The potential at half the step height (called the *half-wave potential*) is used to identify the ion. Most elements can be determined by polarography. The optimum concentrations are in the range 10^{-2}–10^{-4}M; modified techniques allow determinations in the parts per million range.

polar solvent *See* solvent.

pollution Any damaging or unpleasant change in the environment that results from the physical, chemical, or biological side-effects of human industrial or social activities. Pollution can affect the atmosphere, rivers, lakes, seas, and soil.

Air pollution is caused by the domestic and industrial burning of carbonaceous fuels, by industrial processes, and by gases in car exhausts. Among recent problems are industrial emissions of sulfur(IV) oxide, a cause of *acid rain*, and emissions of chlorofluorocarbons (CFCs), previously widely used in refrigeration, aerosols, etc., and linked to the depletion of ozone in the stratosphere. Carbon dioxide, produced by burning fossil fuels, is slowly building up in the atmosphere, which could result in an overall increase in the temperature of the Earth due to its greenhouse effect. Car exhausts also emit carbon monoxide and, if leaded gasoline (petrol) is used, lead. Atmospheric CO has not yet reached dangerous levels, but vegetation near main roads has in the past been found to contain a high proportion of lead, with levels sufficiently high in urban areas to cause concern about the effects on children. Leaded gasoline has thus been banned in many countries. Photochemical smog, caused by the action of sunlight on volatile hydrocarbons and nitrogen oxides from car exhausts, is a problem in several countries. Catalytic converters have helped reduce harmful emissions from vehicle exhausts.

Water pollutants include those that are *biodegradable*, such as sewage effluent, which cause no permanent harm if adequately treated and dispersed, as well as those that are nonbiodegradable, such as certain chlorinated hydrocarbon pesticides (e.g. DDT) and heavy metals, such as lead, copper, and zinc in some industrial effluents, which cause *heavy-metal pollution*. When these pollutants accumulate in the environment they can become very concentrated in food chains. The pesticides DDT, aldrin, and dieldrin are now banned in many countries. Water supplies can become polluted by leaching of nitrates, pesticides, or animal wastes from agricultural land. The discharge of waste heat can cause thermal pollution of the environment, which can be reduced by the use of cooling towers. In lakes, rivers, and the sea, spillage from tankers and the discharge of inadequately treated sewage effluent are the main problems.

Other forms of pollution are noise from aircraft, traffic, and industry and radioactivity from improper disposal of radioactive waste.

polonium A strongly radioactive rare metallic or metalloid element belonging to group 16 (formerly VIA) of the periodic table. It occurs in very minute quantities in uranium ores. Over 30 radioisotopes are known, nearly all alpha-particle emitters. Polonium is a volatile metal and evaporates with time. It is also extremely hazardous; a quantity of polonium quickly reaches a temperature of a few hundred degrees Celsius because of alpha emission. For this reason it has been used as a lightweight heat source in space satellites.

Symbol: Po; m.p. 254°C; b.p. 962°C; r.d. 9.32 (20°C); p.n. 84; stablest isotope ^{209}Po (half-life 102 years).

polyatomic Describing a molecule (or ion or radical) that consists of several atoms (three or more). Examples are calcium carbonate ($CaCO_3$) and methane (CH_4).

polybasic Describing an acid that has two or more replaceable hydrogen atoms. For example, phosphorus(V) acid, H_3PO_4, is tribasic.

polychromic acids *See* chromium(IV) oxide.

polycrystalline Describing a substance composed of very many minute interlocking crystals that have solidified together.

polycyclic Describing a compound that has two or more rings in its molecules.

polydioxoboric(III) acid *See* boric acid.

polymer A compound in which there are very large molecules made up of repeating molecular units called MONOMERS. Polymers do not usually have a definite relative molecular mass, because there are variations in the lengths of different chains. They may be natural substances (e.g. cellulose, starch, silicates, or proteins) or synthetic materials (e.g. nylon or silicone). The two major classes of synthetic polymers are thermosetting and thermoplastic. The former are infusible, and heat may only make them harder, whereas the latter soften on heating.

polymorphism The ability of certain chemical substances to exist in more than one physical form. For elements this is called ALLOTROPY. The existence of two forms is called *dimorphism.* Crystalline structures of compounds can vary with temperature as a result of different packing arrangements of the particles. There is a transition temperature between forms and usually a marked change in density.

polyvalent Describing an atom or group that has a valence of more than one.

porcelain A ceramic traditionally made from feldspar, kaolin (China clay), marble, and quartz.

Porter, George, Baron (1920–2001) British physical chemist. In collaboration with Ronald NORRISH, Porter developed the technique of flash photolysis for investigating fast chemical reactions, starting in the late 1940s. He shared the 1967 Nobel Prize for chemistry with Norrish and Manfred EIGEN.

positive catalyst *See* catalyst.

potash *See* potassium carbonate; potassium hydroxide.

potash alum *See* alum; aluminum potassium sulfate.

potassamide *See* potassium monoxide.

potassium A light soft silvery reactive alkali metal; the third element of group 1 (formerly IA) of the periodic table. It has the argon electronic configuration of argon plus an outer $4s^1$ electron. The element gives a distinct lilac color to flames which, however, is easily masked by the intense yellow coloration resulting from any trace impurity of sodium. Consequently the flames must be viewed through blue cobalt glass. The ionization potential of potassium is low and its chemistry is largely the chemistry of the K^+ ion. Potassium accounts for 2.4% of the lithosphere and occurs in large salt deposits such as carnallite ($KCl.MgCl_2.6H_2O$) and schönite (K_2SO_4), and sylvite (KC1). Large amounts also occur in mineral forms that are not of much use for recovery of potassium, e.g., orthoclase, $K_2Al_2Si_6O_{10}$. The commercial usage of potassium is much less than that of sodium; industrially the element is obtained by electrolysis of the fused hydroxide. Potassium hydroxide is in turn obtained by electrolysis of carnallite solutions (the magnesium is initially precipitated as $Mg(OH)_2$), and both hydrogen and chlorine are recovered as by-products.

The chemistry of potassium and its compounds is very similar to that of the other group 1 elements. Particular ways in which the chemistry of potassium differs from that of sodium can be summarized as follows:

1. Burning in air produces the superoxide KO_2.
2. Because of the larger size of the K^+ ion and hence lower lattice energies, potassium salts are often more soluble than corresponding sodium salts.
3. Potassium salts are usually less heavily hydrated than corresponding sodium salts.

The most abundant isotope of potassium is ^{39}K (93.1%), with ^{41}K also a stable isotope (6.8%). The naturally occurring radioactive isotope ^{40}K (0.11%) has a half-life of 1.28×10^9 years.

Symbol: K; m.p. 63.65°C; b.p. 774°C; r.d. 0.862 (20°C); p.n. 19; r.a.m. 39.0983.

potassium–argon dating A technique for dating rocks. It depends on the radioactive decay of the radioisotope ^{40}K to ^{40}Ar (half-life 1.27×10^{10} years) and is based on the assumption that argon has been trapped in the rock since the time that the rock cooled (i.e. since it formed). If the ratio $^{40}Ar/^{40}K$ is measured it is possible to estimate the age of the rock. *See also* radioactive dating.

potassium bicarbonate *See* potassium hydrogencarbonate.

potassium bromide (KBr) A white or colorless cubic crystalline solid that is extremely soluble in water. It is formed by the action of bromine on hot potassium hydroxide solution or by the neutralization of the carbonate with hydrobromic acid. It is used extensively in the manufacture of photographic plates, films, and papers and was once used as a sedative in medicine.

potassium carbonate (pearl ash; potash; K_2CO_3) A white deliquescent solid prepared by thermal decomposition of potassium hydrogencarbonate ($KHCO_3$). Potassium carbonate is very soluble in water, its solutions being strongly alkaline due to salt hydrolysis. It is used in the laboratory as a drying agent and industrially in the manufacture of soft soap, hard glass, and in dyeing and wool finishing procedures. Potassium carbonate crystallizes out between 10–25°C as $K_2CO_3.3H_2O$; it dehydrates at 100°C to $K_2CO_3.H_2O$ and at 130°C to K_2CO_3.

potassium chlorate ($KClO_3$) A white soluble crystalline solid prepared by the electrolysis of a concentrated solution of potassium chloride. Industrially, it is prepared by the fractional crystallization of a solution containing sodium chlorate (V) and potassium chloride. When heated just above its melting point of 356°C, potassium perchlorate is formed or it can be made by subjecting a hot solution of potassium chloride to electrolysis. When heated further, potassium chlorate decomposes to yield oxygen and potassium chloride. It is a powerful oxidizing agent and is used in explosives, matches, weedkillers, and fireworks, and as a disinfectant. It oxidizes iodide ions to iodine when in an acidic medium.

potassium chloride (KCl) A white or colorless ionic solid prepared by neutralizing hydrochloric acid with potassium hydroxide solution. Potassium chloride occurs naturally as the minerals *sylvite* and carnallite, which both occur in deposits left by evaporated shallow seas. It is more soluble than sodium chloride in hot water but less soluble in cold water. On evaporation of an aqueous solution, colorless cubic crystals similar to those of sodium chloride are produced. Potassium chloride is used as a fertilizer, in the manufacture of potassium hydroxide, and in photography.

potassium chromate (K_2CrO_4) A bright yellow toxic solid prepared by adding potassium hydroxide solution to a solution of potassium dichromate. It is extremely soluble in water. Addition of an acid to an aqueous solution of potassium chromate converts the chromate ions into dichromate ions. The salt is used as an indicator in silver nitrate titrations and in pigments and enamels. The crystals are isomorphous with potassium sulfate.

potassium cyanide (KCN) A white ionic solid that is very soluble in water and extremely poisonous. It smells of bitter almonds. It is made industrially by combining potassium hydroxide with hydrogen cyanide. Potassium cyanide is used to recover gold and silver, in electroplating, and in fumigants. It is also used as a reducing agent and in chemical analysis. Aqueous solutions of potassium cyanide are strongly hydrolyzed, the solutions being alkaline. On standing, these solutions slowly evolve hydrocyanic acid.

potassium dichromate ($K_2Cr_2O_7$) An orange-red crystalline toxic solid prepared by adding potassium chloride solution to a concentrated solution of sodium dichromate and crystallizing out or by acidifying a solution of potassium chromate and evaporating the same. It is less soluble than sodium dichromate in cold water but more soluble in hot water. Potassium dichromate is used as an oxidizing agent in volumetric analysis and organic chemistry, and also in the manufacture of dyes, glass, glues, and ceramics. It also finds use in fireworks, photography, and lithography. The crystals are triclinic, anhydrous, nondeliquescent, and also a fire risk due to their strong oxidative capacity. Addition of an alkali to a solution of the dichromate yields the chromate.

potassium hydride (KH) A white or light gray ionic solid prepared by passing hydrogen over heated potassium, the metal being suspended in an inert medium. It is an excellent reducing agent but also a fire hazard because it reacts with water to release hydrogen.

potassium hydrogencarbonate (potassium bicarbonate; $KHCO_3$) A white crystalline solid prepared by passing carbon dioxide through a saturated solution of potassium carbonate. Potassium hydrogencarbonate is more soluble in water than sodium hydrogencarbonate. When heated, it undergoes thermal decomposition to give the carbonate, water, and carbon dioxide. The salt forms monoclinic crystals. Its solutions are strongly alkaline as a result of salt hydrolysis.

potassium hydrogentartrate (cream of tartar; $HOOC(CHOH)_2COO^-K^+$) A white crystalline solid, an acid salt of tartaric acid that occurs in grape juice. It is used as the acid ingredient of BAKING POWDER.

potassium hydroxide (caustic potash; lye; KOH) A highly corrosive white solid manufactured by the electrolysis of potassium chloride solution in a mercury cell. It can also be prepared by heating either potassium carbonate or potassium sulfate with slaked lime. In the laboratory, it can be made by reacting potassium, potassium monoxide, or potassium superoxide with water. Potassium hydroxide resembles sodium hydroxide but is more soluble in water and alcohol. It is preferred over sodium hydroxide for the absorption of carbon dioxide and sulfur(IV) oxide, due to its greater solubility. Potassium hydroxide is used as an electrolyte in Ni–Cd, alkaline (Zn–Mn), mercury (Zn–Hg), and silver (Zn–Ag) storage batteries and in the production of soft soaps. It forms crystalline hydrates with 1, 1.5, and 2 molecules of water.

potassium iodate (KIO_3) A white solid formed either by adding iodine to a hot concentrated solution of potassium hydroxide or by the electrolysis of potassium iodide solution. No hydrates are known. It is a source of iodide and iodic acid. When treated with a dilute acid and a reducing agent, the iodate ions are reduced to iodine. It is used in chemical analysis, and as a food additive.

potassium iodide (KI) A white ionic solid readily soluble in water. It is prepared by dissolving iodine in hot concentrated potassium hydroxide solution. Both the iodide and iodate are formed but the latter is removed by fractional crystallization. Potassium iodide has a sodium chloride cubic lattice. In solution with iodine it forms potassium triiodide PI_3. Potassium iodide is used in medicine, particularly in the treatment of goiter resulting from iodine deficiency. It is also added to table salt to help prevent goiter, and is used in photography. Dilute acidified solutions of potassium iodide can act as reducing agents, manganate(VII) ions being reduced to manganese(II) ions, copper(II) ions to copper(I) ions, and iodate ions to iodine.

potassium manganate(VII) (potassium permanganate; $KMnO_4$) A purple solid soluble in water. It is prepared by oxidizing potassium manganate(VI) with chlorine. Potassium permanganate is used in volumetric analysis as an oxidizing agent, as a bactericide, and as a disinfectant. In aqueous solution its behavior as an oxidizing agent depends on the pH of the solution.

potassium monoxide (K_2O) An ionic solid that is white when cold and yellow when hot. It is prepared by heating potassium with potassium nitrate. Potassium monoxide dissolves violently in water to form potassium hydroxide solution. The hydrate $K_2O.3H_2O$ is known. Potassium monoxide dissolves in liquid ammonia with the formation of potassium hydroxide and *potassamide* (KNH_2).

potassium nitrate (saltpeter; niter or nitre; KNO_3) A white solid, soluble in water, formed by fractional crystallization of sodium nitrate and potassium chloride solutions. It occurs naturally as *niter* or *nitre* (*saltpeter*) in rocks in Hungary, Spain, India, South Africa, and Brazil. When heated it decomposes to give the nitrite and oxygen. Unlike sodium nitrate it is nondeliquescent. Potassium nitrate is a strong oxidizing agent and is used in gunpowder, fireworks, fertilizers, and in the laboratory preparation of nitric acid.

potassium nitrite (KNO_2) A creamy deliquescent solid that is readily soluble in water. It reacts with cold dilute mineral acids to produce solutions of nitrous acid. Potassium nitrite is used in organic chemistry.

potassium permanganate *See* potassium manganate(VII).

potassium sulfate (K_2SO_4) A white solid prepared by the neutralization of either potassium hydroxide or potassium carbonate with dilute sulfuric acid. It occurs naturally as schönite in deposits at Stassfurt, Germany. Potassium sulfate is soluble in water, forming a neutral solution. It is used as a fertilizer and in the chemical industry in the preparation of alums. It is also used in cement and in glass-making. The anhydrous salt crystallizes in the rhombic form.

potassium sulfide (K_2S) A yellowish-brown deliquescent solid prepared by saturating an aqueous solution of potassium hydroxide with hydrogen sulfide and then adding an equal volume of potassium hydroxide. Industrially it is produced by heating potassium sulfate and carbon at high temperature. The sulfide crystallizes out from aqueous solution as $K_2S.5H_2O$. Its aqueous solutions undergo hydrolysis, the solution being strongly alkaline. Potassium sulfide dust is a fire risk.

potassium superoxide (KO_2) A yellow paramagnetic solid prepared by burning potassium in excess oxygen. When treated with cold water or dilute mineral acids, hydrogen peroxide is produced. If heated strongly, it yields oxygen and potassium monoxide. Potassium superoxide is a powerful oxidizing agent.

potassium thiocyanate (KSCN) A colorless hygroscopic solid. Its solution is used in a test for iron(III) compounds, with which it turns a blood-red color.

potential energy *See* energy.

potential energy curve A curve of the potential energy of electrons in a diatomic molecule in which the potential energy of an electronic state is plotted vertically and the interatomic distance is plotted horizontally, with the minimum of the curve being the average internuclear distance.

potentiometric titration A titration in which an electrode is used in the reaction mixture. The end point can be found by monitoring the electric potential during the titration.

praseodymium A soft ductile malleable silvery-yellow element of the lanthanoid series of metals. It occurs in association with other lanthanoids in minerals such as bastnaesite and monazite. Praseodymium is used in several alloys, as a catalyst, in carbon arc lamp electrodes, in misch metal, in enamel, and in yellow glass for eye protection.

Symbol: Pr; m.p. 931°C; b.p. 3512°C; r.d. 6.773 (20°C); p.n. 59; most common isotope ^{141}Pr; r.a.m. 140.91.

precipitate A suspension of small solid particles formed in a liquid as a result of a chemical reaction.

precursor A substance from which another substance is formed in a chemical reaction.

pressure Symbol: p The pressure on a surface due to forces from another surface or from a fluid acting at 90° to unit area of the surface:

pressure = force/area

The SI derived unit by which pressure is measured is the PASCAL (Pa).

Priestley, Joseph (1733–1804) English chemist and minister. Priestley was one of the most eminent chemists of the 18th century. In 1774 he discovered oxygen, but because he believed in the phlogiston theory of combustion he called it 'dephlogisticated air'. He also discovered and isolated a number of other compounds. Priestley was also interested in electricity and optics.

Prigogine, Ilya (1917–) Russian-born Belgian chemist. Prigogine is best known for extending thermodynamics and statistical mechanics to systems that are far from equilibrium. He called such systems 'dissipative structures'. Prigogine has been particularly interested in applying his ideas to biological and social systems. He won the 1977 Nobel Prize for chemistry for his investigations of non-equilibrium systems.

primary cell A voltaic cell in which the chemical reaction that produces the e.m.f. is not reversible. *Compare* accumulator.

primary standard A substance that can be used directly for the preparation of standard solutions without reference to some other concentration standard. Primary standards should be easy to purify, dry, capable of preservation in a pure state, unaffected by air or CO_2, of a high molecular mass (to reduce the significance of mass determination errors), stoichiometric, and readily soluble. Any likely impurities should be easily identifiable.

principal quantum number *See* atom.

producer gas (air gas) A combustible mixture of carbon monoxide (25–30%), nitrogen (50–55%), and hydrogen (10–15%), prepared by passing air with a little steam in it through a thick layer of white-hot coke in a furnace or 'producer'. Producer gas is used as a fuel while still hot to prevent heat loss for such purposes as industrial heating, the firing of retorts, and in glass furnaces. *Compare* water gas.

product *See* chemical reaction.

promethium A soft silvery radioactive element of the lanthanoid series of metals. It occurs naturally in trace amounts in the mineral pitchblende as the result of radioactive decay of such elements as uranium, thorium, and plutonium. It can also be produced artificially by the fission of uranium. Promethium-147 is used in some miniature nuclear-powered batteries for use on spacecraft.

Symbol: Pm; m.p. 1168°C; b.p. 2730°C (approx.); r.d. 7.22 (20°C); p.n. 61; most stable isotope ^{145}Pm (half-life 18 years).

promoter (activator) A substance that

improves the efficiency of a CATALYST. It does not itself catalyze the reaction but assists the catalytic activity. For example, alumina or molybdenum promotes the catalytic activity of finely divided iron in the Haber process. The manner in which a promoter functions is not fully understood; no one theory explains all the examples.

protactinium A toxic radioactive element of the actinoid series of metals. It occurs in minute quantities in uranium ores such as autunite, tobernite, and pitchblende as a radioactive decay product of actinium. It is also synthesized by bombarding thorium nuclei with neutrons.

Symbol: Pa; m.p. 1840°C; b.p. 4000°C (approx.); r.d. 15.4 (calc.); p.n. 91; most stable isotope ^{231}Pa (half-life 32 500 years).

protic acid *See* acid.

proton An elementary particle with a positive charge ($+1.602\ 192 \times 10^{-19}$ C) and rest mass $1.672\ 614 \times 10^{-27}$ kg. Protons are nucleons, found in all nuclides.

proton number (atomic number) Symbol: Z The number of protons in the nucleus of an atom. In a neutral atom it is also equal to the number of electrons orbiting the nucleus. The proton number determines the chemical properties of an element because the element's electron structure, which determines chemical bonding, depends on electrostatic attraction to the positively charged nucleus. An element's proton number determines its position in the periodic table.

Prussian blue *See* cyanoferrate.

prussic acid See hydrocyanic acid.

pseudo-first order Describing a reaction that appears to exhibit first-order kinetics under special conditions, even though the 'true' order is greater than one. For example, in the hydrolysis of a substance in the presence of a large volume of water, the concentration of water remains approximately constant. The rate of reaction is thus found experimentally to be proportional to the concentration of the substance, even though it also depends on the amount of water present. Such a reaction is described as 'bimolecular of the first order'.

pseudohalogens A small group of simple inorganic compounds with symmetrical molecules that resemble the halogens in some reactions and compounds. For example, cyanogen (C_2N_2) has compounds analogous to halogen compounds (e.g. HCN, KCN, CH_3CN, etc.). Thiocyanogen ($S_2C_2N_2$), is another example.

p-type semiconductor See semiconductor.

Pyrex (*Trademark*) A particularly strong, heat resistant, chemically inert borosilicate glass commonly used for laboratory glassware.

pyrites A mineral sulfide of a metal; e.g. iron pyrites, FeS_2.

pyrolusite See manganese(IV) oxide.

pyrolysis The decomposition of chemical compounds by subjecting them to very high temperature.

pyrometer An instrument used in the chemical industry to measure high temperature, e.g. in reactor vessels.

pyrophoric **1.** Describing a compound that ignites spontaneously in air.
2. Describing a metal or alloy (misch metal) that sparks when struck. Lighter flints are made of pyrophoric alloy.

pyrophosphoric acid See phosphoric(V) acid.

pyrosulfuric acid *See* oleum.

Q

quadrivalent (tetravalent) Having a valence of four.

qualitative analysis Analysis carried out with the purpose of identifying the components of a sample. Classical methods involved simple preliminary tests followed by a carefully devised scheme of systematic tests and procedures. Modern methods include the use of such techniques as infrared spectroscopy and emission spectrography. *Compare* quantitative analysis.

quantitative analysis Analysis carried out with the purpose of determining the concentration of one or more of the known components of a sample. Classical wet methods include volumetric and gravimetric analysis. A wide range of more modern instrumental techniques are also used, including polarography and various types of chromatography and spectroscopy. *Compare* qualitative analysis.

quantized Describing a physical quantity that can take only certain discrete values, and not a continuous range of values.

quantum (plural *quanta*) A definite amount of energy released or absorbed in a process. Energy often behaves as if it were 'quantized' in this way. The quantum of electromagnetic radiation is the photon.

quantum electrodynamics The use of quantum mechanics to describe how particles and electromagnetic radiation interact.

quantum mechanics *See* quantum theory.

quantum number An integer or half integer that specifies the value of a quantized physical quantity (e.g. energy, angular momentum, etc.). *See* atom; Bohr theory; spin.

quantum states States of an atom, electron, particle, etc., specified by a unique set of quantum numbers.

quantum theory A mathematical theory originally introduced by Max Planck (1900) to explain the radiation emitted from hot bodies. Quantum theory is based on the idea that energy (or certain other physical quantities) can be changed only in certain discrete amounts for a given system. Other early applications were the explanations of the photoelectric effect and the Bohr theory of the atom.

Quantum mechanics is a system of mechanics that developed from quantum theory and is used to explain the behavior of atoms, molecules, etc. In one form it is based on de Broglie's idea that particles can have wavelike properties – this branch of quantum mechanics is called *wave mechanics*. *See* orbital.

quartz *See* silicon(IV) oxide.

quasicrystal A type of crystal structure in which there is long-range order but in which the symmetry of the repeating unit is not one allowed in normal crystals. There are examples of solids, such as the compound AlMn, that have icosahedral symmetry and a long-range repeating order in the stucture. These are known as quasicrystals.

quicklime *See* calcium oxide.

quinquevalent *See* pentavalent.

R

racemate *See* optical activity.

racemic mixture *See* optical activity.

racemization The conversion of an optical isomer into a racemic mixture, which is not optically active.

rad An m.k.s. unit formerly used to measure absorbed dose of ionizing radiation, defined as being equivalent to an absorption of 10^{-2} joule of energy in one kilogram of material. In SI units it is equal to 10^{-2} gray.

radian Symbol: rad The SI derived unit of plane angle, equal to that subtended at the center of a circle by an arc whose length is equal to the circle's radius. One complete revolution (360°) is 2π radian.

radiation In general, the emission of energy from a source, either as waves (light, sound, etc.) or as moving particles (beta rays or alpha rays).

radical A group of atoms in a molecule. *See also* free radical.

radioactive Describing an element or nuclide that exhibits natural radioactivity.

radioactive dating (radiometric dating) A technique for dating archaeological specimens, rocks, etc., by measuring the extent to which some radionuclide has decayed to give a product. *See* carbon; potassium–argon dating; rubidium–strontium dating; uranium–lead dating.

radioactive decay series The series of definite isotopes into which a given radioactive element is transformed as it decays.

radioactive isotope *See* radioisotope.

radioactivity The disintegration or decay of certain unstable nuclides with emission of radiation. The emission of ALPHA PARTICLES, BETA PARTICLES, and GAMMA RAYS are the three most important forms of radiation that occur during decay.

radiocarbon dating *See* carbon.

radiochemistry The chemistry of radioactive isotopes of elements. Radiochemistry involves such topics as the preparation of radioactive compounds, the separation of isotopes by chemical reactions, the use of radioactive labels in studies of mechanisms, and experiments on the chemical reactions and compounds of radioactive elements.

radioisotope (radioactive isotope) A radioactive isotope of an element. Tritium, for instance, is a radioisotope of hydrogen. Radioisotopes are extensively used in research as souces of radiation and as tracers in studies of chemical reactions. Thus, if an atom in a compound is replaced by a radioactive nuclide of the element (a *label*) it is possible to follow the course of the chemical reaction. Radioisotopes are also used in medicine for diagnosis and treatment.

radiolysis Chemical decomposition produced by high-energy radiation, i.e. x-rays, gamma rays, or particles.

radiometric dating *See* radioactive dating.

radio waves A form of electromagnetic radiation with wavelengths ranging from approximately 10^{-1} to 10^3 meters. *See also* electromagnetic radiation.

radium A soft white radioactive luminescent metallic element of the alkaline-earth group, the sixth element of group 2 (formerly IIA) of the periodic table. Radium is found in uranium ores, especially pitchblende and carnotite, from which it is ultimately obtained via electrolysis using a mercury electrode. It was formerly used in luminous paints and radiotherapy, and continues to be used as a radiation source in research laboratories. It has several short-lived radioisotopes and one long-lived isotope, radium-226 (half-life 1602 years).

Symbol: Ra; m.p. 700°C; b.p. 1140°C; r.d. 5 (approx. 20°C); p.n. 88; r.a.m. 226.0254.

radon A colorless odorless monatomic radioactive gas, the sixth member of the rare gases; i.e. group 18 (formerly VIIIA) of the periodic table. It occurs naturally as a product of the uranium radioactive decay series. It has 19 short-lived radioisotopes; the most stable, radon-222, is a decay product of radium-226 and itself disintegrates into an isotope of polonium with a half-life of 3.82 days. ^{222}Rn is sometimes used in radiotherapy. Radon accumulates in uranium mines and can also seep into the basements of buildings constructed over rocks or sands containing uranium or radium ores, where it may pose a health risk.

Symbol: Rn; m.p. –71°C; b.p. –61.8°C; mass density 9.73 kg m^{-3} (0°C); p.n. 86.

raffinate The liquid remaining after the solvent extraction of a dissolved substance. *See* solvent extraction.

r.a.m. *See* relative atomic mass.

Raman effect A change in the frequency of electromagnetic radiation, such as light, that occurs when a photon of radiation undergoes an inelastic collision with a molecule. This type of scattering is called *Raman scattering*, in contrast to normal (Rayleigh) scattering. The Raman effect was first observed by the Indian physicists Sir Chandrasekhara Venkata Raman (1888–1970) and his colleague Sir Kariamanikkam Srinivasa Krishnan in 1928. The effect has been used extensively in *Raman spectroscopy* for the determination of molecular structure, particularly since the advent of the laser. In some cases it is possible to draw conclusions about molecular structure by observing whether certain lines are present or absent.

Ramsay, Sir William (1852–1916) British chemist, famous for his discovery of the rare gases. After a talk by Lord Rayleigh in 1892 on some discrepancies in the density of nitrogen, Ramsay predicted that these discrepancies were due to a hitherto undiscovered gas. After careful experiments he was able to isolate a new element which Rayleigh and Ramsay named argon. Ramsay was also able to identify helium, in collaboration with the spectroscopic observations of Sir William Crookes. (Helium had previously been discovered in the Sun.) These discoveries led Ramsay to speculate in his book *The Gases of the Atmosphere* (1896) that there is a whole group of inert elements in the periodic table. Ramsay and his colleagues found the missing elements between 1894 and 1901.

Raney nickel A catalytic form of nickel produced by treating a nickel–aluminum alloy with sodium hydroxide. The aluminum dissolves (as aluminate) in the sodium hydroxide, leaving the Raney nickel as a spongy mass, which is pyrophoric when dry. It is used especially for catalyzing hydrogenation reactions.

Raoult's law A relationship between the pressure exerted by the vapor of a solution and the presence of a solute. It states that the partial vapor pressure of a solvent above a solution (p) is proportional to the mole fraction of the solvent in the solution (X) and that the proportionality constant is the vapor pressure of pure solvent, (p_0), at the given temperature: i.e. $p = p_0X$. It was discovered by François Raoult (1830–1901), a French chemist. Solutions that obey Raoult's law are said to be *ideal solutions*. Because of solvation forces this behavior is rare and in general Raoult's law holds only for dilute solutions.

rare earths *See* lanthanoids.

rare gases (noble gases; inert gases; group 0 elements; group 18 elements) A group of monatomic low-boiling gases belonging to group 18 of the periodic table and occupying a position between the highly electronegative group 17 elements and the highly electropositive group 1 elements. They are helium (He), neon (Ne), argon (Ar), krypton (Kr), xenon (Xe), and radon (Rn). Their completely filled s and p levels gives a closed-shell structure that is associated with a general lack of chemical reactivity and very high ionization energies. It is the acquisition of a stable rare-gas configuration that is associated with determining the valence of many covalent and ionic compounds.

Prior to 1962 the rare gases were frequently called *inert gases* because chemical compounds with them were not known (there were a few clathrates and supposed 'hydrates'), but the realization that the ionization potential of xenon was sufficiently low to be accessible to chemical reaction led to the preparation of several fluorides, oxides, oxyfluorides, and a hexafluoroplatinate of xenon. Several unstable krypton and radon compounds have also been synthesized.

Argon forms about 1% of the atmosphere but the other gases are present in air in only minute traces. They are obtained by fractional distillation of liquid air or of natural gas. Considerable amounts of argon are produced as a by-product from ammonia-plant tail gas. Applications for the rare gases depend heavily on their inertness; for example, welding (Ar), filling light bulbs (Ar), gas-stirring in high temperature metallurgy (Ar), powder technology (Ar), discharge tubes (Ne), and chemical research as inert carriers (He, Ar). Neon has been proposed as an oxygen diluent for deep-sea diving because of its lower solubility in blood. Liquid helium is extensively used in low-temperature work, and radon is used therapeutically as a source of alpha-particles.

The gases helium and argon are formed by radioactive decay in various minerals, for example argon by radioactivity decay of ^{40}K. This mechanism can be applied to age-determination of minerals, a technique known as potassium-argon dating. Radon is generally obtained as ^{222}Rn from the decay of radium. Its half-life is only about 3.8 days.

rate constant (velocity constant; specific reaction rate) Symbol: k The constant of proportionality in the rate expression for a chemical reaction. For example, in a reaction A + B → C, the rate may be proportional to the concentration of A multiplied by that of B; i.e.

$$\text{rate} = k[A][B]$$

where k is the rate constant for this particular reaction. The constant is independent of the concentrations of the reactants but depends on temperature; consequently the temperature at which k is recorded must be stated. The units of k vary depending on the number of terms in the rate expression, but are easily determined remembering that rate has the unit s^{-1}.

rate-determining step (limiting step) The slowest step in a multistep reaction. Many chemical reactions are made up of a number of steps in which the one with the lowest rate is the one that determines the rate of the overall process.

The overall rate of a reaction cannot exceed the rate of the slowest step. For example, the first step in the reaction between acidified potassium iodide solution and hydrogen peroxide is the rate-determining step:

$$H_2O_2 + I^- \rightarrow H_2O + OI^- \text{ (slow)}$$
$$H^+ + OI^- \rightarrow HOI \text{ (fast)}$$
$$HOI + H^+ + I^- \rightarrow I_2 + H_2O \text{ (fast)}$$

rate of reaction A measure of the amount of reactant consumed in a chemical reaction in unit time. It is thus a measure of the number of effective collisions between reactant molecules. The rate at which a reaction proceeds can be measured by the rate the reactants disappear or by the rate at which the products are formed. The principal factors affecting the rate of reaction are temperature, pressure, concentration of reactants, light, and the action of a catalyst. The units usually used to measure the rate of a reaction are mol dm^{-3} s^{-1}. *See also* mass action.

rationalized units A system of units in which the equations defining the system have a logical form related to the shape of the system. SI units form a rationalized system of units. For example, SI formulae concerned with circular symmetry contain a factor of 2π; those concerned with radial symmetry contain a factor of 4π.

raw material A substance from which other substances are made. In the chemical industry it may be simple (such as nitrogen from air, used to make ammonia) or complex (such as coal and petroleum, used to make a wide range of products).

reactant *See* chemical reaction.

reaction *See* chemical reaction.

reaction coordinate *See* energy profile.

reagent A compound that reacts with another. The term is usually used for common laboratory chemicals, e.g. sodium hydroxide, hydrochloric acid, etc. used for experiment and analysis.

realgar The mineral form of arsenic(II) sulfide As_4S_4, one of the chief ores of arsenic. It is bright red in color and was formerly used as a pigment, in tanning, and in fireworks. Its use for these purposes has been largely discontinued due to its toxicity.

real gas *See* gas laws.

rearrangement A reaction in which the groups of a compound rearrange themselves to form a different compound.

reciprocal proportions, law of *See* equivalent proportions.

recrystallization The repeated crystallization of a compound in order to ensure purity of the sample or crystals of better form.

red lead *See* dilead(II) lead(IV) oxide.

redox (oxidation–reduction) Relating to the process of oxidation and reduction, which are intimately connected in that during oxidation by chemical agents the oxidizing agent itself becomes reduced, and vice versa. Thus an oxidation process is always accompanied by a reduction process. In electrochemical processes this is equally true, oxidation taking place at the anode and reduction at the cathode. These systems are often called *redox systems*, particularly when the interest centers on both compounds.

Oxidizing and reducing power is indicated quantitatively by the *redox potential* or standard electrode potential, $E^{\ominus}$. Redox potentials are normally expressed as reduction potentials. They are obtained by electrochemical measurements and the values are referred to the H^+/H_2 couple for which $E^{\ominus}$ is set equal to zero. Thus increasingly negative potentials indicate increasing ease of oxidation or difficulty of reduction. Thus in a redox reaction the half reaction with the most positive value of $E^{\ominus}$ is the reduction half and the half reaction with the least value of $E^{\ominus}$ (or most highly negative) becomes the oxidation half. *See also* electrode potential; oxidation; reduction.

redox potential *See* redox.

redox systems *See* redox.

red phosphorus *See* phosphorus.

reducing agent *See* reduction.

reduction The gain of electrons by such species as atoms, molecules, or ions. It often involves the loss of oxygen from a compound, or addition of hydrogen. Reduction can be effected chemically, i.e. by the use of *reducing agents* (*electron donors*), or electrically, in which case the reduction process occurs at the cathode. For example,

$$2Fe^{3+} + Cu \rightarrow 2Fe^{2+} + Cu^{2+}$$

where Cu is the reducing agent and Fe^{3+} is reduced, and

$$2H_2O + SO_2 + 2Cu^{2+} \rightarrow 4H^+ + SO_4^{2-} + 2Cu^+$$

where SO_2 is the reducing agent and Cu^{2+} is reduced.

Tin(II) chloride, sulfur(IV) oxide, and the hydrogensulfite ion are common reducing agents. *See also* redox.

reduction potential *See* electrode potential.

refining The process of removing impurities from a substance or of extracting a substance from a mixture.

refractory Describing a compound (e.g. an inorganic oxide) that has a very high melting point. Refractory materials are used as furnace linings.

relative atomic mass (r.a.m.; average atomic mass) Symbol: A_r The ratio of the average mass per atom of the naturally occurring isotopes or, if given in parentheses, the most stable isotope of an element to 1/12 of the mass of an atom of nuclide ^{12}C. It was formerly called *atomic weight.*

relative density The ratio of the density of a given substance to the density of some reference substance. The relative densities of liquids are usually measured with reference to the density of water at 4°C. Relative densities are also specified for gases; usually with respect to air at STP. The temperature of the substance is either stated or is understood to be 20°C. Relative density was formerly called *specific gravity.*

relative molecular mass Symbol: M_r The ratio of the average mass per molecule of the naturally occurring form of an element or compound to 1/12 of the mass of an atom of nuclide ^{12}C. This ratio was formerly called *molecular weight.* It does not have to be used only for compounds that have discrete molecules; for ionic compounds (e.g. NaCl) and giant-molecular structures (e.g. BN) the formula unit is used.

relaxation The process by which an excited species loses energy and falls to a lower energy level (such as the ground state).

rem An m.k.s. unit formerly used for measuring equivalent ionizing radiation doses absorbed by living tissue. One rem is equivalent to a dosage of one roentgen of x-rays or gamma rays absorbed by an average adult male. In SI units it is equal to 10^{-2} sievert.

resin A yellowish insoluble organic compound exuded by trees as a viscous liquid that gradually hardens on exposure to air to form a brittle amorphous solid. Synthetic resins are artificial polymers used in making adhesives, insulators, and paints.

resolution (of racemates) The separation of a racemate into the two optical isomers. This cannot be done by normal methods, such as crystallization or distillation, because the isomers have identical physical properties. The main methods involve mechanical, chemical, or biochemical techniques.

resonance (mesomerism) The behavior of many molecules or ions cannot be adequately explained by a single structure using simple single and double bonds. The bonding electrons of the molecule or ion have instead a different distribution, one in which the actual bonds can be regarded as a hybrid of two or more conventional forms called *resonance forms* or *canonical forms*. The result is a resonance hybrid. For example, three equivalent LEWIS STRUCTURES can be written for sulfur(VI) oxide, depending on which of the three oxygen atoms is double bonded to the sulfur atom. However, experiments reveal that all the bonds in sulfur(VI) oxide are of the same length and strength. This means that the bonds to the oxygen atoms must be neither double nor single, but a hybrid of the possible Lewis structures.

resonance forms *See* resonance.

resonance ionization spectroscopy (RIS) A type of spectroscopy that detects specific types of atoms using lasers. The laser ionizes the atoms of interest. The frequency of the laser is chosen so that only the atoms of interest in a sample are excited by the laser. This method is very selective because ionization occurs only for those atoms whose energy levels fit in with the frequency of the laser light. This selectivity has led to many practical uses for this technique.

retort A piece of laboratory apparatus consisting of a glass bulb with a long narrow neck. In industrial chemistry, various metallic vessels in which distillations or reactions take place are called retorts.

reverbatory furnace A furnace for smelting metals. It has a curved roof so that heat is reflected downward, the fuel being in one part of the furnace and the ore in the other.

reverse osmosis A process by which water containing a salt is separated via a semipermeable membrane into pure water and salt water streams by subjecting the incoming water to pressures significantly greater than its OSMOTIC PRESSURE. This action causes pure water to be forced through the membrane, leaving the salt behind. *See* osmosis.

reversible change A change in the pressure, volume, or other properties of a system, in which the system remains at equilibrium throughout the change. Such processes could be reversed; i.e. returned to the original starting position through the same series of stages. They are never realized in practice. An isothermal reversible compression of a gas, for example, would have to be carried out infinitely slowly and involve no friction, etc. Ideal energy transfer would have to take place between the gas and the surroundings to maintain a constant temperature.

In practice, all real processes are *irreversible changes* in which there is not an equilibrium throughout the change. In an irreversible change, the system can still be returned to its original state, but not through the same series of stages. For a closed system, there is always an entropy increase involved in an irreversible change.

reversible reaction A chemical reaction that can proceed in either direction. For example, the reaction:

$$N_2 + 3H_2 \rightleftharpoons 2NH_3$$

is reversible. In general, there will be an equilibrium mixture of reactants and products.

rhenium A dense silvery rare transition metal, the third element of group 7 (formerly VIIB) of the periodic table. It usually occurs naturally with molybdenum and platinum ores, and is usually extracted from flue dust in molybdenum smelters. The metal is chemically similar to manganese. It has the second highest melting point of all the elements, making it especially useful for alloys for electrical contacts, thermocouples, and other high-temperature environments. It is also used as a catalyst.

Symbol: Re; m.p. 3180°C; b.p. 5630°C; r.d. 21.02 (20°C); p.n. 75; most common isotope ^{187}Re; r.a.m. 186.207.

rheology The study of the ways in which matter can flow. This topic is of particular interest in the study of polymers.

rhodium A rare silvery hard transition metal, the second element of group 9 (formerly part of subgroup VIIIB) of the periodic table. It is difficult to work and highly resistant to corrosion. Rhodium occurs native in the natural alloy osmiridium but is most often obtained from copper and nickel ore refinery residues. It is used in protective finishes, high-temperature alloys, and as a catalyst. Rhodium compounds are characteristically pink.

Symbol: Rh; m.p. 1966°C; b.p. 3730°C; r.d. 12.41 (20°C); p.n. 45; most common isotope ^{103}Rh; r.a.m. 102.90550.

rhombic crystal *See* crystal system.

ring A closed loop of atoms in a molecule, as in the crown-shaped allotropes of sulfur (S_8) and selenium (Se_8). A *fused ring* is one joined to another ring in such a way that they share two atoms.

RIS *See* resonance ionization spectroscopy.

rock A definable part of the Earth's crust made up of a mixture of minerals, some of which usually predominate in abundance over other, accessory minerals. Rock may be hard (for example, granite), soft (chalk), consolidated (clay), or loose (sand). *See also* mineral.

rock salt (halite) A transparent naturally occurring mineral form of sodium chloride.

roentgen Symbol: R A c.g.s. unit of radiation exposure, formerly used for x-rays and gamma rays, defined in terms of the ionizing effect of one electrostatic unit of electricity on a cubic centimeter of air. In SI units one roentgen is equal to 2.58×10^{-4} coulomb per kilogram of dry air.

Rose's metal A fusible alloy containing 50% bismuth, 25–28% lead, and the balance in tin. Its low melting point (about 100°C) leads to its use in fire-protection devices.

rotary dryers Devices commonly used in the chemical industry for the drying, mixing, and sintering of solids. They consist essentially of a heated rotating inclined cylinder, which is longer in length than in diameter. Gases flow through the cylinder in either a countercurrent or co-current direction to regulate the flow of solids, which are fed into the end of the cylinder. Rotary dryers can be applied to both batch and continuous processes.

rotational spectroscopy The spectroscopic study of the rotational motion of molecules. Rotational spectroscopy gives information about interatomic distances. The transitions between different rotational energy levels in molecules correspond to the microwave and far infrared regions of the electromagnetic spectrum. It is possible for there to be transitions between rotational energy levels in pure rotational spectra only if the molecule has a permanent dipole moment. In the near infrared region rotational transitions are superimposed on vibrational transitions, resulting in a *vibrational–rotational* spectrum. This type of spectrum is considerably more complicated than a purely rotational spectrum.

rubidium A soft light silvery highly reactive alkali metal; the fourth element of group 1 (formerly IA) of the periodic table. It is found in small amounts in several complex silicate minerals, including lepidolite, and in some brines. Naturally occurring rubidium comprises two isotopes, one of which, ^{87}Rb, is radioactive (half-life 5×10^{10} years). Rubidium is used in vacuum tubes, photocells, and in making special glass. Rubidium–strontium dating relies on the decay rate of ^{87}Rb.

Symbol: Rb; m.p. 39.05°C; b.p. 688°C; r.d. 1.532 (20°C); p.n. 37; most common isotope ^{85}Rb; r.a.m. 85.4678.

rubidium–strontium dating A technique for dating rocks. It depends on the radioactive decay of the radioisotope ^{87}Rb to ^{87}Sr (half-life 4.7×10^{10} years). If the ratio $^{87}Sr/^{87}Rb$ is measured and compared to the implied original quantity of ^{87}Rb, it is possible to estimate the age of the rock. *See also* radioactive dating.

ruby A rare and highly valued gemstone variety of the mineral corundum (aluminum oxide, Al_2O_3) colored red by chromium impurities. Rubies can also be made synthetically. They are used in jewelry, as bearings in watches, and in some lasers.

Russell–Saunders coupling The type

of coupling of electronic angular momentum vectors that describes light atoms. Russell–Saunders coupling is applicable if the energies of electrostatic Coulomb repulsion are considerably greater than the energies of SPIN–ORBIT COUPLING. The concept of Russell–Saunders coupling was pioneered by the American astronomer Henry Norris Russell (1877–1957) and American physicist Frederick Albert Saunders (1875–1963) in 1925.

rusting The corrosion of iron in air to form an oxide. Both oxygen and moisture must be present for rusting to occur. The process is electrolytic, involving electrochemical reactions in which different areas of the wet iron serve as anode and cathode:

$$Fe(s) \rightarrow Fe^{2+}(aq) + 2e^{-}$$

$$2H_2Ol + O_2(aq) + 4e^{-} \rightarrow 4OH^{-}(aq)$$

Iron(II) hydroxide precipitates and is oxidized to the red hydrated iron(III) oxide $Fe_2O_3.H_2O$. Rusting is greatly speeded up by impurities in the iron and electrolytes in the water.

ruthenium A hard brittle silver-white transition metal, the second element of group 8 (formerly part of subgroup VIIIB) of the periodic table. It occurs naturally with platinum in the natural alloy osmiridium and in some sulfide ores. Its relatively high melting point and hardness makes it useful in hard heat-resistant alloys, such as those used for electrical contacts. It is also used as a catalyst, and as a pigment in decorative glasses and enamels. Ruthenium is also used in jewelry alloyed with palladium.

Symbol: Ru; m.p. 2310°C; b.p. 3900°C; r.d. 12.37 (20°C); p.n. 44; most common isotope ^{102}Ru; r.a.m. 101.07.

Rutherford, Ernest, Lord Rutherford (1871–1937) New Zealand-born physicist. Ernest Rutherford is generally regarded as one of the greatest experimental physicists ever. He conducted fundamental research on radioactivity and nuclear physics. In 1899–1900 he found that there are three types of radioactivity. He called the different types of radiation alpha rays, beta rays, and gamma rays. He showed that radioactivity is characterized by a half-life. Following a suggestion of Rutherford, Hans Geiger and Ernest Marsden performed an experiment involving the scattering of alpha particles which Rutherford interpreted in 1911 as showing that an atom consists of a nucleus at the center surrounded by electrons. In 1919 he showed that artificial disintegration of the nucleus could be induced by bombarding it with alpha particles. Rutherford also pioneered the use of radioactivity in dating rocks. He won the 1908 Nobel Prize for chemistry.

rutherfordium A radioactive synthetic transactinide metallic element, the first transactinide. Atoms of rutherfordium are produced by bombarding ^{249}Cf with ^{12}C or by bombarding ^{248}Cm with ^{18}O. Only a very few atoms have ever been created.

Symbol: Rf; m.p. 2100°C (est.); b.p. 5200°C (est.); r.d. 23 (est.); p.n. 104; most stable isotope ^{261}Rf (half-life 65 s).

rutile A naturally occurring reddish-brown to black form of titanium(IV) oxide, TiO_2. It is an ore of titanium and is also used in making ceramics and to coat welding rods. Clear varieties are used as a semiprecious gemstone.

Rydberg constant A constant that occurs in formula for the frequencies of spectral lines in atomic spectra. For the hydrogen atom it has the value $1.0968 \times 10^{7}m^{-1}$. The value of the Rydberg constant can be calculated from the BOHR THEORY of the hydrogen atom and from quantum mechanics. These calculations showed that:

$$R = M_0^2 me^4 c^3 / 8h^3$$

where M_0 is the magnetic constant, m and e are the mass and charge respectively of an electron, h is the Planck constant and c is the speed of light in a vacuum. The Rydberg constant is named for the Swedish physicist Johannes Robert Rydberg (1854–1919) for his work on atomic spectra. *See also* hydrogen atom spectrum.

S

sacrificial protection A method of protection against electrolytic corrosion (especially rusting). In protecting steel pipelines, for instance, zinc or magnesium rods are buried in the ground at points along the line and connected to the pipeline. Rusting of iron is an electrochemical process in which Fe^{2+} ions dissolve in the water in contact with the surface. A more electropositive element, such as zinc, protects against this because Zn^{2+} ions dissolve in preference. In other words, the zinc rod is 'sacrificed' for the steel pipeline. The same effect occurs with galvanized iron. If the zinc coating is scratched, the exposed iron does not rust until all the zinc has dissolved away. (Note that a tin coating, being less electropositive than iron, has the opposite effect.)

safety lamp (Davy lamp) A type of oil lamp for use in coal mines designed so that it will not ignite any methane (firedamp) present and cause an explosion. The flame is surrounded by a fine metal gauze which dissipates the heat from the flame but never reaches the ignition temperature of methane. If methane is present it burns inside the gauze; the lamp can be used as a test for pockets of methane.

sal ammoniac *See* ammonium chloride.

saline Containing a salt, especially an alkali-metal halide such as sodium chloride.

salt A compound with an acidic and a basic radical, or a compound formed by total or partial replacement of the hydrogen in an acid by a metal; the product of neutralizing an acid with a base. In general terms a salt is a material that has identifiable cationic and anionic components.

salt bridge An electrical contact between two half cells, used to prevent mixing. A glass U-tube filled with potassium chloride in agar is often used.

saltcake An impure form of SODIUM SULFATE, formerly produced industrially as a by-product of the LEBLANC PROCESS.

salt hydrate A metallic salt in which the metal ions are surrounded by a fixed number of water molecules. The water molecules are bound to the metal ions by ion–dipole interactions. In the solid state the water molecules form part of the crystal structure. It is thus a clathrate.

Salt hydrates that contain several molecules of water for each metal ion can lose them progressively (effloresce) if the pressure of water vapor is kept below the dissociation pressure of the system; eventually an anhydrous salt is formed. Alternatively, if the vapor pressure is continuously raised the system will add molecules of water (deliquesce) until a saturated solution forms.

salting out The addition of an ionic salt to a solution in order to precipitate a solute (or cause it to be evolved as a gas). The dissolved substance is more soluble in pure water than in the ionic solution. Certain colloids may also be precipitated in this way.

saltpeter *See* potassium nitrate.

sal volatile *See* ammonium carbonate.

samarium A hard brittle silver-gray element of the lanthanoid series of metals. It occurs in association with other lanthanoids in such minerals as allanite, bastnaesite, gadolinite, monazeite, and samarskite. Samarium is alloyed with cobalt to produce powerful permanent magnets. It is also used in alloys for nuclear reactor parts, as a catalyst in organic reactions, and as an ingredient in optical glass.

Symbol: Sm; m.p. 1077°C; b.p. 1791°C; r.d. 7.52 (20°C); p.n. 62; most common isotope ^{152}Sm; r.a.m. 150.36.

sand A mineral form of silicon(IV) oxide (silica, SiO_2) consisting chiefly of separate grains of quartz. It results from erosion of quartz-containing rocks; the grains may be rounded by the tumbling action of water or wind. Sand is used in concrete and mortar and in making glass and other ceramics.

sandstone A sedimentary rock consisting of consolidated sand grains cemented together with calcium carbonate, clay, or oxides of iron. It is used mainly as a building stone.

sandwich compound A type of organometallic compound formed between transition metal ions and aromatic compounds, in which the metal ion is 'sandwiched' between two rings. Four, five, six, seven, and eight-membered rings are known to form complexes with a number of elements including V, Cr, Mn, Co, Ni, and Fe. The bonding in such compounds is not between the metal ion and individual atoms on the ring; rather it involves bonding of d electrons of the ion with the pi-electron system of the ring as a whole. FERROCENE is the original example, having two parallel cyclopentadienyl rings sandwiching the iron ion. There are a number of related types of compound. Thus, *half sandwich compounds* (sometimes known as *piano-stool compounds*) have only one ring bound to the central ion. *Bent sandwich compounds* have two rings that are not parallel, and the ion is attached to other ligands. *Multidecker sandwich compounds* have more than two parallel rings. *See illustration.*

sapphire A valuable gemstone variety of the mineral corundum (aluminum oxide, Al_2O_3) colored blue, green, pink, violet, or other colors by impurities. Sapphires can also be made synthetically. They are used in jewelry and certain kinds of lasers.

saturated solution A solution that contains the maximum equilibrium amount of solute at a given temperature. A solution is saturated if it is in equilibrium with its solute. If a saturated solution of a solid is cooled slowly, the solid may stay temporarily in solution; i.e. the solution may contain more than the equilibrium amount of solute. Such solutions are said to be *supersaturated solutions*.

saturated vapor A vapor that is in equilibrium with its solid or liquid phase. A saturated vapor is at the maximum pressure (the saturated vapor pressure) at a given temperature. If the temperature of a saturated vapor is lowered, the vapor condenses. Under certain circumstances, the substance may stay temporarily in the vapor phase; i.e. the vapor contains more than the equilibrium concentration of the substance. The vapor is then said to be a *supersaturated vapor*.

s-block elements The elements of the first two groups of the periodic table; i.e. group 1 (H, Li, Na, K, Rb, Cs, Fr) and group 2 (Be, Mg, Ca, Sr, Ba, Ra). They are so called because their outer shells have the electronic configurations ns^1 or ns^2. The s-block excludes those with inner $(n - 1)$d levels occupied (i.e. it excludes transition elements, which also have s^2 and occasionally s^1 configurations).

scandium A light soft reactive silvery transition metal, the first member of group 3 (formerly IIIB) of the periodic table. It is found in minute amounts in over 800 minerals, especially thortveitite ($(Sc,Y)_2Si_2O_7$) and minerals often associated with lanthanoids. Scandium is used in high-intensity lights, in electronic devices, and in titanium carbide alloys. Scandium has several radioactive isotopes, some of which are used as tracers in medicine.

Symbol: Sc; m.p. 1541°C; b.p. 2831°C; r.d. 2.989 (0°C); p.n. 21; most common isotope ^{45}Sc; r.a.m. 44.955910.

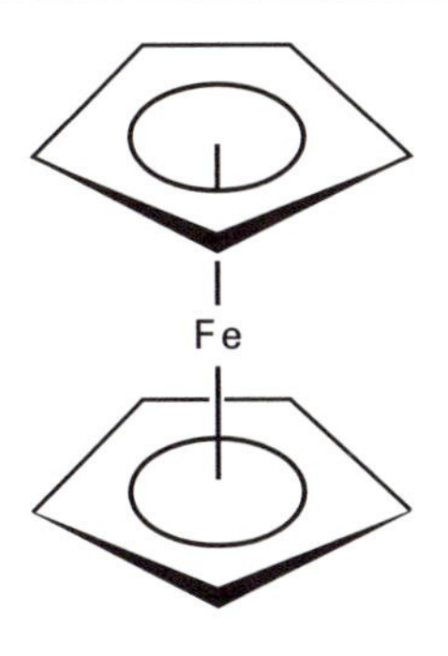

parallel sandwich

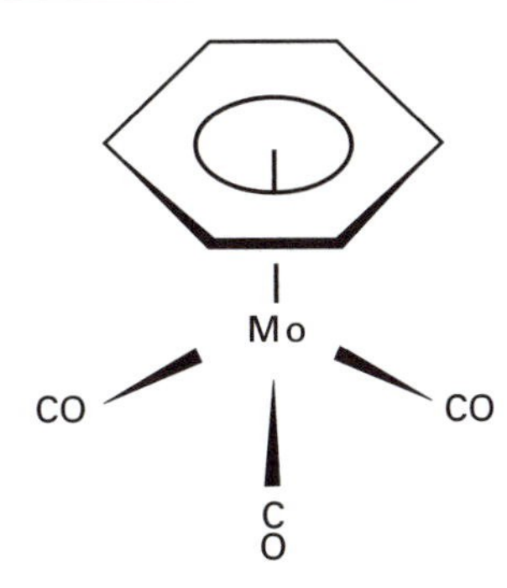

half sandwich (piano stool)

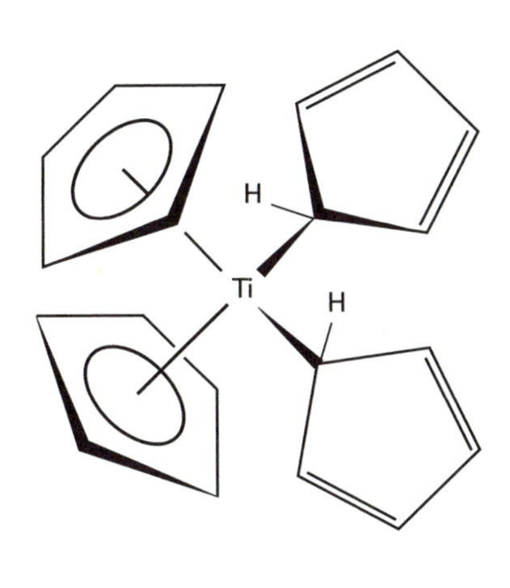

bent sandwich

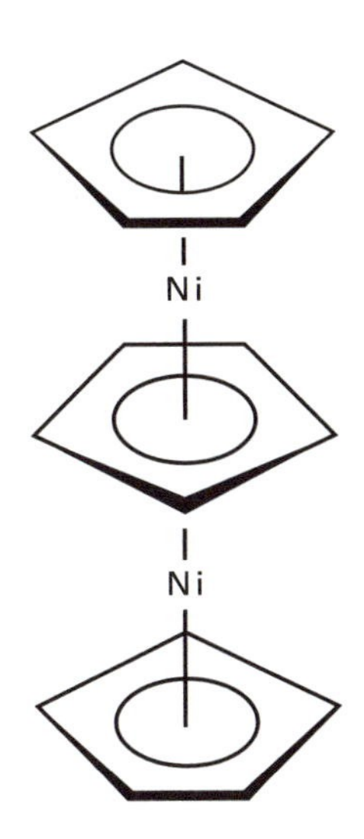

multidecker sandwich

Sandwich compound

scanning tunneling microscope (STM) A type of microscope that makes use of quantum-mechanical tunneling to investigate the atomic structure of a sample of material. In this technique a sharp conducting tip is brought near a surface. This causes electrons to tunnel from the surface to the tip, with the probability of this occurring depending on both the distance between the surface and the tip and the electron density at the surface. The tunneling produces an electric current, which is kept constant by moving the tip up or down as it is moved over the surface; a computer converts the movements into an image. Single atoms can be observed using STM.

Scheele, Karl Wilhelm (1742–86) Swedish chemist. Scheele was one of the greatest chemists of the 18th century. He discovered a large number of elements and compounds including oxygen (independently of Joseph PRIESTLEY), nitrogen, chlorine (although he did not recognize it as an element), citric acid, glycerol, hydrogen cyanide, hydrogen fluoride, hydrogen sulfide, lactic acid, and uric acid. He also discovered that light affects certain silver salts. He wrote a book entitled *A Chemical Treatise on Air and Fire* (1777) in which he described his experiments.

schönite *See* potassium sulfate.

Schottky defect *See* defect.

Schrödinger, Erwin (1887–1961) Austrian physicist. Schrödinger developed the branch of quantum mechanics known as wave mechanics. In this work Schrödinger formulated what is known as the *Schrödinger equation*. He solved his equation for simple systems such as the hydrogen atom. He also showed how wave mechanics and the matrix mechanics of Werner HEISENBERG are equivalent. Schrödinger shared the 1933 Nobel Prize for physics with Paul DIRAC for his work on quantum mechanics. Schrödinger wrote a book entitled *What is Life?* (1944) which was influential in stimulating interest in molecular biology.

Schrödinger equation The fundamental equation in the WAVE MECHANICS formulation of quantum mechanics. For calculating energy levels of electrons in atoms, molecules, and solids, the *time-independent Schrödinger equation* is used. This equation can be written:

$$\nabla^2\psi + 8\pi^2 m(E - U)\psi/h^2 = 0$$

where ∇^2 is an operator, U the potential energy, h the Plank constant, and ψ the wavefunction. E is the energy of the system. The operator ∇^2 is:

$$\partial/\partial x^2 + \partial/\partial y^2 + \partial/\partial z^2$$

scrubber The part of a chemical plant that removes impurities from a gas by passing it through a liquid.

seaborgium A radioactive synthetic transactinide metallic element created by bombarding ^{249}Cf with ^{18}O nuclei using a cyclotron. Six isotopes are known.

Symbol: Sg; m.p., b.p. and r.d. unknown; p.n. 106; most stable isotope ^{266}Sg (half-life 27.3s).

second Symbol: s The base unit of time in all metric systems, including the SI. In the SI it is defined as the duration of 9 192 631 770 periods of a particular wavelength of radiation corresponding to a transition between two hyperfine levels in the ground state of the cesium-133 atom, an unvarying physical constant. Earlier metric systems defined the second as the fraction 1/86400 of a mean solar day. Irregularities in the Earth's rotation, however, make the mean solar day less than constant.

secondary cell *See* accumulator.

second-order reaction A reaction in which the rate of reaction is proportional to the product of the concentrations of two of the reactants or to the square of the concentration of one of the reactants; i.e.

$$\text{rate} = k[\text{A}][\text{B}]$$

or

$$\text{rate} = k[\text{A}]^2$$

For example, the hydrolysis by dilute alkali of an ester is a second-order reaction:

$$\text{rate} = k[\text{ester}][\text{alkali}]$$

The rate constant for a second-order reaction has the units mol^{-1} dm^3 s^{-1}. Unlike a first-order reaction, the time for a definite fraction of the reactants to be consumed is dependent on the original concentrations.

sedimentation The settling of a suspension, either under gravity or in a centrifuge. The speed of sedimentation can be used to estimate the average size of the particles. This technique is used with an ultracentrifuge to find the relative molecular masses of macromolecules.

seed A small crystal added to a gas or liquid to assist solidification or precipitation of a substance from a solution. The seed, usually a crystal of the substance to be formed, enables particles to pack into predetermined positions so that a larger crystal can form.

selenium A metalloid or nonmetallic element; the third member of group 16 (formerly VIA) of the periodic table. It exists in several allotropic forms. It occurs in minute quantities in sulfide ores and is also recovered from anode sludges formed during electrolytic refining. The common gray metallic allotrope is both a semiconductor and very light-sensitive, and thus finds use in photocells, solar cells, and xerography. It is also used in some glasses and enamels, and in trace amounts is an essential dietary mineral. The red allotrope is unstable and reverts to the gray form under normal conditions.

Symbol: Se; m.p. 217°C (gray); b.p. 684.9°C (gray); r.d. 4.79 (gray); p.n. 34; most common isotope ^{80}Se; r.a.m. 78.96.

self-organization The order that can arise in a system that is far from thermodynamic equilibrium. The existence of self-organization does not contradict the second law of thermodynamics since self-organization occurs in systems that are open and through which there is an energy flow. Self-organization can be analyzed in terms of chaos theory and nonequilibrium statistical mechanics and thermodynamics.

Semenov, Nikolay Nikolaevich (1896–1986) Russian physical chemist. Semenov studied chemical chain reactions in the 1920s. He showed how such reactions could lead to combustion and violent explosions when branching occurs in the chain. He gave an account of his work in the influential book *Chemical Kinetics and Chain Reactions* (1934), the English translation of which was published in 1935. He shared the 1956 Nobel Prize for chemistry with Sir Cyril HINSHELWOOD.

semiconductor A crystalline material that conducts only under certain conditions. Its conductivity, unlike that of metals, increases with temperature because the highest occupied and lowest unoccupied energy levels are very close. A semiconductor can be formed by introducing small amounts of an impurity to an ultrapure material. If these impurities can withdraw electrons from the occupied energy level, leaving 'holes' that permit conduction, the result is known as a *p-type semiconductor*. Alternatively, these impurities may donate electrons that enter the unoccupied energy level, thus forming an *n-type semiconductor*. Semiconductors are used extensively by the electronics industry in such devices as integrated circuits.

semipermeable membrane A membrane that, when separating a solution from a pure solvent, permits the solvent molecules to pass through it but does not allow the transfer of solute molecules. Synthetic semipermeable membranes are generally supported on a porous material, such as unglazed porcelain or fine wire screens, and are commonly formed of cellulose or related materials. They are used in osmotic studies, gas separations, reverse osmosis water systems for homes and beverage industries, and in medical applications.

Equilibrium is reached at a semipermeable membrane if the chemical potentials on both sides become identical; migration of solvent molecules towards the solution is an attempt by the system to reach equilibrium. The pressure required to halt this migration is the OSMOTIC PRESSURE.

semipolar bond *See* coordinate bond.

septivalent *See* heptavalent.

sequestration The formation of a complex with an ion in solution, so that the ion does not have its normal activity. Sequestering agents are often chelating agents.

sesqui- Prefix indicating a 2/3 ratio in a compound. A sesquioxide, for example, would have the formula M_2O_3. A sesquicarbonate is a mixed carbonate and hydrogencarbonate of the type $Na_2CO_3.NaHCO_3.2H_2O$, which contains 2 CO_3 units and 3 sodium ions.

shell A group of electrons that share the

same principal quantum number n. Early work on x-ray emission studies used the terms K, L, M, and these are still sometimes used for the first three shells: $n = 1$, K-shell; $n = 2$, L-shell; $n = 3$, M-shell. *See also* atom.

sheradizing A technique used to obtain corrosion-resistant articles of iron or steel by heating with zinc dust in a sealed rotating drum for several hours at about 375°C. At this temperature the two metals combine, forming internal layers of zinc–iron and zinc–steel alloys and an external layer of pure zinc. Sheradizing is principally used for small parts, such as springs, washers, nuts, and bolts. It is named for its inventor, English metallurgist Sherard Cowper-Coles (1866–1936).

short period *See* period.

SI *See* SI units.

side reaction A chemical reaction that takes place to a limited extent at the same time as a main reaction. Thus the main product of a reaction may contain small amounts of other compounds.

Sidgwick, Nevil Vincent (1873–1952) British chemist. The first notable contribution by Sidgwick to chemistry was his book *The Organic Chemistry of Nitrogen* (1910). He subsequently became interested in the electronic structure of atoms, including work on coordination compounds and the hydrogen bond. He summarized his research in an influential book *The Electronic Theory of Valency* (1927). He spent the rest of his career trying to understand compounds in terms of valence theory. This culminated in his monumental two-volume book *The Chemical Elements and their Compounds* (1950).

siemens Symbol: S The SI derived unit of electrical conductance, equal to a conductance of one ampere per volt.1S = 1A v^{-1}. In earlier metric systems the equivalent of the siemens was known as the *mho*. It is named for German electrical engineer and inventor Ernst von Siemens (1816–92).

sievert Symbol: Sv The SI derived unit of absorbed dose equivalent in living tissue, equal to that produced by ionizing radiation which, when multiplied by certain dimensionless factors, gives a value of one joule per kilogram (1 J kg^{-1}). The dimensionless factors are used to modify the absorbed dose to take account of the fact that different types of radiation cause different biological effects. The sievert is named for the Swedish radiologist and physicist Rolf Sievert (1896–1966).

sigma bond *See* orbital.

sigma orbital *See* orbital.

silane Any of a small number of hydrides of silicon: SiH_4, Si_2H_6, Si_3H_8, etc. Silanes can be made by the action of acids on magnesium silicide (Mg_2Si). They are unstable compounds and ignite spontaneously in air. Only the first few members of the series are known and there is not the range of 'hydrosilicon' compounds to compare with the hydrocarbons.

silica *See* silicon(IV) oxide.

silica gel A gel made by coagulating sodium silicate sol. The gel is dried by heating and used as a catalyst support and as a drying agent. The silica gel used in desiccators and in packaging to remove moisture is often colored with a cobalt salt to indicate whether it is still active (blue = dry; pink = moist).

silicates A large number of compounds containing metal ions and complex silicon–oxygen combinations. The negative ions in silicates are of the form SiO_4^{4-}, $Si_2O_7^{6-}$, etc., and many are polymeric, containing SiO_4 units linked in long chains, sheets, or three-dimensional arrays. *Aluminosilicates* and *borosilicates* are similar materials containing aluminum or boron atoms in the structure. Silicates occur naturally in many rocks and minerals.

silicic acid A colloidal or jellylike hydrated form of silicon(IV) oxide (SiO_2), made by adding an acid (such as hy-

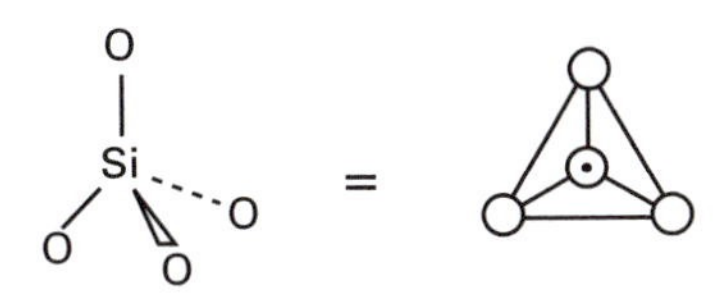

Si_4O^{4-} as in Be_2SiO_4 (phenacite)

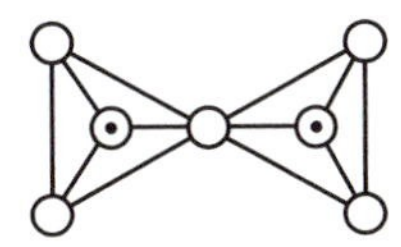

$Si_2O_5^{2-}$ as in $Sc_2Si_2O_7$ (thortveitite)

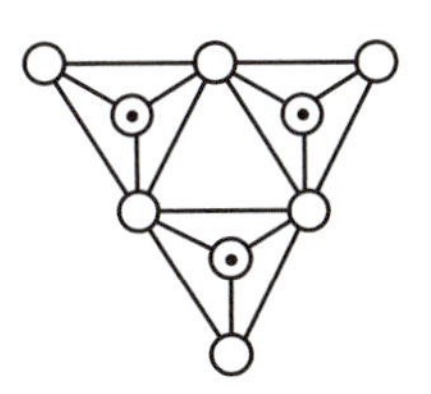

$Si_3O_9^{6-}$ as in $BaTiSi_3O_9$ (bentonite)

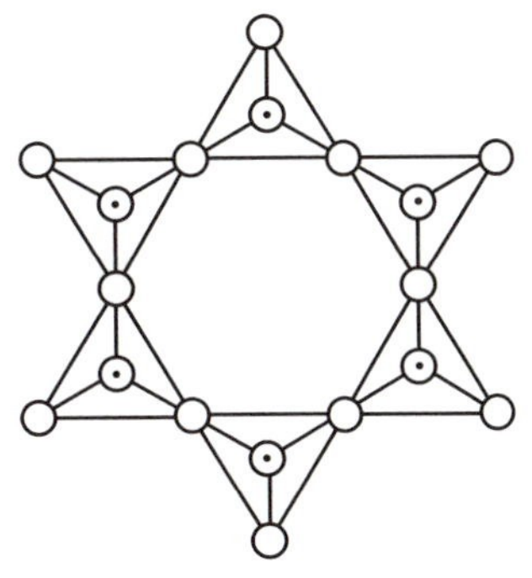

$Si_6O_{18}^{12-}$ as in $Be_3Al_2Si_6O_{18}$ (beryl)

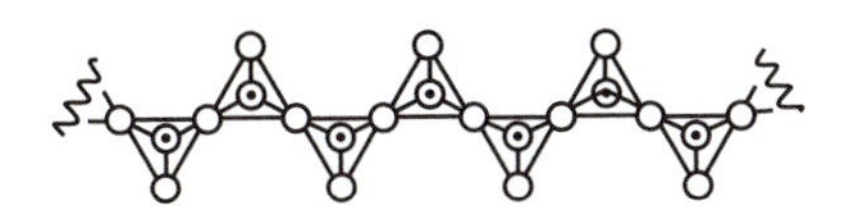

single chain : pyroxenes

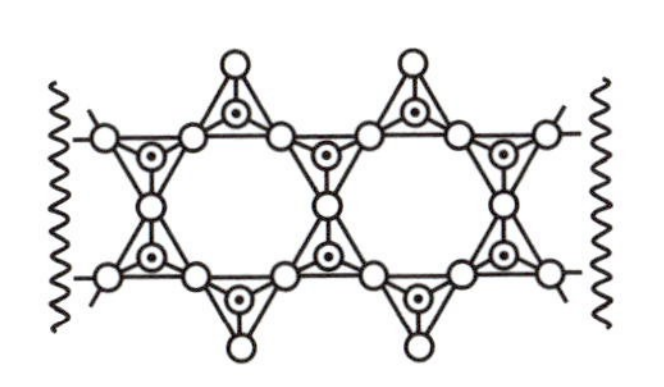

double chain : amphiboles

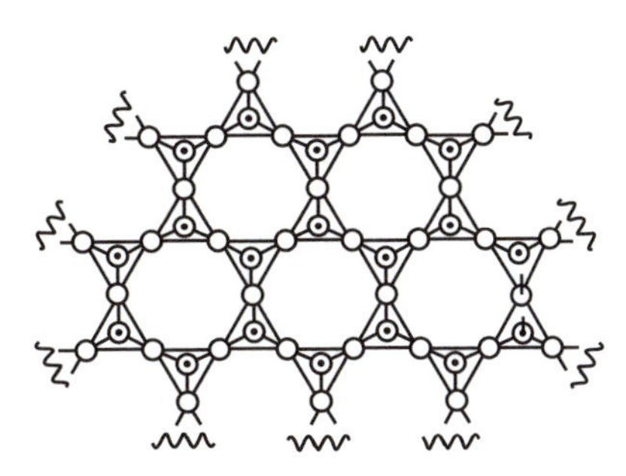

sheet : micas

Silicate: examples of silicate structures

drochloric acid) to a soluble SILICATE.

silicide A compound of silicon with a more electropositive element.

silicon A hard brittle gray semiconducting metalloid; the second element of group 14 (formerly IVA) of the periodic table. It has the electronic configuration of neon with four additional outer electrons; i.e., $[Ne]3s^23p^2$.

Silicon accounts for 27.7% of the mass of the Earth's crust and occurs in a wide variety of silicates with metals, and in clays, micas, and sand, which is largely silicon(IV) oxide (SiO_2). The element is obtained on a small scale by the reduction of SiO_2 by carbon or calcium carbide. It is employed in certain alloys, but its major use by far is in semiconducting electronic components and chips. For semiconductor applications very pure silicon is produced by direct reaction of silicon with an HCl/Cl_2 mixture to give silicon tetrachloride ($SiCl_4$), which can then be purified by fractional distillation. This compound is then decomposed on a hot wire in an atmosphere of hydrogen. For ultrapure samples zone refining is used. Silicon has two allotropes: the familiar gray one has a diamond lattice; the other is an amorphous brownish powder.

Silicon does not react with hydrogen except under extreme conditions in the presence of chlorine, in which case compounds such as trichlorosilane ($HSiCl_3$) are obtained. The hydride SiH_4 itself may be prepared in the laboratory by hydrolysis of magnesium silicide, in which case a number of other hydrides up to Si_6H_{14} are also obtained, or more conveniently by reduction of silicon tetrachloride with lithium tetrahydridoaluminate:

$$SiCl_4 + LiAlH_4 \rightarrow SiH_4 + LiCl + AlCl_3$$

The silanes are mildly explosive.

Silicon combines with oxygen when heated in air to form silicon(IV) oxide (SiO_2). The mineral silicates constitute a vast collection of materials with a wide range of structures including chains, rings, lattices, twisted chains, sheets, and varyingly cross-linked structures all based on different assemblies of $[SiO_4]^{2-}$ tetrahedra. Being a metalloid, silicon is somewhat amphoteric and consequently silica is weakly acidic, dissolving in fused alkalis to form the appropriate silicate.

Symbol: Si; m.p. 1410°C; b.p. 2355°C; r.d. 2.329 (20°C); p.n. 14; most common isotope ^{28}Si; r.a.m. 28.0855

silicon bronze *See* bronze.

silicon carbide (carborundum; SiC) A black very hard crystalline solid made by heating silicon(IV) oxide with carbon in an electric furnace. Sand and coke may be employed for lower grades. It is used as an abrasive and as a refractory material.

silicon(IV) chloride (silicon tetrachloride; $SiCl_4$) A colorless fuming liquid which rapidly hydrolyzes in moist air or water. It is used to produce pure silica for making special kinds of glass.

silicon dioxide *See* silicon(IV) oxide.

silicones Polymeric synthetic silicon compounds containing chains of alternating silicon and oxygen atoms, with organic groups bound to the silicon atoms. Silicones are used as lubricants and water repellants and in waxes and varnishes. Silicone rubbers are superior to natural rubbers in their resistance to both high and low temperatures and to chemicals. *See also* siloxanes.

silicon(IV) oxide (silicon dioxide; silica; SiO_2) A hard crystalline compound occurring naturally chiefly in three crystalline forms: *quartz* (hexagonal), *tridymite* (rhombic), and *crystobalite* (tetragonal or cubic). Sand is mostly fine particles of quartz. Fused silica is a glassy substance used in laboratory apparatus. Silicon(IV) oxide is used in the manufacture of glass, in refractory materials, and in metallurgical furnaces.

silicon tetrachloride *See* silicon(IV) chloride.

siloxanes Compounds containing Si–O–Si groups with organic groups bound to the silicon atoms. The silicones are polymers of siloxanes.

silver A soft ductile malleable brilliant white precious transition metal, the second element of group 11 (formerly IB) of the periodic table. Silver occurs native and in such minerals as argentine (silver sulfide, Ag_2S) and chlorargyrite (AgCl). It is also found in gold, copper, and lead ores, from which it is extracted as a by-product during refining. Silver is not as reactive as copper, but is more reactive than gold. It darkens in air due to the formation of silver sulfide. It is used in jewelry, dental amalgams, electrical conductors, and batteries. Until recently it was also used in coinage alloys, tableware, and as a reflective surface for mirrors. Silver compounds are extensively used in photography.

Symbol: Ag; m.p. 961.93°C; b.p. 2212°C; r.d. 10.5 (20°C); p.n. 47; most common isotope ^{107}Ag; r.a.m. 107.8682.

silver(I) bromide (AgBr) A pale yellow precipitate obtained by adding a soluble bromide solution to a solution of silver(I) nitrate. The halide dissolves in concentrated ammonia solution but not in dilute ammonia solution. It is used extensively in photography for making light-sensitive emulsions. Silver(I) bromide crystallizes in a similar manner to sodium(I) chloride. Unlike silver(I) chloride and silver(I) iodide, silver(I) bromide does not absorb ammonia gas.

silver(I) chloride (AgCl) A compound prepared as a curdy white precipitate by the addition of a soluble chloride solution to a solution of silver(I) nitrate. It occurs in nature as the mineral *chlorargyrite*. Silver(I) chloride is insoluble in water but dissolves in concentrated hydrochloric acid and in concentrated ammonia solution. It is rather unreactive but is affected by sunlight, and is thus used extensively in photography to make light-sensitive emulsions. Silver(I) chloride crystallizes in the sodium chloride lattice pattern.

silver(I) iodide (AgI) A pale yellow precipitate prepared by the addition of a soluble iodide solution to a solution of silver(I) nitrate. Silver(I) iodide occurs in nature as *iodargyrite* in the form of hexagonal crystals. It is virtually insoluble in ammonia solution, and is used in the photographic industry. Silver(I) iodide will absorb ammonia gas. It is trimorphic, contracting when heated and expanding when cooled.

silver(I) nitrate ($AgNO_3$) A white crystalline poisonous solid prepared by dissolving silver in dilute nitric acid and crystallizing the solution. Silver(I) nitrate is dimorphous, crystallizing in the orthorhombic and hexagonal systems. It undergoes thermal decomposition to yield silver, dinitrogen oxide, and oxygen. Probably the most important silver salt, it is used in the medical industry as an antiseptic and for the treatment of warts, and in the photographic industry to make light-sensitive emulsions. In the laboratory, it is used both as an analytical and volumetric reagent.

silver(I) oxide (argentous oxide; Ag_2O) A brown amorphous solid formed by the addition of sodium or potassium hydroxide solution to a solution of silver(I) nitrate. Silver(I) oxide turns moist red litmus blue and decomposes (at 160°C) to give silver and oxygen. Silver(I) oxide dissolves in concentrated ammonia solution, forming the complex ion $[Ag(NH_3)_2]^+$. It is used in organic chemistry.

silver(II) oxide (argentic oxide; AgO) A black diamagnetic substance prepared by the oxidation of either silver or silver(I) oxide using ozone. Alternatively it can be prepared as a precipitate by the addition of a solution of potassium persulfate ($K_2S_2O_8$) to one of silver(I) nitrate.

simple cubic crystal *See* cubic crystal.

single bond A covalent bond between two elements that involves one pair of electrons only. It is represented by a single line, for example H–Br, and is usually a *sigma bond*, although it can be a *pi* bond. *Compare* multiple bond. *See also* orbital.

sintering A process in which certain

powdered substances (e.g. metals, ceramics) coagulate into a single mass when heated to a temperature below the substance's melting point. Sintered glass is a porous material used for laboratory filtration.

SI units (SI; *S*ystème *I*nternational d'Unités) The modern coherent rationalized internationally adopted metric system of units. It has seven BASE UNITS and two dimensionless units, formerly called supplementary units. DERIVED UNITS are formed by multiplication and/or division of base units. Standard prefixes are used for decimal multiples and submultiples of SI units, along with standard symbols for both units and prefixes.

slag (basic slag) Glasslike compounds of comparatively low melting point formed during the extraction of metals when the impurities in an ore react with a flux. The fact that the slag is a liquid of comparatively low density means that it can be separated from the liquid metal on which it floats. In a blast furnace, the flux used is limestone ($CaCO_3$) and the main impurity is silica (SiO_2). The slag formed is mainly calcium silicate:

$$CaCO_3 + SiO_2 \rightarrow CaSiO_3$$

It is used as railroad ballast and in fertilizers and concrete.

slaked lime *See* calcium hydroxide.

slaking *See* calcium hydroxide.

slip planes The planes of weakness in a crystal, e.g. the boundaries between the octahedral layers in graphite.

slurry A thin paste of suspended solid particles in a liquid.

Smalley, Richard Errett (1943–) American chemist. In 1981 Smalley devised a procedure to produce microclusters of a hundred or so atoms, by vaporizing a metal using a laser. The released atoms are cooled by a jet of helium and condense into variously sized clusters. In 1985 a visiting British chemist, Harold KROTO, persuaded Smalley to direct his laser beam at a graphite target, leading to the discovery of a new alloprope of carbon, now known as BUCKMINSTERFULLERENE.

smelting An industrial process for extracting metals from their ores at high temperatures. Generally the ore is reduced with carbon (for zinc and tin) or carbon monoxide (for iron). Copper and lead are obtained by reduction of the oxide with the sulfide; for example:

$$2Cu_2O + Cu_2S \rightarrow 6Cu + SO_2$$

A flux is also used to combine with impurities and form a SLAG on top of the molten metal.

smithsonite *See* calamine.

soap *See* detergents.

soapstone A soft type of TALC which has a greasy feel and which is easy to carve to make ornaments. It was formerly known as steatite.

soda *See* sodium carbonate; sodium hydroxide.

soda ash *See* sodium carbonate.

soda glass *See* glass.

soda lime A gray solid produced by adding sodium hydroxide solution to calcium oxide, to give a mixture of $Ca(OH)_2$ and NaOH on evaporation. It is used in the laboratory as a drying agent and as an absorbent for carbon dioxide.

sodamide ($NaNH_2$) A white ionic solid formed by passing dry ammonia over sodium at 300–400°C. The compound reacts with water to give sodium hydroxide and ammonia. It reacts with red-hot carbon to form sodium cyanide and with nitrogen(I) oxide to form sodium azide. Sodamide is used in the manufacture of sodium hydroxide and in the explosives industry.

soda water A solution of carbon dioxide dissolved in water under pressure, used

BASE AND DIMENSIONLESS SI UNITS

Physical quantity	*Name of SI unit*	*Symbol for SI unit*
length	meter	m
mass	kilogram(me)	kg
time	second	s
electric current	ampere	A
thermodynamic temperature	kelvin	K
luminous intensity	candela	cd
amount of substance	mole	mol
*plane angle	radian	rad
*solid angle	steradian	sr

*supplementary units

DERIVED SI UNITS WITH SPECIAL NAMES

Physical quantity	*Name of SI unit*	*Symbol for SI unit*
frequency	hertz	Hz
energy	joule	J
force	newton	N
power	watt	W
pressure	pascal	Pa
electric charge	coulomb	C
electric potential difference	volt	V
electric resistance	ohm	Ω
electric conductance	siemens	S
electric capacitance	farad	F
magnetic flux	weber	Wb
inductance	henry	H
magnetic flux density	tesla	T
luminous flux	lumen	lm
illuminance (illumination)	lux	lx
absorbed dose	gray	Gy
activity	becquerel	Bq
dose equivalent	sievert	Sv

DECIMAL MULTIPLES AND SUBMULTIPLES USED WITH SI UNITS

Submultiple	*Prefix*	*Symbol*	*Multiple*	*Prefix*	*Symbol*
10^{-1}	deci-	d	10^{1}	deca-	da
10^{-2}	centi-	c	10^{2}	hecto-	h
10^{-3}	milli-	m	10^{3}	kilo-	k
10^{-6}	micro-	μ	10^{6}	mega-	M
10^{-9}	nano-	n	10^{9}	giga-	G
10^{-12}	pico-	p	10^{12}	tera-	T
10^{-15}	femto-	f	10^{15}	peta-	P
10^{-18}	atto-	a	10^{18}	exa-	E
10^{-21}	zepto-	z	10^{21}	zetta-	Z
10^{-24}	yocto-	y	10^{24}	yotta-	Y

in fizzy drinks. When the pressure is released, the liquid effervesces with bubbles of carbon dioxide gas. *See also* carbonic acid.

sodium A soft silver-white reactive alkali metal; the second element of group 1 (formerly IA) of the periodic table. It has the electronic configuration of neon plus one additional outer 3s electron. The element undergoes electronic excitation in flames and in sodium-vapor lamps to give a distinctive yellow color arising from intense emission at the so called 'sodium-D' line pair. The element's ionization potential is rather low and the sodium atom thus readily loses its electron (i.e. the metal is strongly reducing). The chemistry of sodium is largely the chemistry of the monovalent Na^+ ion.

Sodium occurs widely as NaCl in seawater and as deposits of halite in dried-up lakes etc. (2.6% of the Earth's crust). The element is obtained commercially via the Downs process by electrolysis of NaCl melts in which the melting point is reduced by the addition of calcium chloride; sodium is produced at the steel cathode. The metal is extremely reactive, vigorously so with the halogens and also with water, in the latter case to give hydrogen and sodium hydroxide. It is used as a coolant in fast-breeder nuclear reactors. The chemistry of sodium is very similar to that of the other members of group 1.

Nearly all sodium compounds are soluble in water. Sodium hydroxide is produced commercially by the electrolysis of brine using diaphragm cells or mercury-cathode cells; chlorine is a coproduct.

Solutions of sodium metal in liquid ammonia are blue and have high electrical conductivities; the main current carrier of such solutions is the solvated electron. Such solutions are used in both organic and inorganic chemistry as efficient reducing agents.
The metal itself has a body-centered crystal structure.

Symbol: Na; m.p. 97.81°C; b.p. 883°C; r.d. 0.971 (20°C); p.n. 11; most common isotope ^{23}Na; r.a.m. 22.989768.

sodium acetate *See* sodium ethanoate.

sodium aluminate ($NaAlO_2$) A white solid produced in the laboratory by adding excess aluminum to a hot concentrated solution of sodium hydroxide. Once the reaction has been initiated, sufficient heat is liberated to keep the reaction going; hydrogen is also produced. In solution the aluminate ions have the structure $Al(OH)_4^-$ and $NaAl(OH)_4$ is known as sodium(I) tetrahydroxoaluminate(III). Addition of sodium hydroxide to an aluminum salt solution produces a white gelatinous precipitate of aluminum hydroxide, which dissolves in excess alkali to give a solution of sodium(I) tetrahydroxoaluminate(III). Sodium aluminate is used in cleaning products, in the treatment of waste effluent, and in zeolite production. It is also used as a mordant.

sodium azide (NaN_3) A white solid prepared by passing nitrogen(I) oxide over heated sodamide. On heating, sodium azide decomposes to give sodium and nitrogen. It is used as a reagent in organic chemistry and in the manufacture of explosives (detonators).

sodium bicarbonate *See* sodium hydrogencarbonate.

sodium bisulfate *See* sodium hydrogensulfate.

sodium bisulfite *See* sodium hydrogensulfite.

sodium borate *See* disodium tetraborate decahydrate.

sodium bromide (NaBr) A white solid formed by the action of bromine on hot sodium hydroxide solution. Alternatively, it can be made by the action of dilute hydrobromic acid on sodium carbonate or hydroxide. Sodium bromide is soluble in water and forms crystals similar in shape to sodium chloride. It is used in medicine, analytical chemistry, and photography. Sodium bromide forms a hydrate, $NaBr.2H_2O$.

sodium carbonate (soda ash; Na_2CO_3) A white amorphous powder, which aggregates on exposure to air owing to the formation of hydrates. Industrially it is prepared by the SOLVAY PROCESS. On crystallizing from aqueous solution large translucent crystals of the decahydrate ($Na_2CO_3.10H_2O$) (*washing soda*) are formed. These effloresce quickly in air, forming the monohydrate $Na_2CO_3.H_2O$. Large quantities of sodium carbonate are used in the manufacture of sodium hydroxide. Sodium carbonate is also used in photography, as a food additive, and in the textile trade. Washing soda is used as a domestic cleanser. The salt produces an alkaline solution in water by hydrolysis:

$$Na_2CO_3 + 2H_2O \rightarrow 2NaOH + H_2CO_3$$

It is used in volumetric analysis to standardize strong acids.

sodium chlorate(V) ($NaClO_3$) A white solid formed by the action of chlorine on hot concentrated sodium hydroxide solution, or by the electrolysis of a concentrated sodium chloride solution. It is soluble in water. Sodium chlorate is a powerful oxidizing agent. If heated it yields oxygen and sodium chloride. The chlorate is used in explosives, in matches, as a weed killer, and in the textile industry.

sodium chloride (common salt; salt; NaCl) A white solid prepared by the neutralization of hydrochloric acid with aqueous sodium hydroxide. It occurs in sea water and natural brines. Natural solid deposits of *rock salt* (halite) are also found in certain places. It is an essential dietary mineral. The compound dissolves in water with the absorption of heat, its solubility changing very little with temperature. It is used to season and preserve food. Industrially, it is used as a raw material in the manufacture of sodium carbonate via the SOLVAY PROCESS, sodium hydroxide via electrolysis, and soap. Sodium chloride is almost insoluble in alcohol.

sodium-chloride structure A form of crystal structure that consists of a face-centered cubic arrangement of sodium ions with chloride ions situated at the middle of

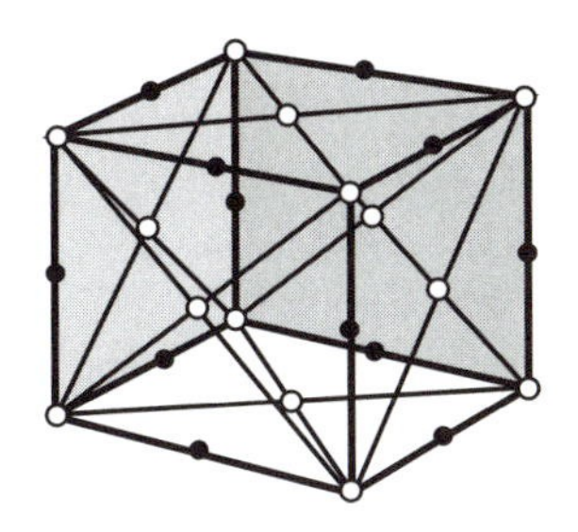

Sodium-chloride structure

each edge and in the center of the cube. Electrostatic attraction holds the oppositely charged ions together.

The lattice can also be thought of as two interpenetrating face-centered cubes, one composed of sodium ions and the other of chloride ions. Other compounds (e.g. sodium bromide and potassium chloride) having their ions arranged in the same positions are also described as having the sodium-chloride structure.

sodium cyanide (NaCN) An extremely poisonous white solid formed either by the action of carbon on sodamide at high temperature or by passing ammonia over sodium at 300–400°C to form sodamide and then reacting it with carbon. The industrial salt is prepared by absorbing hydrogen cyanide into a sodium hydroxide solution and purifying the solid by recrystallization from liquid ammonia. It is used as a source of cyanide and hydrocyanic acid. In the cyanide process it is used in the extraction of silver and gold. Its aqueous solutions are alkaline due to salt hydrolysis.

sodium dichromate(VI) ($Na_2Cr_2O_7$) An orange deliquescent solid that is very soluble in water. It is made from finely powdered chromite, which is heated with calcium oxide and sodium carbonate in a furnace. If an alkali is added to a solution of the dichromate, the chromate is produced. At high temperatures, sodium dichromate decomposes to give the chromate, chromic oxide, and oxygen. It is used as an oxidizing agent, particularly in organic chemistry, and in volumetric analysis to estimate iron(II) ions and iodide ions. It is also used as a mordant.

sodium dihydrogen orthophosphate *See* sodium dihydrogen phosphate(V).

sodium dihydrogen phosphate(V) (sodium dihydrogen orthophosphate; NaH_2PO_4) A white solid prepared by titrating phosphoric acid with sodium hydroxide solution using methyl orange as the indicator. On evaporation white crystals of the monohydrate are formed. It is used in some baking powders. The monohydrate crystallizes in the orthorhombic system.

sodium dioxide *See* sodium superoxide.

sodium ethanoate (sodium acetate; CH_3COONa) A white solid prepared by the neutralization of ethanoic acid with sodium carbonate or sodium hydroxide. Sodium ethanoate reacts with sulfuric acid to form sodium hydrogensulfate and ethanoic acid; with sodium hydroxide it gives rise to sodium carbonate and methane. Sodium ethanoate is used in the dyeing industry.

sodium fluoride (NaF) A toxic white solid formed by the action of hydrofluoric acid on sodium carbonate or sodium hydroxide (the reaction between metallic sodium and fluorine is too violent). Sodium fluoride is soluble in water and reacts with concentrated sulfuric acid to yield hydrogen fluoride. It is used as a constituent of ceramic enamels, as an antiseptic, as an agent to prevent fermentation, and in minute amounts to fluoridate water supplies.

sodium hexafluoroaluminate (Na_3AlF_6) A compound found in nature as the mineral *cryolite*. It is usually white or colorless, but may be reddish or brown because of impurities. It is used to make enamels, opaque glass, and glazes for ceramics. It crystallizes in the monoclinic system but in forms that closely resemble cubes and isometric octahedrals. Sodium hexafluoroaluminate is also used as a flux in the manufacture of aluminum.

sodium hydride (NaH) A white crystalline solid prepared by passing a stream of pure dry hydrogen over sodium at 350°C; the sodium is usually suspended in an inert medium. Electrolysis of fused sodium hydride yields hydrogen at the anode, suggesting the presence of the H^- ion. Sodium hydride reacts vigorously with water to form sodium hydroxide solution and hydrogen. It is used as a powerful reducing agent to convert water to hydrogen, concentrated sulfuric acid to hydrogen sulfide, and iron(III) oxide to iron. Sodium hydride bursts into flames spontaneously on contact with the halogens at room temperature. It dissolves in liquid ammonia to give sodamide, $NaNH_2$.

sodium hydrogencarbonate (sodium bicarbonate; baking soda; $NaHCO_3$) A white solid formed either by passing an excess of carbon dioxide through sodium carbonate or hydroxide solution, or by precipitation when cold concentrated solutions of sodium chloride and ammonium hydrogencarbonate are mixed. Sodium hydrogencarbonate decomposes on heating to give sodium carbonate, carbon dioxide, and water. With dilute acids, it yields carbon dioxide. It is used as an antacid and as a constituent of baking powder, and in effervescent beverages and fire extinguishers. It is also used in the textile, ceramics, and paper-making industries. Its aqueous solutions are alkaline as a result of salt hydrolysis. Sodium hydrogencarbonate forms monoclinic crystals.

sodium hydrogensulfate (sodium bisulfate; $NaHSO_4$) A white solid formed either by the partial neutralization of sulfuric acid with sodium hydroxide solution, by the action of concentrated sulfuric acid on sodium nitrate, or by heating equimolar quantities of sodium chloride and concentrated sulfuric acid. In aqueous solution sodium hydrogensulfate is strongly acidic. It crystallizes as the monohydrate

($NaHSO_4.H_2O$), which dehydrates on warming. If heated strongly the pyrosulfate ($Na_2S_2O_7$) is formed, which decomposes to give the sulfate and sulfur(VI) oxide. Sodium hydrogensulfate is used as a cheap source of sulfuric acid and in the dyeing industry.

sodium hydrogensulfite (sodium bisulfite; $NaHSO_3$) A white powder prepared by saturating a solution of sodium carbonate with sulfur(IV) oxide. It is isolated from the aqueous solution by precipitation with alcohol. If heated it undergoes thermal decomposition to give sodium sulfate, sulfur(VI) oxide, and sulfur. Sodium hydrogensulfite is used to sterilize equipment used in the brewing and winemaking industries, and also in medicine as an antiseptic.

sodium hydroxide (caustic soda; NaOH) A white deliquescent slightly translucent solid. In solution it is a strong alkali and electrolyte. In the laboratory it can be prepared by reacting sodium, sodium monoxide, or sodium peroxide with water. Industrially it can be prepared by the electrolysis of sodium chloride using a cell having a mercury cathode or a cell in which the anode and cathode are separated by a diaphragm. Sodium hydroxide dissolves readily in water with the evolution of heat. Its solutions have a soapy feel and are very corrosive. It is used to absorb acidic gases, such as carbon dioxide and sulfur(IV) oxide, and to remove heavy metals from sewage and industrial effluents. Industrially it is also used in soap and paper manufacture and in the purification of bauxite.

sodium iodide (NaI) A white solid formed by reacting sodium carbonate or sodium hydroxide with hydroiodic acid; the solution is then evaporated. Sodium iodide forms colorless crystals having a cubic shape. It is used as a source of iodine and in medicine and photography. Acidified solutions of sodium iodide exhibit reducing properties owing to the formation of hydroiodic acid.

sodium metasilicate *See* sodium silicate.

sodium monoxide (Na_2O) A whitish deliquescent solid formed by burning sodium in a deficiency of oxygen or alternatively by reducing sodium peroxide or sodium hydroxide with the requisite amount of sodium. It reacts violently with water to form sodium hydroxide and with acids to form solutions of their salts. Sodium monoxide forms cubic crystals. It dissolves in liquid ammonia to give a mixture of sodamide and sodium hydroxide.

sodium nitrate (Chile saltpeter; $NaNO_3$) A white solid formed by the neutralization of nitric acid with either sodium carbonate or sodium hydroxide. It occurs naturally in large quantities in South America in impure mineral deposits called *caliche*. Sodium nitrate is very soluble in water and on crystallization forms colorless deliquescent crystals. On heating it decomposes to give sodium nitrite and oxygen. When heated with concentrated sulfuric acid, nitric acid is produced. Sodium nitrate is used as a fertilizer and as a source of nitrates and nitric acid. The salt crystallizes in the rhombohedral system and is isomorphous with the iodate.

sodium nitrite ($NaNO_2$) A yellowish-white solid formed by the thermal decomposition of sodium nitrate. It is readily soluble in water. The anhydrous salt forms orthorhombic crystals. When treated with cold dilute hydrochloric acid, sodium nitrite forms nitrous acid. It is used in organic chemistry and industrially as a corrosion inhibitor.

sodium orthophosphate *See* trisodium phosphate(V).

sodium peroxide (Na_2O_2) A yellowish-white ionic solid formed by the direct combination of burning sodium with excess oxygen. It reacts with water to form sodium hydroxide and hydrogen peroxide; the latter decomposes rapidly in the alkaline solution to give oxygen. Sodium peroxide is used as a bleaching agent and an oxidizing agent for such materials as wool and wood pulp. Sodium peroxide can convert nitrogen(II) oxide to sodium nitrate and iodine to sodium iodate.

sodium silicate (sodium metasilicate; $Na_2SiO_3.5H_2O$) A colorless crystalline solid used in detergents and other cleaning compounds, in refractories, and in cements for ceramics. A concentrated aqueous solution is known as *water glass* and is used as a sizing agent and for making precipated silica and silica gel.

sodium sulfate (Na_2SO_4) A white soluble solid often formed in the laboratory by heating a mixture of sodium chloride and concentrated sulfuric acid. Industrially it is more frequently prepared by reacting magnesium sulfate with sodium chloride. It forms two hydrates, the decahydrate ($Na_2SO_4.10H_2O$) and the heptahydrate ($Na_2SO_4.7H_2O$). Sodium decahydrate, found in nature as the mineral *mirabilite*, also known as *Glauber's salt*, is used in the manufacture of glass and as a purgative in medicine. It effloresces on exposure to air to form the anhydrous salt, which is used as a drying agent in organic chemistry. At room temperature sodium sulfate crystallizes in the orthorhombic system; around 250°C there is a transition to the hexagonal system. The decahydrate crystallizes in the monoclinic form.

sodium sulfide (Na_2S) A yellow-red solid formed by the reduction of sodium sulfate using carbon (or carbon monoxide or hydrogen) at a high temperature. It is a corrosive material and deliquesces to release hydrogen sulfide. Sodium sulfide forms hydrates containing 4½, 5, and 9 H_2O. In aqueous solution it is readily hydrolyzed, the solution being strongly alkaline. Its reducing properties make it useful in metallurgy and in the production of dyestuffs and wood pulp.

sodium sulfite (Na_2SO_3) A white solid formed by reacting the exact amount of sulfur(IV) oxide with either sodium carbonate or sodium hydroxide. It is readily soluble in water and crystallizes as colorless crystals of the heptahydrate ($Na_2SO_3.7H_2O$). When treated with dilute mineral acids, sulfur(IV) oxide is evolved. At high temperatures, sodium sulfite undergoes thermal decomposition to give sodium sulfate and sodium sulfide. It is used as a bleaching agent for textiles and paper, and as an antioxidant in some canned goods.

sodium superoxide (sodium dioxide; NaO_2) A pale yellow solid obtained by heating sodium peroxide in oxygen at 490°C. It reacts with water to give hydrogen peroxide, sodium hydroxide solution, and oxygen. Commercial sodium peroxide contains 10% sodium superoxide.

sodium thiosulfate(IV) (hypo; $Na_2S_2O_3$) A white solid prepared either by boiling sodium sulfite with flowers of sulfur or by passing sulfur(IV) oxide into a suspension of sulfur in boiling sodium hydroxide. Sodium thiosulfate is readily soluble in water and crystallizes as large colorless crystals of the pentahydrate ($Na_2S_2O_3.5H_2O$). It reacts with dilute acids to give sulfur and sulfur(IV) oxide. It is used in photography as the fixative 'hypo' (it was formerly known as sodium hyposulfate) and industrially as an antichlor. In volumetric analysis, solutions of sodium thiosulfate are usually prepared from the pentahydrate. On heating it disproportionates to give sodium sulfate and sodium sulfide.

soft iron Iron that has a low carbon content and is unable to form permanent magnets.

soft solder *See* solder.

soft water *See* hardness.

sol A COLLOID consisting of solid particles distributed in a liquid medium. A wide variety of sols are known; the colors often depend markedly on the particle size. The term *aerosol* is used for solid or liquid phases dispersed in a gaseous medium.

solder An alloy used in joining metals.

The molten solder wets the surfaces to be joined, without melting them, and solidifies on cooling to form a hard joint. The surfaces must be clean and free of oxide, a result usually achieved with the aid of a *flux*. *Soft solders* consist of lead with up to 60% tin and melt in the range 183–250°C. Soft-soldering is used, for example, in plumbing and making electrical joints. *Brazing solders* are copper–zinc alloys that have higher melting points and produce stronger joints than soft solders; silver can be added to produce *silver solders*.

solid The state of matter in which the particles occupy fixed positions, giving the substance a definite shape. The particles are held in these positions by bonds. Three kinds of attraction fix the positions of the particles: electrovalent, covalent, and intermolecular. Since these bonds act over short distances the particles in solids are packed closely together. The strengths of these three types of bonds are different and so, therefore, are the mechanical properties of different solids.

solid solution A solid composed of two or more substances mixed together at the molecular level. Atoms, ions, or molecules of one component in the crystal are at lattice positions normally occupied by the other component. Certain alloys are solid SOLUTIONS of one metal in another. Isomorphic salts can also sometimes form solid solutions, as in the case of crystalline alums.

solubility Symbol: S The amount of one substance that can dissolve in another to form a saturated solution under specified conditions of temperature and pressure. Solubilities are stated as moles of solute per kilogram of solvent (*molality*), or as kilograms of solute per cubic meter of solvent (*density*). The solubility of a solid in a liquid generally increases with temperature, whereas that of a gas in a liquid generally decreases. An increase in pressure on a gas above a liquid leads to a proportional increase in the solubility of the gas. *See also* concentration.

solubility product Symbol: K_s If an ionic solid is in contact with its saturated solution, there is a dynamic equilibrium between solid and solution:

$$AB(s) \rightleftharpoons A^+(aq) + B^-(aq)$$

The equilibrium constant for this is given by

$$[A^+][B^-]/[AB]$$

The concentration of undissolved solid [AB] is also constant, so

$$K_s = [A^+][B^-]$$

K_s is the solubility product of the salt (at a given temperature). For a salt A_2B_3, for instance:

$$K_s = [A^+]^2[B^-]^3, \text{ etc.}$$

Solubility products are meaningful only for sparingly soluble salts. If the product of ions exceeds the solubility product, precipitation occurs.

solute A material that is dissolved in a solvent to form a solution.

solution A liquid system of two or more species that are intimately dispersed within each other at a molecular level. The system is therefore totally homogeneous. The major component is called the solvent (generally liquid in the pure state) and the minor component is called the solute (gas, liquid, or solid).

The process occurs because of a direct intermolecular interaction of the solvent with the ions or molecules of the solute. This interaction is called solvation. Part of the energy of this interaction appears as a change in temperature on dissolution. *See also* solid solution; solubility.

solvation The attraction of a solute species (e.g. an ion) for molecules of solvent. In water, for example, a positive ion will be surrounded by water molecules, which tend to associate around the ion because of attraction between the positive charge of the ion and the negative part of the polar water molecule. The energy of this solvation (hydration in the case of water) is the 'force' needed to overcome the attraction between positive and negative ions when an ionic solid dissolves. The attraction of the dissolved ion for solvent molecules may extend for several layers. In the case of transition metal elements, ions may also form complexes by coordination to the nearest layer of molecules.

Solvay process (ammonia–soda process) An industrial process for making sodium carbonate. The raw materials are calcium carbonate and sodium chloride (with ammonia). The calcium carbonate is heated:

$$CaCO_3 \rightarrow CaO + CO_2$$

Carbon dioxide is bubbled into a solution of sodium chloride saturated with ammonia, precipitating sodium hydrogencarbonate and leaving ammonium chloride in solution. The sodium hydrogencarbonate is then heated:

$$2NaHCO_3 \rightarrow Na_2CO_3 + H_2O + CO_2$$

The ammonia is regenerated by heating the ammonium chloride with the calcium oxide:

$$2NH_4Cl + CaO \rightarrow CaCl_2 + 2NH_3 + H_2O$$

solvent A liquid capable of dissolving other materials (solids, liquids, or gases) to form a solution. The solvent is generally the major component of the solution. Solvents can be divided into classes, the most important being the following:

Polar. A solvent in which the molecules possess a moderate to high dipole moment and in which polar and ionic compounds are easily soluble. Polar solvents are usually poor solvents for nonpolar compounds. For example, water is a good solvent for many ionic species, such as sodium chloride or potassium nitrate, and polar molecules, such as the sugars, but does not dissolve paraffin wax.

Nonpolar. A solvent in which the molecules do not possess a permanent dipole moment and consequently will solvate nonpolar species in preference to polar species. For example, some organic compounds are good solvents for iodine and paraffin wax, but do not dissolve sodium chloride.

Amphiprotic. A solvent that undergoes self-ionization and can act both as a proton donor and as an acceptor. Water is a good example and ionizes according to:

$$2H_2O = H_3O^+ + OH^-$$

Aprotic. A solvent that can neither accept nor yield protons. An aprotic solvent is therefore the opposite to an amphiprotic solvent.

solvent extraction (liquid–liquid extraction) A method of removing a substance from solution by shaking it with and dissolving it in a better solvent that is imiscible with the original solvent.

solvolysis A reaction between a compound and the solvent in which it is dissolved. *See also* hydrolysis.

sorption *See* absorption.

specific Denoting a physical quantity per unit mass. For example, volume (V) per unit mass (m) is called *specific volume* (v):

$$V/m = v$$

In certain physical quantities the term does not have this meaning: for example, specific gravity is more properly called relative density.

specific gravity *See* relative density.

specific heat capacity Symbol: c The amount of heat needed to raise a unit mass of a substance by one degree. It is measured in joules per kilogram per Kelvin ($J\ kg^{-1}K^{-1}$) in SI units.

specific rotatory power Symbol: α_m The rotation of plane-polarized light in degrees produced by a 10-centimeter length of solution containing one gram of a given substance per milliliter of stated solvent. The specific rotatory power is a measure of the optical activity of substances in solution. It is measured at 20°C using the D-line of sodium.

specific volume *See* specific.

spectra *See* spectrum.

spectral line A particular wavelength of light emitted or absorbed by an atom, ion, or molecule. *See* line spectrum.

spectral series A group of related lines

in the absorption or emission spectrum of a substance. The lines in a spectral series arise when the transitions all occur between one particular energy level and a set of different levels. *See also* Bohr theory.

spectrograph An instrument for producing a photographic record of a spectrum.

spectrographic analysis A method of analysis in which the sample is excited electrically (by an arc or spark) and emits radiation characteristic of its component atoms. This radiation is passed through a slit, dispersed by a prism or a grating, and recorded as a spectrum, either photographically or photoelectrically. The photographic method was widely used for qualitative and semiquantitative work but photoelectric detection also allows wide quantitative application.

spectrometer **1.** An instrument for examining the different wavelengths present in electromagnetic radiation. Typically, spectrometers have a source of radiation, which is collimated by a system of lenses and/or slits. The radiation is dispersed by a prism or grating, and recorded photographically or by a photocell. There are many types for producing and investigating spectra over the whole range of the electromagnetic spectrum. Often spectrometers are called *spectroscopes*. *See also* spectrophotometer.
2. Any of various other instruments for analyzing the energies, masses, etc., of particles. *See* mass spectrometer.

spectrophotometer A form of spectrometer able to measure the intensity of radiation at different wavelengths in a spectrum, usually in the visible, infrared, or ultraviolet regions.

spectroscope *See* spectrometer.

spectroscopy **1.** The production and analysis of spectra. There are many spectroscopic techniques designed for investigating the electromagnetic radiation emitted or absorbed by substances. Spectroscopy, in various forms, is used for analysis of mixtures, for identifying and determining the structures of chemical compounds, and for investigating energy levels in atoms, ions, and molecules. In the visible and longer wavelength ultraviolet, transitions correspond to electronic energy levels in atoms and molecules. The shorter wavelength ultraviolet corresponds to transitions in ions. In the x-ray region, transitions in the inner shells of atoms or ions are involved. The infrared region corresponds to vibrational changes in molecules, with rotational changes at longer wavelengths.
2. Any of various techniques for analyzing the energy spectra of beams of particles or for determining mass spectra.

spectrum (plural *spectra*) **1.** A range of electromagnetic radiation emitted or absorbed by a substance under particular circumstances. In an *emission spectrum*, light or other radiation emitted by the object is analyzed to determine the particular wavelengths produced. The emission of radiation may be induced by a variety of methods; for example, by high temperature, bombardment by electrons, absorption of higher-frequency radiation, etc. In an *absorption spectrum* a continuous flow of radiation is passed through the sample. The radiation is then analyzed to determine which wavelengths are absorbed. *See also* band spectrum; continuous spectrum; line spectrum.
2. In general, any distribution of a property. For instance, a beam of particles may have a spectrum of energies. A beam of ions may have a mass spectrum (the distribution of masses of ions). *See* mass spectrometer.

spelter Commercial zinc containing about 3% impurities, mainly lead.

spin A property of certain elementary particles whereby the particle acts as if it were spinning on an axis; i.e. it has an angular momentum. Such particles also have a magnetic moment. In a magnetic field the spins line up at an angle to the field direction and precess around this direction. Certain definite orientations to the field direction occur such that $m_s h/2\pi$ is the component of angular momentum along this direction. Here m_s is the spin quantum number, which for an electron has values +1/2 and –1/2, and h is the Planck constant.

spin–orbit coupling The coupling between the spin angular momentum and the orbital angular momentum of an electron. This is a FINE STRUCTURE feature of atomic spectra.

spin paired *See* electron pair.

spin quantum number *See* atom; spin.

square-planar *See* complex.

stabilization energy The difference in energy between the delocalized structure and the conventional structure for a compound. The stabilization energy can be determined by comparing the experimental value for the heat of formation with that calculated for a conventional structure.

stabilizer A substance added to prevent chemical change (i.e. a negative catalyst).

stainless steel *See* steel.

stalactites and stalagmites Pillars of calcium carbonate found hanging from the ceiling (stalactites) and standing on the floor (stalagmites) in limestone caverns. They form when water containing dissolved carbon dioxide forms weak carbonic acid, which is able to react with limestone rock (calcium carbonate) to give a solution of calcium hydrogencarbonate. This solution accumulates in drops on the roof of caves. As it evaporates, the reaction is reversed, causing calcium carbonate to be deposited. Over a very long period, these deposits build up to create a stalactite hanging from the cavern roof. Water that drips onto the floor from the tip of this stalactite also evaporates to leave a deposit of calcium carbonate, and this deposit gradually grows to form a stalagmite. Stalagmites and stalactites can eventually meet to form a single rock pillar.

standard cell A voltaic cell whose e.m.f. is used as a standard. *See* Clark cell; Weston cadmium cell.

standard electrode A half cell used for measuring electrode potentials. The hydrogen electrode (standard hydrogen half cell) is the basic standard but, in practice, calomel electrodes are usually used.

standard hydrogen half cell *See* hydrogen electrode.

standard pressure An internationally agreed value of 101 325 Pa (approximately 100 kPa), equal in non-SI units to a barometric height of 760 millimeters of mercury at 0°C or one ATMOSPHERE.

standard solution A solution that contains a known mass of reagent in a definite volume of solution. A standard flask or volumetric flask is used for this purpose. The solutions may be prepared by direct determination of mass for primary standards. If the reagent is not available in a pure form or is deliquescent the solution must be standardized by titration against another known standard solution. *See* primary standard.

standard state The standard conditions used as a reference system in thermodynamics: pressure is 101 325 Pa; temperature is 25°C (298.15 K); concentration is one mol. The substance under investigation must also be pure and in its usual state, given the above conditions.

standard temperature An internationally agreed value for which many measurements are quoted. It is the melting temperature of water, 0°C (273.15 K). *See also* STP.

stannane (tin(IV) hydride; SnH_4) A colorless poisonous volatile reactive gas prepared by the action of lithium tetrahydroaluminate(III) on tin(II) chloride. It is unstable, decomposing immediately at 150°C. Stannane is used as a reducing agent.

stannate *See* tin.

stannic chloride *See* tin(IV) chloride.

stannic compounds Compounds of tin(IV).

stannite *See* tin.

stannous chloride *See* tin(II) chloride.

stannous compounds Compounds of tin(II).

Stark effect The splitting of atomic spectral lines due to the presence of an external electric field. This phenomenon was discovered by the German physicist Johannes Stark (1874–1957) in 1913 and can be explained using classical electron theory and the BOHR THEORY of the atom.

states of matter The three physical conditions or phases in which substances occur: solid, liquid, and gas. The addition or removal of energy (usually in the form of heat) enables one state to be converted into another.

The major distinctions between the states of matter depend on the kinetic energies of their particles and the distances between them. In solids, the particles have low kinetic energy and are closely packed; in gases they have high kinetic energy and are very loosely packed; kinetic energy and separation of particles in liquids are intermediate.

Solids, for instance, have fixed shapes and volumes, i.e. they do not flow like liquids and gases, and they are difficult to compress. In solids the atoms or molecules occupy fixed positions in space. In most cases there is a regular pattern of atoms, meaning that the solid is crystalline.

Liquids have fixed volumes (i.e. low compressibility) but flow to take up the shape of their container. Their atoms or molecules move about at random, but they are quite close to one another and their motion is hindered.

Gases have no fixed shape or volume. They expand spontaneously to fill the container and are easily compressed. The molecules have almost free random motion.

A PLASMA is sometimes considered to be a fourth state of matter.

stationary phase *See* chromatography.

statistical mechanics The subject that links the microscopic and macroscopic aspects of a system by using statistical methods to relate the properties of the very large number of microscopic particles, such as atoms and molecules, to the thermodynamics of the system. The derivation of the MAXWELL–BOLTZMANN DISTRIBUTION in the second half of the 19th century can be considered to be the start of statistical mechanics.

steam distillation A method of isolating or purifying substances by exploiting Dalton's law of partial pressures to lower the boiling point of the mixture. When two immiscible liquids are distilled, the boiling point will be lower than that of the more volatile component and consequently will be below 100°C if one component is water. The method is particularly useful for recovering materials from tarry mixtures.

steatite *See* soapstone.

steel An alloy of iron with small amounts of carbon and, often, other metals. *Carbon steels* contain 0.05–1.5% carbon – the more carbon, the harder the steel. *Alloy steels* also contain small amounts of other elements, such as chromium, manganese, and vanadium. Their properties depend on the composition. *Stainless steels*, for instance, are resistant to corrosion. The main nonferrous constituent is chromium (10–25%) with up to 0.7% carbon.

step An elementary stage in a chemical reaction, in which energy may be transferred from one molecule to another, bonds may be broken or formed, or electrons may be transferred. For example, in the reaction between hypochlorite ions and iodide ions in aqueous solution there are three steps:

Step 1

$$OCl^-(aq) + H_2O(l) \rightarrow HOCl(aq) + OH^-(aq)$$

Step 2

$$I^-(aq) + HOCl(aq) \rightarrow HOI(aq) + Cl^-(aq)$$

Step 3

$$OH^-(aq) + HOI(aq) \rightarrow H_2O(l) + OI^-(aq)$$

steradian Symbol: sr The SI dimensionless unit of solid angle, equal to that subtended at the center of a sphere by the area of a square on the sphere's surface whose sides are equal in length to the radius of the sphere.

stereochemistry The branch of chemistry concerned with the shapes of molecules and the way these affect the chemical properties.

stereoisomerism *See* isomerism.

stereospecific Describing a chemical reaction involving an asymmetric atom (usually carbon) that results in only one geometric isomer. *See* isomerism; optical activity.

steric effect An effect in which the shape of a molecule influences its reactions. A particular example occurs in molecules containing large groups, which hinder the approach of a reactant (*steric hindrance*).

steric hindrance *See* steric effect.

still An apparatus for distillation.

Stock, Alfred (1876–1946) German chemist. Stock began studying boron hydrides in 1909. Boron had previously been little studied and was thought to react only with strongly electronegative elements, such as oxygen. However Stock found that magnesium boride and acid produced B_4H_{10}, the first of the several boron hydrides that he discovered. It was clear from this and the other hydrides, B_2H_6 and $B_{10}H_{14}$, that the bond between boron and hydrogen could not be the familiar covalent bond which occurs between carbon and hydrogen. Its actual form was not solved until the work of William LIPSCOMB in the 1950s.

stoichiometric coefficient *See* chemical equation.

stoichiometric compound *See* stoichiometry.

stoichiometry The proportions in which elements form compounds. A *stoichiometric compound* is one in which the atoms have combined in small whole numbers.

storage battery *See* accumulator.

STP (NTP) Standard temperature and pressure. Conditions used internationally when measuring quantities that vary with both pressure and temperature (such as the density of a gas). The values are 101 325 pascals (Pa) (approximately 100 kPa) and 0°C (273.15 kelvin). *See also* standard pressure; standard temperature.

straight chain *See* chain.

strong acid An acid that is almost completely dissociated into its component ions in solution. *Compare* weak acid.

strong base A base that is completely or almost completely dissociated into its component ions in solution.

strontia *See* strontium oxide.

strontium A soft reactive alkaline earth metal; the fourth member of group 2 (formerly IIA) of the periodic table. The electronic configuration is that of krypton with two additional outer 5s electrons. Strontium is of low abundance in the Earth's crust, strontium occurring in the minerals strontianite ($SrCO_3$) and celestite ($SrSO_4$).

The element is produced industrially by roasting the carbonate to give the oxide and then reducing the oxide with aluminum:

$$3SrO + 2Al \rightarrow Al_2O_3 + 2Sr$$

Strontium has a low ionization potential, has large atoms, and is therefore very electropositive. Its chemistry is therefore characterized by high reactivity. The properties of strontium fall into sequence with other alkaline earths. Thus it reacts directly with oxygen (finely divided strontium will ignite in air), nitrogen, sulfur, the halogens, and hydrogen to form respectively the oxide SrO, nitride Sr_3N_2, sulfide SrS, halides SrX_2, and hydride SrH_2, all of which are largely ionic in character. The oxide SrO and the metal react readily with water to form the hydroxide $Sr(OH)_2$, which is basic and between $Ca(OH)_2$ and $Ba(OH)_2$ in terms of solubility. The carbonate and sulfate are both insoluble.

As the metal is very electropositive the salts are never much hydrolyzed in solution and the ions are largely solvated as $[Sr(H_2O)_6]^{2+}$.

Strontium is used in special alloys and to help create high vacuums. Its compounds flame bright red and thus find use in fireworks and flares. The fission of uranium in atomic bombs creates the long-lived radioactive isotopes ^{89}Sr and ^{90}Sr, which can accumulate in bone (in preference to calcium) and cause cancer.

Symbol: Sr; m.p. 769°C; b.p. 1384°C; r.d. 2.54 (20°C); p.n. 38; most common isotope ^{88}Sr; r.a.m. 87.62.

strontium bicarbonate *See* strontium hydrogencarbonate.

strontium carbonate ($SrCO_3$) A white insoluble solid that occurs naturally as the mineral *strontianite*. It can be prepared by passing carbon dioxide over strontium oxide or hydroxide or by passing the gas through a solution of a strontium salt. Strontium carbonate is then formed as a precipitate. It is used as a slagging agent in certain metal furnaces and to produce a red color in fireworks, and also as a phosphor for cathode-ray screens.

strontium chloride ($SrCl_2$) A white deliquescent solid obtained directly from strontium and chlorine or by passing chlorine over heated strontium oxide. It is used in fireworks and flares to produce a red color. The hexahydrate ($SrCl_2.6H_2O$) is prepared by the neutralization of hydrochloric acid by strontium hydroxide or carbonate.

strontium hydrogencarbonate (strontium bicarbonate; $Sr(HCO_3)_2$) A compound present in solutions, formed by the action of carbon dioxide on a suspension in cold water of strontium carbonate, to which it reverts on heating:

$$SrCO_3 + CO_2 + H_2O = Sr(HCO_3)_2$$

strontium hydroxide ($Sr(OH)_2$) A solid that normally occurs as the octahydrate ($Sr(OH)_2.8H_2O$), which is prepared by crystallizing an aqueous solution of strontium oxide. Strontium hydroxide is readily soluble in water and its aqueous solutions are strongly basic. It is used in the purification of sugar.

strontium monoxide *See* strontium oxide.

strontium nitrate ($Sr(NO_3)_2$) A colorless crystalline compound, often used in pyrotechnics to produce a bright red flame.

strontium oxide (strontia; strontium monoxide SrO) A grayish-white powder prepared by the thermal decomposition of strontium carbonate, hydroxide, or nitrate. Strontium oxide is soluble in water, forming an alkaline solution because of the attraction between the oxide ions and protons from the water:

$$O^{2-} + H_2O \rightarrow 2OH^-$$

It is used as a drying agent and to make various strontium salts, as well as in pigments, soaps, and lubricants.

strontium sulfate ($SrSO_4$) A white sparingly soluble salt that occurs naturally as the mineral *celestite*. It can be prepared by dissolving strontium oxide, hydroxide, or carbonate in sulfuric acid. It is used as a substitute for barium sulfate in paints. It is also used in fireworks to impart a red color and in paints and glazes.

structural formula The formula of a compound showing the numbers and types of atoms present, together with the way in which these are arranged in the molecule. Often this arrangement can be shown by grouping the atoms, as in the structural formula for hydrogen peroxide (HOOH), or by illustrating their orientation toward each other in space. *Compare* empirical formula; molecular formula.

structural isomerism *See* isomerism.

sublimate A solid formed by sublimation.

sublimation The conversion of a solid into a vapor without the solid first melting. For instance, at standard pressure iodine, solid carbon dioxide, and ammonium chloride sublime. At certain conditions of external pressure and temperature an equilibrium can be established between the solid phase and vapor phase.

subshell A subdivision of an electron shell. It is a division of the orbitals that make up a shell into sets of orbitals that are degenerate (i.e. have the same energy) in the free atom. For example, in the third or M-shell there are the 3s, 3p, and 3d subshells. The 3p subshell has three p orbitals and the 3d sub shell has five d orbitals, indicating that they are 3- and 5-degenerate, respectively.

substitution reaction A reaction in which an atom or group of atoms in an organic molecule is replaced by another atom or group.

substrate A material that is acted on by a catalyst.

sulfate A salt or ester of sulfuric acid.

sulfide A compound of sulfur with a more electropositive element. Simple sulfides contain the ion S^{2-}. *Polysulfides* can also be formed containing ions with chains of sulfur atoms (S_x^{2-}).

sulfinate *See* dithionite.

sulfinic acid *See* dithionous acid.

sulfite A salt or ester of sulfurous acid.

sulfonamide A type of organic compound with the general formula $R.SO_2.NH_2$. Sulfonamides, which are amides of sulfonic acids, are active against bacteria, and some are used in pharmaceuticals ('sulfa drugs').

sulfonate A salt or ester of sulfonic acid. *See* sulfonic acid.

sulfonation A reaction introducing the $-SO_2OH$ (sulfonic acid) group into an organic compound. Sulfonation of aromatic compounds is usually accomplished by refluxing with concentrated sulfuric acid for several hours. The attacking species is SO_3 (sulfur(VI) oxide) and the reaction is an example of electrophilic substitution.

sulfonic acid A type of organic compound containing the $-SO_2.OH$ group. The simplest example is benzenesulfonic acid $C_6H_5SO_2OH$). Sulfonic acids are strong acids. Electrophilic substitution can introduce other groups onto the benzene ring; the $-SO_2.OH$ group directs substituents into the 3-position.

sulfonium compound An organic compound of general formula R_3SX, where R is an organic radical and X is an electronegative radical or element; it contains the ion R_3S^+. An example is diethylmethylsulfonium chloride, $(C_2H_5)_2.CH_3.-S^+Cl^-$, made by reacting diethyl sulfide with chloromethane.

sulfoxide An organic compound of general formula RSOR′, where R and R′ are organic radicals. An example is dimethyl sulfoxide, $(CH_3)_2SO$, commonly used as a solvent.

sulfur A nonmetallic solid element, yellow in its common forms; the second member of group 16 (formerly VIA) of the periodic table. It has the electronic configuration $[Ne]3s^23p^4$.

Sulfur occurs in elemental form in Sicily and some southern states of the USA, and in large quantities in combined forms such as sulfide ores (e.g. iron pyrite, FeS_2) and sulfate minerals (e.g. anhydrite, ($CaSO_4$)). It forms about 0.5% of the Earth's crust. Elemental sulfur is extracted commercially by the *Frasch process*, in which superheated water is forced down the outer of three concentric tubes leading into the deposit. This causes the sulfur to melt. Air is then blown down the central tube, causing the molten sulfur to be forced out of the well, where it is collected and cooled at the well head. In contrast, the large amount of sulfur obtained from the smelting of sulfide ores or roasting of sulfates is not obtained as elemental sulfur but is rather used directly as SO_2/SO_3 for conversion to sulfuric acid via the contact process, which has many uses.

Sulfur exhibits allotropy and its structure in all its phases is quite complex. The common crystalline modification, rhombic sulfur, is in equilibrium with a triclinic modification above 96°C. Both have structures based on S_8 rings, but the crystals are quite different. If molten sulfur is poured into water a dark red 'plastic' variety is obtained in a semielastic form. Its structure appears to be a helical chain of S atoms. Unlike selenium and tellurium, sulfur does not appear to have a gray 'metal-like' allotrope.

Sulfur reacts directly with hydrogen but the position of the equilibrium at normal temperatures precludes the use of this reaction as a preparative method for hydrogen sulfide. The element burns readily in oxygen with a characteristic blue flame to form sulfur(IV) oxide (SO_2) and traces of sulfur(VI) oxide (SO_3). Sulfur(IV) oxide is used as a food preservative, and to sterilize fruit juices and barrels used to age wines and spirits. Sulfur also forms a large number of oxyacid species, some containing peroxide groups and some with two or more S atoms. It forms four distinct fluorides, S_2F_2, SF_4, SF_6, S_2F_{10}; three chlorides, S_2Cl_2, SCl_2, SCl_4; a bromide S_2Br_2, but no iodide. Apart from SF_6, which is surprisingly stable and inert, the halides are generally susceptible to hydrolysis, commonly to give SO_2, sometimes H_2S, and the hydrogen halide. The halides SF_4, S_2Cl_2, and SCl_2 are frequently used as selective fluorinating agents and chlorinating agents in organic chemistry. The chlorides are also industrially important for hardening rubbers. Sulfur hexafluoride may be prepared by direct reaction of the elements but SF_4 is prepared by fluorinating SCl_2 with NaF. In a limited supply of chlorine, sulfur combines to give sulfur monochloride, S_2Cl_2; in excess chlorine at higher temperatures the dichloride is formed,

$$2S + Cl_2 \rightarrow S_2Cl_2$$
$$S_2Cl_2 + Cl_2 \rightarrow 2SCl_2$$

In addition to these binary halides, sulfur forms two groups of oxyhalides; the sulfur dihalide oxides (thionyl halides), SOX_2 (F, Cl, and Br, but not I), and the sulfur dihalide dioxides (sulfuryl halides), SO_2X_2 (F and Cl, but not Br and I).

Most metals react with sulfur to form sulfides, of which there are a large number of structural types. With electropositive elements of groups 1 and 2 the sulfides are largely ionic (S^{2-}) in the solid phase. Hydrolysis occurs in solution thus:

$$S^{2-} + H_2O \rightarrow SH^- + OH^-$$

Most transition metals form sulfides, which although formally treated as FeS, CoS, NiS, etc., are largely covalent and frequently nonstoichiometric.

In addition to the complexities of the binary compounds with metals, sulfur forms a range of binary compounds with elements such as Si, As, and P, which may be polymeric and generally of rather complex structure, e.g. $(SiS_4)_n$, $(SbS_3)_n$, N_4S_4, As_4S_4, P_4S_3.

Sulfur is used in fungicides and in the vulcanizing process for rubber. It also forms a wide range of organic sulfur compounds, most of which have the typically revolting smell of H_2S.

Symbol: S; m.p. 112.8°C; b.p. 444.6°C; r.d. 2.07; p.n. 16; most common isotope ^{32}S; r.a.m. 32.066.

sulfur dichloride dioxide (sulfuryl chloride; SO_2Cl_2) A colorless fuming liquid formed by the reaction of chlorine with sulfur(IV) oxide in sunlight. It is used as a chlorinating agent.

sulfur dioxide *See* sulfur(IV) oxide.

sulfuretted hydrogen *See* hydrogen sulfide.

sulfuric(IV) acid *See* sulfurous acid.

sulfuric acid (oil of vitriol; H_2SO_4) A colorless oily liquid manufactured by the CONTACT PROCESS. The concentrated acid is diluted by adding it slowly to water, with careful stirring. Concentrated sulfuric acid acts as an oxidizing agent, giving sulfur(IV) oxide as the main product, and also as a dehydrating agent. The diluted acid acts as a strong dibasic acid, neutralizing bases and reacting with active metals and carbonates to form sulfates.

Sulfuric acid is used in the laboratory to dry gases (except ammonia) and to prepare nitric acid. In industry, it is used to manufacture fertilizers (e.g. ammonium sulfate), explosives, chemicals, rayon, pigments, and detergents. It is also used to clean metals, and as the electrolyte in vehicle batteries.

sulfur monochloride *See* disulfur dichloride.

sulfurous acid (sulfuric(IV) acid; trioxosulfuric(IV) acid; H_2SO_3) A weak acid found only in solution, made by passing sulfur(IV) oxide into water. The solution is unstable and smells of sulfur(IV) oxide. It is used as a reducing agent, converting iron(III) ions to iron(II) ions, chlorine to chloride ions, and orange dichromate(VI) ions to green chromium(III) ions.

sulfur(IV) oxide (sulfur dioxide; SO_2) A colorless choking gas prepared by burning sulfur or by heating metal sulfides in air, or by treating a sulfite with an acid. It is a powerful reducing agent, used as a bleach. It is also used for sterilizing and as a food preservative. It dissolves in water to form sulfurous acid (sulfuric(IV) acid) and combines with oxygen, in the presence of a catalyst, to form sulfur(VI) oxide. This latter reaction is important in the manufacture of sulfuric(VI) acid.

sulfur(VI) oxide (sulfur trioxide; SO_3) A fuming volatile white solid prepared by passing sulfur(IV) oxide and oxygen over hot vanadium(V) oxide, which acts as a catalyst, and then cooling the product in ice. Sulfur(VI) oxide reacts vigorously with water to form sulfuric acid. *See also* contact process.

sulfur trioxide *See* sulfur(VI) oxide.

sulfuryl Describing a compound containing the group $=SO_2$.

sulfuryl chloride *See* sulfur dichloride dioxide.

superconductivity The property of zero electrical resistance exhibited by certain metals and metallic compounds at low temperatures. Metals superconduct at temperatures close to absolute zero but a number of ceramic metal oxides are now known that superconduct at higher (94K or better) temperatures. Synthetic organic superconductors have also been produced.

supercooling The cooling of a liquid at a given pressure to a temperature below its melting temperature at that pressure without solidifying it. The liquid particles lose energy but do not spontaneously fall into the regular geometrical pattern of the solid. A supercooled liquid is in a metastable state and will usually solidify if a small crystal of the solid is introduced to act as a 'seed' for the formation of crystals. As soon as this happens, the temperature returns to the melting temperature until the substance has completely solidified.

superfluidity A property of liquid helium at very low temperatures (2.186 K) when it makes a transition to a superfluid state in which it has a high thermal conductivity and flows without friction. *See also* helium.

superheating The raising of a liquid's temperature above its boiling temperature, accomplished by increasing the pressure.

supernatant Denoting a clear liquid that lies above a sediment or a precipitate.

superoxide An inorganic compound containing the O_2^- ion.

superphosphate A mixture – mainly calcium hydrogen phosphate and calcium sulfate – used as a fertilizer. It is made from calcium phosphate and either sulfuric acid or phosphoric(V) acid.

supersaturated solution *See* saturated solution.

supersaturated vapor *See* saturated vapor.

supramolecular chemistry A branch of chemistry concerned with the synthesis and study of large structures consisting of molecules assembled together in a definite pattern. In a *supramolecule* the molecular units (which are sometimes known as *synthons*) are joined by intermolecular bonds – i.e. by hydrogen bonds or by ionic attractions. A particular interest in supramolecular research is the idea of 'self-assembly' – i.e. that the molecules form well-defined structures spontaneously as a result of their geometry and chemical properties. In this way, supramolecular chemistry is 'chemistry beyond molecules'.

One type of supramolecule is a *helicate*, which has a double helix made of two chains held by copper ions along its axis. The structure is analogous to the double helix of DNA. *See also* host–guest chemistry.

supramolecule *See* supramolecular chemistry.

surfactant A substance that lowers surface tension and has properties of wetting, foaming, detergency, dispersion, and emulsification, e.g. a soap or other detergent.

suspension A system in which small particles of a solid or liquid are dispersed in a liquid or gas.

sylvite *See* potassium chloride.

symbol, chemical *See* chemical symbol.

symmetry The property of an object that certain operations, called *symmetry operations*, change the object to a state in which it is indistinguishable from its original state. Examples of symmetry operations are rotation about an axis, reflection through a plane, and inversion through a point. A *symmetry element* is a geometrical feature of the object with respect to which the symmetry operation is carried out. For example, an axis of rotation, a plane of reflection, and a point through which inversion can be performed are symmetry elements. A symmetry operation has to be associated with a symmetry element and vice versa. *See also* molecular symmetry.

syn-anti isomerism *See* isomerism.

synthesis The preparation of chemical compounds from simpler compounds.

synthesis gas A mixture of carbon monoxide and hydrogen produced by steam reforming of natural gas.

$$CH_4 + H_2O \rightarrow CO + 3H_2$$

Synthesis gas is a useful starting material for the manufacture of a number of organic compounds.

synthon *See* supramolecular chemistry.

Système International d'Unités *See* SI units.

T

talc (French chalk) An extremely soft greenish or white mineral, a hydrous silicate of magnesium ($Mg_3Si_4O_{10}(OH)_2$). Purified it is sold as talcum powder. It is also used as a lubricant, an electrical insulator, a filler (in paint, rubber, and paper) and an ingredient of some ceramics.

tantalum A hard ductile rare silvery transition metal, the third member of group 5 (formerly VB) of the periodic table. It occurs chiefly in the minerals colombite and tantalite, in association with niobium. It is strong, highly resistant to corrosion, and easily worked. Tantalum is used in turbine blades and cutting tools, and in surgical and dental work. It is also used in electronic components and in filaments for light bulbs subject to shock and vibration.

Symbol: Ta; m.p. 2996°C; b.p. 5425 ± 100°C; r.d. 16.654 (20°C); p.n. 73; most common isotope ^{181}Ta; r.a.m. 180.9479.

tartar emetic (potassium antimonyl tartrate; $KSbO(C_4H_4O_6).\frac{1}{4}H_2O$) A poisonous substance used as an emetic in medicine and as a mordant in dyeing and as an insecticide.

Taube, Henry (1915–) American inorganic chemist. Taube has developed a number of experimental techniques for studying the kinetics and mechanism of inorganic reactions, in particular electron-transfer reactions. Transition metals such as iron, copper, cobalt, and molybdenum form coordination compounds of a type first described by Alfred WERNER. In a typical coordination compound a metal ion is attached to a number of ligands, such as water or ammonia. It was thought that the ligands would keep the ions apart and inhibit electron transfer between ions. Taube showed experimentally that ligand bridges form between interacting complexes, thus allowing electrons to be transferred. For his work in this field Taube was awarded the 1983 Nobel Prize for chemistry.

tautomerism Isomerism in which each isomer can convert into the other, so that the two isomers are in equilibrium. The isomers are called *tautomers*. Tautomerism often results from the migration of a hydrogen atom.

technetium A silver-gray radioactive transition metal, the second element of group 7 (formerly VIIB) of the periodic table. It does not occur naturally on Earth although it has been detected in the spectra of stars. It is produced artificially by bombarding molybdenum with deuterium nuclei or neutrons, and also during the fission of uranium. It is used in medicine as a radioactive label.

Symbol: Tc; m.p. 2172°C; b.p. 4877°C; r.d. 11.5 (est.); p.n. 43; r.a.m. 98.9063 (^{99}Tc); most stable isotope ^{98}Tc (half-life 4.2×10^6 years).

telluride A compound of tellurium and another element. *See* tellurium.

tellurium A brittle silvery semiconducting metalloid element; the fourth member of group 16 (formerly VIA) of the periodic table. It is found chiefly in combination with metals in minerals known as *tellurides*. Commercially it is recovered primarily during the refining of copper, lead, and gold. Tellurium is used mainly as an additive to improve the qualities of stainless steel and various metals. It is also used in semiconductors, and as a catalyst in the petroleum industry. Tellurium compounds are poisonous.

Symbol: Te; m.p. 449.5°C; b.p. 989.8°C; r.d. 6.24 (20°C); p.n. 52; most common isotope ^{128}Te; r.a.m. 127.6.

temperature scale A practical scale against which temperature can be measured. A temperature scale is determined by one or more *fixed points*, which are reproducible and are assigned an agreed temperature. On the Celsius scale, for example, the fixed points are the temperature of pure melting ice (the *ice temperature*) and the temperature of pure boiling water (the *steam temperature*). The difference between these fixed points is subdivided into temperature interval units. The International Practical Temperature Scale has 11 fixed points that cover the range 13.81 kelvin (K) to 1337.58 K.

temporary hardness A type of water hardness caused by the presence of dissolved calcium, iron, and magnesium hydrogencarbonates. This form of hardness can be removed by boiling the water. Temporary hardness arises because rainwater combines with carbon dioxide from the atmosphere to form weak carbonic acid. This acid reacts with carbonate rocks, such as limestone, to produce calcium hydrogencarbonate, which then goes into solution. When this solution is heated, the complete reaction sequence is reversed so that calcium carbonate is precipitated. If this compound is not removed, it will accumulate as scale in boilers and hot-water pipes and reduce their efficiency. *Compare* permanent hardness. *See also* water softening.

tera- Symbol: T A prefix used with SI units denoting 10^{12}. For example, 1 terawatt (TW) = 10^{12} watts (W).

terbium A soft ductile malleable silver-gray rare element of the lanthanoid series of metals. It occurs in association with other lanthanoids in minerals such as gadolinite, monazite, and xenotime. It is also found in apatite. One of its few uses is as a dopant in solid-state devices. Terbium compounds are also used in lasers and in color television sets.

Symbol: Tb; m.p. 1356°C; b.p. 3123°C; r.d. 8.229 (20°C); p.n. 65; only natural isotope ^{159}Tb; r.a.m. 158.92534.

ternary compound A chemical compound formed from three elements; e.g. Na_2SO_4 or $LiAlH_4$.

tervalent (trivalent) Having a valence of three.

tesla Symbol: T The SI derived unit of magnetic flux density, equal to a flux density of one weber of magnetic flux per square meter. 1 T = 1 Wb m^{-2}. It is named for the Croatian-born electrical engineer and inventor Nicola Tesla (1857–1943).

tetracarbonyl nickel(0) *See* nickel carbonyl.

tetraethyl lead *See* lead tetraethyl.

tetragonal crystal *See* crystal system.

tetrahedral compound A compound such as methane in which an atom has four bonds directed toward the corners of a regular tetrahedron. The angles between the bonds in such a compound are about 109°.

tetrahydrate A crystalline hydrated compound containing four molecules of water of crystallization per molecule of compound. *See also* complex.

tetravalent (quadrivalent) Having a valence of four.

thallium A soft malleable bluish-white metallic element belonging to group 13 (formerly IIIA) of the periodic table. It is found in lead, zinc, and cadmium ores, and in pyrites (e.g. FeS_2). Thallium and its compounds are highly toxic and were used previously as a rodent and insect poison. Various compounds are now used in photocells, infrared detectors, and low-melting glasses.

Symbol: Tl; m.p. 303.5°C; b.p. 1457°C; r.d. 11.85 (20°C); p.n. 81; most common isotope ^{205}Tl; r.a.m. 204.3833.

thermal dissociation The decomposition of a chemical compound into component atoms or molecules by the action of heat. Often it is temporary and reversible.

thermite (thermit) A mixture of aluminum powder and (usually) iron(III) oxide. When ignited the oxide is reduced; the reaction is strongly exothermic and molten iron is formed:

$$2Al + Fe_2O_3 \rightarrow Al_2O_3 + 2Fe$$

Thermite is used in incendiary bombs and for welding steel. *See also* Goldschmidt process.

thermochemistry The branch of chemistry concerned with heats of reaction, solvation, etc.

thermodynamic engine *See* heat engine.

thermodynamics The study of heat and other forms of energy and the various related changes in physical quantities such as temperature, pressure, density, etc.

The *first law of thermodynamics* states that the total energy in a closed system is conserved (constant). In all processes energy is simply converted from one form to another, or transferred from one system to another.

A mathematical statement of the first law is:

$$\Delta Q = \Delta U + \Delta W$$

Here, ΔQ is the heat transferred to the system, ΔU the change in internal energy (resulting in a rise or fall of temperature), and ΔW is the external work done *by* the system.

The *second law of thermodynamics* can be stated in a number of ways, all of which are equivalent. One is that heat cannot pass from a cooler to a hotter body without some other process occurring. Another is the statement that heat cannot be totally converted into mechanical work, i.e. a heat engine cannot be 100% efficient.

The *third law of thermodynamics* states that the entropy of a substance tends to zero as its thermodynamic temperature approaches zero.

Often a zeroth law of thermodynamics is given: that if two bodies are each in thermal equilibrium with a third body, then they are in thermal equilibrium with each other. This is considered to be more fundamental than the other laws because they assume it. *See also* Carnot cycle; entropy.

thermodynamic temperature Symbol: T A temperature that measures changes in the entropy of a closed system with respect to heat energy. The thermodynamic temperature interval unit is the kelvin. *See also* absolute temperature.

thin-layer chromatography A technique widely used for the analysis of mixtures. Thin-layer chromatography employs a solid stationary phase, such as alumina or silica gel, spread evenly as a thin layer on a glass plate. A base line is carefully scratched near the bottom of the plate using a needle, and a small sample of the mixture is spotted onto the base line using a capillary tube. The plate is then stood upright in solvent, which rises up to the base line and beyond by capillary action. The components of the spot of the sample will dissolve in the solvent and tend to be carried up the plate. However, some of the components will cling more readily to the solid phase than others and will not move up the plate so rapidly. In this way, different fractions of the mixture eventually become separated. When the solvent has almost reached the top, the plate is removed and quickly dried. The plate is developed to locate the positions of colorless fractions by spraying with a suitable chemical or by exposure to ultraviolet radiation. The components are identified by comparing the distance they have traveled up the plate with standard solutions that have been run simultaneously, or by computing an R_F value.

thionyl Describing a compound containing the group =SO.

thiosulfate A salt containing the ion $S_2O_3^{2-}$. *See* sodium thiosulfate(IV).

thixotropy The change of viscosity with movement. Fluids that undergo a decrease in viscosity with increasing velocity are said to be thixotropic. Nondrip paint is an example of a thixotropic fluid.

thoria *See* thorium dioxide.

thorium A soft ductile gray toxic radioactive element of the actinoid series of metals. It has several long-lived radioisotopes and is found in a variety of minerals, including monazite and pitchblende. Thorium is used in magnesium alloys, incandescent gas mantles, refractory materials, and nuclear fuel elements.

Symbol: Th; m.p. 1750°C; b.p. 4790°C (approx.); r.d. 11.72 (20°C); p.n. 90; r.a.m. 232.0381; most stable isotope ^{232}Th (half-life 1.41×10^{10} years).

thorium dioxide (thoria; ThO_2) A white insoluble compound, used as a refractory, in incandescent gas mantles, and as a replacement for silica in some types of optical glass.

thulium A very rare soft malleable ductile silvery-gray element of the lanthanoid series of metals. It occurs in association with other lanthanoids in minerals such as gadolinite and xenotime. It is used in arc lighting. Thulium-170, an artificially created radioactive isotope, is used as a power source in portable x-ray machines.

Symbol: Tm; m.p. 1545°C; b.p. 1947°C; r.d. 9.321 (20°C); p.n. 69; only natural isotope ^{169}Tm; r.a.m. 168.93421.

tin A white lustrous malleable metal; the fourth member of group 14 (formerly IVA) of the periodic table. Tin is the first distinctly metallic element of the group even though it retains some amphoteric properties. Its electronic structure has outer s^2p^2 electrons ($[Kr]4d^{10}5s^25p^2$). The element is of low abundance in the Earth's crust (0.004%) but is widely distributed, largely as the mineral cassiterite (SnO_2). The metal has been known since early Bronze Age civilizations when the ores used were relatively rich. Currently worked ores are as low as 1–2% tin and considerable concentration must be carried out before roasting. The metal itself is obtained by reduction using carbon:

$$SnO_2 + C \rightarrow Sn + CO_2$$

Tin is an expensive metal and several processes are therefore used to recover it from scrap tin-plate. These may involve dry chlorination to give the volatile $SnCl_4$, or electrolytic methods using an alkaline electrolyte:

$$Sn + 4OH^- \rightarrow Sn(OH)_4^{2-} + 2e^- \text{ (anode)}$$

$$Sn(OH)_4^{2-} \rightarrow Sn^{2+} + 4OH^- \text{ (cathode)}$$

$$Sn^{2+} + 2e \rightarrow Sn \text{ (cathode)}$$

Tin does not react directly with hydrogen but an unstable hydride, SnH_4, can be prepared by reduction of $SnCl_4$. The low stability is due to the rather poor overlap of the diffuse orbitals of the tin atom with the small H-orbitals. Tin forms both tin(II) oxide and tin(IV) oxide. Both are amphoteric, dissolving in acids to give tin(II) and tin(IV) salts, and in bases to form *stannites* and *stannates*,

$$SnO + 4OH^- \rightarrow [SnO_3]^{4-} + 2H_2O$$

stannite (relatively unstable)

$$SnO_2 + 4OH^- \rightarrow [SnO_4]^{4-} + 2H_2O$$

stannate

The halides, SnX_2, may be prepared by dissolving tin metal in the hydrogen halide or by the action of heat on SnO plus the hydrogen halide. Tin(IV) halides may be prepared by direct reaction of halogen with the metal. Although tin(II) halides are ionized in solution their melting points are all low, suggesting considerable covalence in all but the fluoride. The tin(IV) halides are volatile and essentially covalent with slight polarization of the bonds. Tin(II) compounds are readily oxidized to tin(IV) compounds and are therefore good reducing agents for general laboratory use.

Tin has three crystalline modifications or allotropes, α-tin or 'gray tin' (diamond structure), stable below 13°C; β-tin or 'white tin', stable between 13°C and 160°C; and γ-tin, stable only above 160°C. The latter two are metallic, with close-packed structures. Tin also has several isotopes. It is used in a large number of alloys including Babbit metal, bell metal, Britannia metal, bronze, gun metal, tin plate, and pewter as well as several special solders.

Symbol: Sn; m.p. 232°C; b.p. 2270°C; r.d. 7.31 (20°C); p.n. 50; most common isotope ^{118}Sn; r.a.m. 118.710.

tincal A naturally occurring form of borax. *See* disodium tetraborate decahydrate.

tin(II) chloride (stannous chloride; $SnCl_2$) A transparent solid made by dissolving tin in hydrochloric acid. Tin(II) chloride is a reducing agent, it combines with ammonia, and forms hydrates with water. It is used as a mordant.

tin(IV) chloride (stannic chloride; $SnCl_4$) A colorless fuming liquid. Tin(IV) chloride is soluble in organic solvents but is hydrolyzed by water. It dissolves sulfur, phosphorus, bromine, and iodine, and it dissolves in concentrated hydrochloric acid to give the anion $SnCl_6^{2-}$.

tin(IV) hydride *See* stannane.

tin(II) oxide (stannous oxide; SnO) A dark green or black solid. It can also be obtained in an unstable red form, which turns black on exposure to air. Tin(II) oxide can be made by precipitating the hydrated oxide from a solution containing tin(II) ions and dehydrating the product at 100°C. It is used in the manufacture of cosmetics and ceramics.

tin(IV) oxide (stannic oxide; tin dioxide; SnO_2) A colorless crystalline solid, which is usually discolored owing to the presence of impurities. It occurs in nature as the mineral *cassiterite.* It exists as hexagonal or rhombic crystals or in an amorphous form. Tin(IV) oxide is insoluble in water. It is used for polishing glass and metal and in cosmetics.

tin(II) sulfide (SnS) A gray solid that can be prepared from tin and sulfur. Above 265°C tin(II) sulfide slowly turns into tin(IV) sulfide and tin:

$$2SnS(s) \rightarrow SnS_2(s) + Sn(s)$$

tin(IV) sulfide (SnS_2) A yellowish solid that can be precipitated by reacting hydrogen sulfide with a solution of a soluble tin(IV) salt. A crystalline form sometimes known as *mosaic gold* is obtained by heating a mixture of tin filings, sulfur, and ammonium chloride. Tin(IV) sulfide is used as a pigment.

titania *See* titanium(IV) oxide.

titanic chloride *See* titanium(IV) chloride.

titanic oxide *See* titanium(IV) oxide.

titanium A silvery lustrous light strong transition metal, the first element of group 4 (formerly subgroup IVB) of the periodic table. It occurs in various ores as titanium(IV) oxide and also in combination with iron and oxygen. The most commercially important minerals are rutile (TiO_2) and ilmenite $(FeTi)_3$). The mineral titanite ($(CaTiO)SlO_4$) is often used for gemstones. Titanium is extracted by conversion of titanium(IV) oxide to the chloride, which is reduced to the metal by heating with sodium. Titanium is chiefly reactive only at high temperatures. Its strength, low density, and resistance to corrosion make it especially useful for alloys used in the aviation and aerospace industries. Titanium pins, plates, and joints are also used in dentistry and medicine to replace damaged teeth or bones. It forms compounds with oxidation states +4, +3, and +2, the +4 state being the most stable. Titanium compounds are used in pigments and cosmetics.

Symbol: Ti; m.p. 1660°C; b.p. 3287°C; r.d. 4.54 (20°C); p.n. 22; most common isotope ^{48}Ti; r.a.m. 47.867.

titanium(IV) chloride (titanic chloride; titanium tetrachloride; $TiCl_4$) A volatile colorless liquid formed by heating titanium(IV) oxide with carbon in a stream of dry chlorine at 700°C. The product is purified by fractional distillation. Titanium(IV) chloride fumes in moist air forming the oxychlorides of titanium. It undergoes hydrolysis in water but this can be prevented by the presence of excess hydrochloric acid. Crystallization of such a solution gives crystals of the di- and pentahydrates. Titanium(IV) chloride is used in the preparation of pure titanium and as an intermediate in the production of titanium compounds from minerals.

titanium dioxide *See* titanium(IV) oxide.

titanium(IV) oxide (titania; titanic oxide; titanium dioxide; TiO_2) A white ionic solid that occurs naturally in three crystalline forms: rutile (tetragonal), brookite (orthorhombic), and anatase (tetragonal). It reacts slowly with acids and is attacked by the halogens at high temperatures. On heating it decomposes to give titanium(III) oxide and oxygen. Titanium(IV) oxide is amphoteric: with hot concentrated sulfuric acid, it forms titanyl sulfate, $TiOSO_4$; when fused with alkalis, it forms titanates. It is used as a white pigment.

titanium tetrachloride *See* titanium(IV) chloride.

titrant *See* titration.

titration A procedure in volumetric analysis in which a solution of known concentration (called the *titrant*) is added to a solution of unknown concentration from a burette until the equivalence point or end point of the titration is reached. *See* volumetric analysis.

tonne Symbol: t An m.k.s. unit of mass equal to 10^3 kilograms in SI units (i.e. one megagram).

topaz A hydrous aluminosilicate mineral containing some fluoride ions, used for making refractories, glasses, and glazes. A clear pale yellow or brown form is used as a gemstone.

torr A unit of pressure equal to one mmHg. In SI units it is equal to 133.322 pascals. It was named for Italian physicist Evangelista Torricelli (1609–47).

tracer An isotope of an element used to investigate chemical reactions or physical processes (e.g. diffusion). *See* isotopes.

transactinide elements Those elements following ^{103}Lw (i.e. following the actinide series). To date, element 104 (rutherfordium) through to element 112 and element 114 have been synthesized; so far, elements 111, 112, and 114 are unnamed. All the transactinides are unstable and have very short half-lives. There is speculation that islands of stable nuclei with very high mass numbers exist, because theoretical calculations predict unusual stability for elements 114 and 118.

trans-isomer *See* isomerism.

transition elements (group 3–12 elements) A class of elements occurring in the periodic table in three series: from scandium to zinc; from yttrium to cadmium; and from lanthanum to mercury. The transition elements are all metals. They owe their properties to the presence of d electrons in the atoms. In the first transition series, calcium has the configuration $3s^23p^64s^2$. The next element scandium has $3s^23p^63d^14s^2$, and the series is formed by filling the d levels up to zinc at $3d^{10}4s^2$. The other two transition series occur by filling of the 4d and 5d levels. Often the elements scandium, yttrium, and lanthanum are considered with the lanthanoids rather than the transition elements, and zinc, cadmium, and mercury are also treated as a separate group. Transition elements have certain characteristic properties resulting from the existence of d levels:

1. They have variable valences – i.e. they can form compounds with the metal in different oxidation states (Fe^{2+} and Fe^{3+}, etc.).
2. They form a vast number of inorganic complexes, in which coordinate bonds are formed to the metal atom or ion.
3. Compounds of transition elements are often colored.

See also d-block elements; lanthanoids; metals; zinc group.

transition state (activated complex) Symbol: ‡ A short-lived high-energy molecule, radical, or ion formed during a reaction between molecules possessing the necessary activation energy. The transition state decomposes at a definite rate to yield either the reactants again or the final products. The transition state can be considered to be at the top of the potential energy profile.

For the reaction,

$$X + YZ = X...Y...Z^{\ddagger} \rightarrow XY + Z$$

the sequence of events is as follows. X approaches YZ and when it is close enough the electrons are rearranged producing a weakening of the bond between Y and Z. A partial bond is now formed between X and Y producing the transition state. Depending on the experimental conditions, the transition state either then breaks down to form the products or reverts back to the reactants.

transition state theory An alternative name for ACTIVATED COMPLEX THEORY.

transition temperature A temperature at which some definite physical change occurs in a substance. Examples of such transitions are change of state, change of crystal structure, and change of magnetic behavior.

transmutation A change of one element into another by radioactive decay or by bombardment of the nuclei with particles.

transport number Symbol: *t* In an electrolyte, the transport number of an ion is the fraction of the total charge carried by that type of ion in conduction.

transuranic elements Those elements of the ACTINOID series that have higher atomic numbers than uranium, i.e. neptunium through lawrencium. The transuranic elements are formed by adding neutrons to the lower actinoids by high-energy bombardment. They generally have such short half-lives that macroscopic isolation is impossible. Neptunium, plutonium, americium, and possibly curium are also formed in trace quantities by natural neutron bombardment from spontaneous fission of ^{235}U.

triammonium phosphate *See* ammonium phosphate.

triatomic Denoting a molecule, radical, or ion consisting of three atoms. For example, O_3 and H_2O are triatomic molecules.

tribasic acid An acid with three replaceable hydrogen atoms (such as phosphoric(V) acid, H_3PO_4). *See* acid.

triclinic crystal *See* crystal system.

tridymite *See* silicon(IV) oxide.

trigonal bipyramid *See* complex.

trigonal crystal *See* crystal system.

trihydrate A crystalline hydrated compound that contains three molecules of water of crystallization per molecule of compound.

triiron tetroxide (ferrosoferric oxide; magnetic iron oxide; Fe_3O_4) A black solid prepared by passing either steam or carbon dioxide over red-hot iron. It may also be prepared by passing steam over heated iron(II) sulfide. Triiron tetroxide occurs in nature as the mineral magnetite. It is insoluble in water but will dissolve in acids to give a mixture of iron(II) and iron(III) salts in the ratio 1:2. Generally it is chemically unreactive; it is, however, a fairly good conductor of electricity. *See also* iron(II) oxide; iron(III) oxide.

trimer A molecule (or compound) formed by addition of three identical molecules.

trimethylaluminum (aluminum trimethyl; $(CH_3)_3Al$) A colorless liquid produced by the sodium reduction of dimethyl aluminum chloride. It ignites spontaneously on contact with air and reacts violently with water, acids, halogens, alcohols, and amines. Aluminum alkyls are used in the Ziegler process for the manufacture of high-density polyethene.

trimolecular Describing a reaction or

step that involves three molecules interacting simultaneously with the formation of a product. For example, the final step in reaction between hydrogen peroxide and acidified potassium iodide is trimolecular:

$$HOI + H^+ + I^- \rightarrow I_2 + H_2O$$

It is uncommon for reactions to take place involving trimolecular steps. The oxidation of nitrogen(II) oxide to nitrogen(IV) oxide,

$$2NO + O_2 \rightarrow 2NO_2$$

is often classified as a trimolecular reaction but many chemists believe it to involve two bimolecular reactions.

trioxoborix(III) acid *See* boric acid.

trioxosulfuric(IV) acid *See* sulfurous acid.

trioxygen *See* ozone.

triple bond A covalent bond formed between two atoms in which three pairs of electrons contribute to the bond. One pair forms a sigma bond (equivalent to a single bond) and two pairs give rise to two pi bonds. It is conventionally represented as three lines, thus H–C≡C–H. *See* multiple bond.

triple point The only point at which the gas, solid, and liquid phases of a substance can coexist in equilibrium. The triple point of water (273.16 K at 101 325 Pa) is used to define the kelvin.

trisodium phosphate(V) (sodium orthophosphate; Na_3PO_4) A white solid prepared by adding sodium hydroxide to disodium hydrogenphosphate. On evaporation white hexagonal crystals of the dodecahydrate ($Na_3PO_4.12H_2O$) may be obtained. These crystals do not effloresce or deliquesce. They dissolve readily in water to produce alkaline solutions owing to salt hydrolysis. Trisodium phosphate is used as a water softener.

tritiated *See* tritium.

tritium Symbol: T, ^3H A radioactive isotope of hydrogen of mass number 3. The nucleus contains 1 proton and 2 neutrons. Tritium decays with emission of low-energy beta radiation to give ^3He. The half-life is 12.3 years. It is useful as a tracer in studies of chemical reactions. Compounds in which ^3H atoms replace the usual ^1H atoms are said to be *tritiated*. A positive tritium ion, T^+, is a *triton*.

triton *See* tritium.

trivalent (tervalent) Having a valence of three.

tungsten A hard gray malleable transition metal, the third element of group 6 (formerly VIB) of the periodic table. It occurs naturally in the minerals wolframite ($(Fe,Mn)WO_4$) and scheelite ($CaWO_4$). It was formerly called *wolfram*. It is used as the filaments in electric lamps and in various steel alloys. Tungsten has the highest melting point of all the metals. Tungsten salts are used in paints and in tanning.

Symbol: W; m.p. 3410 ± 20°C; b.p. 5650°C; r.d. 19.3 (20°C); p.n. 74; most common isotope ^{184}W; r.a.m. 183.84.

tungsten carbide Either of two carbides (W_2C and WC) produced by heating powdered tungsten with carbon. The carbides are extremely hard and are used in industry to make cutting tools or as an abrasive. WC has a very high melting point (2770°C) and will conduct electricity. W_2C also has a very high melting point (2780°C) but is a less efficient conductor of electricity. It is very resistant to chemical attack and behaves in a manner very similar to that of tungsten. It is strongly attacked by chlorine to give tungsten hexachloride, WCl_6.

turquoise A naturally occurring hydrated copper aluminum phosphate, used as a green-blue gemstone.

U

ultracentrifuge A high-speed centrifuge used for separating out very small particles. Because the sedimentation rate depends on the particle size, the ultracentrifuge can be used to measure the molecular mass of colloidal particles and large molecules (e.g. proteins).

ultrahigh vacuum *See* vacuum.

ultraviolet (UV) A form of electromagnetic radiation, shorter in wavelength than visible light. Ultraviolet wavelengths range between about 1 nm and 400 nm. Ordinary glasses are not transparent to these waves; quartz is a much more effective material for making lenses and prisms for use with ultraviolet. Like light, ultraviolet radiation is produced by electronic transitions between the outer energy levels of atoms. However, having a higher frequency, ultraviolet photons carry more energy than those of light and can induce photolysis of compounds. *See also* electromagnetic radiation.

unimolecular Describing a reaction (or step) in which only one molecule is involved. For example, radioactive decay is a unimolecular reaction:

$$Ra \rightarrow Rn + \alpha$$

Only one atom is involved in each disintegration.

In a unimolecular chemical reaction, the molecule acquires the necessary energy to become activated and then decomposes. The majority of reactions involve only uni- or BIMOLECULAR steps. The following reactions are all unimolecular:

$$N_2O_4 \rightarrow 2NO_2$$
$$PCl_5 \rightarrow PCl_3 + Cl_2$$
$$CH_3CH_2Cl \rightarrow C_2H_4 + HCl$$

unit A reference value of a quantity, used to express other values of the same quantity. *See also* SI units.

unit cell The smallest group of atoms, ions, or molecules that, when repeated at regular intervals in three dimensions, will produce the lattice of a crystal system.

unit processes (chemical conversions) The recognized steps used in chemical processes, e.g. alkylation, distillation, hydrogenation, pyrolysis, nitration, etc.

univalent (monovalent) Having a valence of one.

universal gas constant *See* gas constant.

universal indicator (multiple-range indicator) A mixture of indicator dyestuffs that shows a gradual change in color over a wide pH range. A typical formulation contains methyl orange, methyl red, bromothymol blue, and phenolphthalein and changes through a red, orange, yellow, green, blue, and violet sequence between pH 3 and pH 10. Several commercial preparations are available as both solutions and test papers.

unsaturated solution *See* saturated solution.

unsaturated vapor *See* saturated vapor.

unun- A prefix used in the names of new chemical elements. Synthetic elements with high proton numbers were given temporary names based on the proton number, pending official agreement on the actual

Element	*Number*	*Symbol*
nil	0	n
un	1	u
bi	2	b
tri	3	t
quad	4	q
pent	5	p
hex	6	h
sept	7	s
oct	8	0
enn	9	e

name to be used. These names were formed from the word elements in the table above. The suffix -ium was also used (because the elements were metals). So element 104, for example, had the temporary systematic name unnilquadium (un + nil + quad + ium). The names for elements using this system are shown in the table below.

uraninite *See* pitchblende.

uranium A toxic radioactive silvery element of the actinoid series of metals. Its three naturally occurring radioisotopes, ^{238}U (99.283% in abundance), ^{235}U (0.711%), and ^{234}U (0.005%), are found in numerous minerals including the uranium oxides pitchblende, uraninite, and carnotite. The readily fissionable ^{235}U is a major nuclear fuel and nuclear explosive, while ^{238}U is a source of fissionable ^{239}Pu. Spent uranium fuel rods have been used to make armor-piercing shells.

Symbol: U; m.p. 1132.5°C; b.p. 3745°C; r.d. 18.95 (20°C); p.n. 92; most stable isotope U^{238}; r.a.m. 238.0289.

uranium hexafluoride (UF_6) A crystalline volatile compound, used in separating uranium isotopes by differences in the rates of gas diffusion.

uranium–lead dating A technique for dating certain rocks depending on the decay of the radioisotope ^{238}U to ^{206}Pb (half-life 4.5×10^9 years) or of ^{235}U to ^{207}Pb (half-life 7.1×10^8 years). The decay of ^{238}U releases alpha particles and an estimate of the age of the rock can be made by measuring the amount of helium present. Another method is to measure the ratio of radioactive lead present to the amount of nonradioactive lead. *See also* radioactive dating.

uranium(IV) oxide (uranium dioxide; urania; UO_2) An extremely poisonous radioactive black crystalline solid. It occurs in pitchblende (uraninite) and is used in ceramics, photographic chemicals, pigments, and as a nuclear fuel.

uranium(VI) oxide (uranium trioxide; UO_3) An extremely poisonous radioactive orange solid, used as a pigment in ceramics and in uranium refining.

uranyl The radical UO_2^{2+}, as in uranyl nitrate $UO_2(NO_3)_2$.

P.N.	*Temporary name*	*Symbol*	*Name*	*Symbol*
101	unnilunium	Unu	mendelevium	Md
102	unnilbiium	Unb	nobelium	No
103	unniltriium	Unt	lawrencium	Lr
104	unnilquadium	Unq	rutherfordium	Rf
105	unnilpentium	Unq	dubnium	Db
106	unnilhexium	Unh	seaborgium	Sg
107	unnilseptium	Uns	bohrium	Bh
108	unniloctium	Uno	hassium	Hs
109	unnilennium	Une	meitnerium	Mt
110	ununnilium	Uun	darmstadtium	Ds
111	unununium	Uuu	unnamed	
112	ununbiium	Uut	unnamed	

V

vacancy *See* defect.

vacuum A space containing gas below atmospheric pressure. A perfect vacuum contains no matter at all, but for practical purposes *soft* (*low*) *vacuum* is usually defined as down to about 10^{-2} pascal, and *hard* (*high*) *vacuum* as below this. *Ultra-high vacuum* is lower than 10^{-7} pascal.

vacuum distillation The distillation of liquids under a reduced pressure, so that the boiling point is lowered. Vacuum distillation is a common laboratory technique for purifying or separating compounds that would decompose at their 'normal' boiling point.

valence (valency) The combining power of an element or radical, equal to the number of hydrogen atoms that will combine with or displace one atom of the element. For simple covalent molecules the valence is obtained directly, for example C in CH_4 is tetravalent; N in NH_3 is trivalent. For ions the valence is regarded as equivalent to the magnitude of the charge; for example Ca^{2+} is divalent, CO_3^{2-} is a divalent radical. The rare gases are zero-valent because they do not form compounds under normal conditions. Because the valence for many elements is constant, the valence of some elements can be deduced without reference to compounds formed with hydrogen. Thus, because the valence of chlorine in HCl is 1, the valence of aluminum in $AlCl_3$ is 3; because oxygen is divalent (H_2O), silicon in SiO_2 is tetravalent. The product of the valence and the number of atoms of each element in a compound must be equal. For example, in Al_2O_3 for the two aluminum atoms (valence 3) the product is 6 and for the three oxygen atoms (valence 2) the product is also 6.

The valence of an element is generally equal to either the number of valence electrons or eight minus the number of valence electrons. Transition metal ions display variable valence.

valence band The highest energy level in a semiconductor or nonconductor that can be occupied by electrons. *See* energy level; semiconductor.

valence-bond theory A technique for calculating the electronic structure of molecules. In valence-bond theory the starting point is to consider the atoms in a molecule and the ways in which valence electrons in the atoms can pair up in chemical bonds. Each way of pairing up the electrons is called a *valence bond structure*. A WAVEFUNCTION is written for each valence bond structure. The total wavefunction Ψ for the molecule is a linear combination of the wavefunctions for the valence-bond structures. The wavefunction incorporates RESONANCE between the various valence bond structures.

valence electron An outer electron in an atom that can participate in forming chemical bonds.

valency *See* valence.

vanadium A strong silvery toxic transition metal, the first element of group 5 (formerly VB) of the periodic table. It occurs in complex ores, especially those of iron, lead, and uranium, in small quantities. It is used in alloy steels to give them added strength and heat resistance. Vanadium forms compounds with oxidation states +5, +4, +3, and +2. It forms colored ions.

Symbol: V; m.p. 1890°C; b.p. 3380°C; r.d. 6.1 (20°C); p.n. 23; most common isotope ^{51}V; r.a.m. 50.94.

vanadium(V) oxide (vanadium pentoxide; V_2O_5) An oxide of vanadium extensively used as a catalyst in oxidation processes, as in the contact process.

van der Waals, Johannes Diderik (1837–1923) Dutch physicist. In his doctoral thesis of 1873 entitled *On the Continuity of the Liquid and Gaseous States* van der Waals studied the kinetic theory of gases and liquids. This work provided the foundation for most of his subsequent work. He derived an equation of state for gases which is more realistic than the ideal gas laws discovered by Robert BOYLE and others because it takes into account both the non-zero sizes of the molecules and the existence of attractive intermolecular forces. The equation of state he found is called the *van der Waals equation* and the intermolecular forces are called *van der Waals forces*. This work resulted in van der Waals winning the 1910 Nobel Prize for physics.

van der Waals equation An equation of state for real gases. For n moles of gas the equation is

$$(p + n^2a/V^2)(V - nb) = nRT$$

where p is the pressure, V the volume, and T the thermodynamic temperature. a and b are constants for a given substance and R is the gas constant. The equation gives a better description of the behavior of real gases than the perfect gas equation ($pV = nRT$).

The equation contains two corrections: b is a correction for the non-negligible size of the molecules; a/V^2 corrects for the fact that there are attractive forces between the molecules, thus slightly reducing the pressure from that of an ideal gas. *See also* gas laws; kinetic theory.

van der Waals force An intermolecular force of attraction arising from weak electrostatic interactions between molecules. Such forces are considerably weaker than those of chemical bonds and typically have energies of less than one joule per mole.

The van der Waals force, named for Dutch physicist Johannes van der Waals (1837–1923), arises because of three effects: permanent dipole–dipole interactions found for any polar molecule; dipole–induced dipole interactions, where one dipole causes a slight charge separation in bonds that have a high polarizability; and *dispersion forces*, which result from temporary polarity arising from an asymmetrical distribution of electrons around the nucleus. Even atoms of the rare gases exhibit dispersion forces.

van't Hoff factor Symbol: i The ratio of the number of particles present in a solution to the number of undissociated molecules added. It is used in studies of colligative properties, which depend on the number of entities present. For example, if n moles of a compound are dissolved and dissociation into ions occurs, then the number of particles present will be in. Osmotic pressure (π), for instance, will be given by the equation

$$\pi V = inRT$$

It is named for the Dutch chemist Jacobus van't Hoff (1852–1911).

van't Hoff isochore The equation:

$$d(\log_e K)/dT = \Delta H/RT^2$$

showing how the equilibrium constant, K, of a reaction varies with thermodynamic temperature, T. ΔH is the enthalpy of reaction and R is the gas constant.

vapor A gas formed by the vaporization of a solid or liquid. Some particles near the surface of a liquid acquire sufficient energy in collisions with other particles to escape from the liquid and enter the vapor; some particles in the vapor lose energy in collisions and re-enter the liquid. At a given temperature an equilibrium is established, which determines the vapor pressure of the liquid at that temperature.

vapor density The ratio of the mass of a certain volume of a vapor to the mass of an equal volume of hydrogen (measured at the same temperature and pressure). Determination of vapor densities is one method of finding the relative molecular mass of a compound (equal to twice the vapor density). Victor Meyer's method, Dumas' method, or Hofmann's method can be used.

vaporization The process by which a liquid or solid is converted into a gas or vapor by heat. Unlike boiling, which occurs at a fixed temperature, vaporization can occur at any temperature. Its rate increases as the temperature rises.

vapor pressure The pressure exerted by a vapor. The *saturated vapor pressure* is the pressure of a vapor in equilibrium with its liquid or solid. It depends on the nature of the liquid or solid and the temperature.

verdigris Any of various greenish basic salts of copper. True verdigris is basic copper acetate, a green or blue solid of variable composition used as a paint pigment. The term is often applied to green patinas formed on metallic copper. Copper cooking vessels can form a coating of basic copper carbonate, $CuCO_3.Cu(OH)_2$. The verdigris formed on copper roofs and domes is usually basic copper sulfate, $CuSO_4.3Cu(OH)_2.H_2O$, while basic copper chloride, $CuCl_2.Cu(OH)_2$, may form in regions near the sea.

vermilion *See* mercury(II) sulfide.

vesicant A substance that causes blistering of the skin. Mustard gas is an example.

vibrational spectroscopy The spectroscopic investigation of the vibrational energy levels of molecules. In the infrared region of the electromagnetic spectrum vibrational transitions are accompanied by rotational transitions. Infrared spectra of molecules are series of *bands*, with each band being associated with a vibrational transition and every line in that band being associated with a rotational transition that accompanies the vibrational transition.

Some features of vibrational spectra can be analyzed by regarding the vibrations as simple harmonic motion but a realistic account of molecular vibrations requires that anharmonicity is taken into account. A diatomic molecule can have a vibrational-rotational spectrum only if it has a permanent dipole moment. A polyatomic molecule can have a vibrational-rotational spectrum only if the normal modes of vibration cause the molecule to have an oscillating dipole moment. The analysis of vibrational–rotational spectra is greatly facilitated by the use of group theory. This type of spectroscopy provides information about interatomic distances and the force constants of chemical bonds.

vicinal positions Positions in a molecule at adjacent atoms. For example, in 1,2-dichloroethane the chlorine atoms are in vicinal positions, and this compound can thus be named *vic*-dichloroethane.

viscosity Symbol: η The resistance to flow of a fluid.

vitreous Resembling glass, or having the structure of a glass.

volatile Easily converted into a vapor.

volt Symbol: V The SI derived unit of electrical potential, potential difference, and e.m.f., defined as the potential difference between two points in a circuit between which a constant current of one ampere flows when the power dissipated is one watt. $1\ V = 1\ W\ A^{-1}$. It is named for the Italian physicist Alessandro Volta (1745–1827).

voltaic cell *See* cell.

voltameter *See* coulombmeter.

volume strength *See* hydrogen peroxide.

volumetric analysis One of the classical wet methods of quantitative analysis. It involves measuring the volume of a solution of accurately known concentration that is required to react with a solution of the substance being determined. The solution of known concentration (the standard solution) is added in small portions from a burette. The process is called a titration and the observed point of complete reaction is called the *end point*. End points are observed with the aid of indicators or by instrumental methods, such as via conduction or light absorption. Volumetric analysis can also be applied to gases. The gas is typically held over mercury in a graduated tube, and volume changes are measured on reaction or after absorption of components of a mixture.

VSEPR (valence-shell electron-pair repulsion) A method of predicting the shape of molecules. In this theory one atom is taken to be the central atom and pairs of valence electrons are drawn round the central atom. The shape of the molecule is determined by minimizing the Coulomb repulsion between pairs and the repulsion between pairs due to the Pauli exclusion principle. Thus, for three pairs of electrons an equilateral triangle is favored and for four pairs of electrons a tetrahedron is favored. The VSEPR theory has had considerable success in predicting molecular shapes. *See* lone pair.

W

washing soda *See* sodium carbonate.

water (H_2O) A colorless liquid that freezes at 0°C and, at atmospheric pressure, boils at 100°C. In the gaseous state water consists of single H_2O molecules. Due to the presence of two lone pairs the atoms do not lie in a straight line, the angle between the central oxygen atom and the two hydrogen atoms being 105°; the distance between each hydrogen atom and the oxygen atom is 0.099 nm. When ice forms, hydrogen bonds some 0.177 nm long develop between the hydrogen atom and oxygen atoms in adjacent molecules, giving ice its tetrahedral crystalline structure with a density of 916.8 kg m^{-3} at STP. Different ice structures develop under higher pressures. When ice melts to form liquid water, the tetrahedral structure breaks down, but some hydrogen bonds continue to exist; liquid water consists of groups of associated water molecules, $(H_2O)_n$, mixed with some monomers and some dimers. This mixture of molecular species has a higher density than the open-structured crystals. The maximum density of water, 999.97 kg m^{-3}, occurs at 3.98°C. This accounts for the ability of ice to float on water and for the fact that water pipes burst as ice expands on freezing.

Although water is predominantly a covalent compound, a very small amount of ionic dissociation occurs ($H_2O \rightleftharpoons H^+ + OH^-$). In every liter of water at STP there is approximately 10^{-7} mole of each ionic species. It is for this reason that, on the pH scale, a neutral solution has a value of 7.

As a polar liquid, water is the most powerful solvent known. This is partly a result of its high dielectric constant and partly its ability to hydrate ions. This latter property also accounts for the incorporation of water molecules into some ionic crystals as water of crystallization.

Water is decomposed by reactive metals (e.g. sodium) when cold and by less active metals (e.g. iron) when steam is passed over the hot metal. It is also decomposed by electrolysis.

water gas A mixture of carbon monoxide and hydrogen produced when steam is passed over red-hot coke or made to combine with hydrocarbons, e.g.

$$C(s) + H_2O(g) \rightarrow CO(g) + H_2(g)$$

$$CH_4(g) + H_2O(g) \rightarrow CO(g) + 3H_2(g)$$

The production of water gas using methane is an important step in the preparation of hydrogen for ammonia synthesis. *Compare* producer gas.

water glass *See* sodium silicate.

water of crystallization Water present in definite proportions in crystalline compounds. Compounds containing water of crystallization are called HYDRATES. Examples are copper(II) sulfate pentahydrate ($CuSO_4.5H_2O$) and sodium carbonate decahydrate ($Na_2CO_3.10H_2O$). The water can be removed by heating.

When hydrated crystals are heated the water molecules may be lost in stages. For example, copper(II) sulfate pentahydrate changes to the monohydrate ($CuSO_4.H_2O$) at 100°C, and to the anhydrous salt ($CuSO_4$) at 250°C. The water molecules in crystalline hydrates may be held by hydrogen bonds (as in $CuSO_4.H_2O$) or, alternatively, may be coordinated to the metal ion as a complex aquo ion.

water softening The removal from water of dissolved calcium, magnesium, and iron compounds, thus reducing the

HARDNESS of the water. The compounds are potentially damaging because they can accumulate in pipes and boilers. They also react with, and therefore waste, soap. Temporary hardness can be removed by boiling the water. Permanent hardness can be removed in a number of ways: by distillation; by the addition of sodium carbonate (which causes dissolved calcium, for example, to precipitate out as calcium carbonate); and by the use of ion-exchange products such as Permutit (utilizes zeolites) and Calgon (utilizes polyphosphates).

watt Symbol: W The SI unit of power, defined as a power of one joule per second. $1\ \mathrm{W} = 1\ \mathrm{J\ s^{-1}}$.

wavefunction Symbol ψ. A function that specifies the state of a system in the WAVE MECHANICS formulation of quantum mechanics. Thus, a wavefunction appears in the SCHRÖDINGER EQUATION, with each wavefunction which is a solution to the Schrödinger equation for a given quantum mechanical system being an EIGENFUNCTION of the equation. The physical interpretation of ψ, known as the *Born interpretation* since it was put forward by Max Born in 1926, is that the square of the wavefunction is a measure of the probability of finding a particle in a small volume element at a given point.

waveguide *See* microwaves.

wavelength Symbol: λ The distance between the ends of one complete cycle of a wave. Wavelength is related to the speed (c) and frequency (ν) thus:

$$c = \nu\lambda$$

wave mechanics *See* quantum theory.

wave number Symbol: σ The reciprocal of the wavelength of a wave. It is the number of wave cycles in unit distance, and is often used in spectroscopy. The unit is the $\mathrm{meter^{-1}}$ ($\mathrm{m^{-1}}$). The circular wave number (symbol: k) is given by:

$$k = 2\pi\sigma$$

weak acid An acid that is not fully dissociated in solution.

weak base A base that is only partly or incompletely dissociated into its component ions in solution.

weber Symbol: Wb The SI unit of magnetic flux, equal to the magnetic flux that, linking a circuit of one turn, produces an e.m.f. of one volt when reduced to zero at uniform rate in one second. $1\ \mathrm{Wb} = 1\ \mathrm{V\ s}$. It is named for German physicist Wilhelm Weber (1804–91).

Werner, Alfred (1866–1919) French-born Swiss chemist. Werner's main contribution to chemistry consists of his elucidation of structure and valence in inorganic molecules. In a series of papers Werner put forward a theory of what he called 'coordination compounds'. He distinguished between the primary and secondary valence of a metal, with the secondary valence being associated with how many ligands can surround a metal atom. Werner put forward his ideas in *New Ideas on Inorganic Chemistry* (1911). He won the 1913 Nobel Prize for chemistry for his work on coordination compounds.

Weston cadmium cell (cadmium cell) A standard cell that produces a constant e.m.f. of 1.0186 volts at 20°C. It consists of an H-shaped glass vessel containing a negative cadmium-mercury amalgam electrode in one leg and a positive mercury electrode in the other. The electrolyte, a saturated cadmium sulfate solution, fills the horizontal bar of the vessel to connect the two electrodes. The e.m.f. of the cell varies very little with temperature, being given by the equation $E = 1.0186 - 0.000\,037\,(T - 293)$, where T is the thermodynamic temperature.

wet cell A type of cell, such as a car battery, in which the electrolyte is a liquid solution.

white arsenic *See* arsenic(III) oxide.

white cast iron *See* cast iron.

white gold *See* palladium.

white lead *See* lead(II) carbonate hydroxide.

white phosphorus *See* phosphorus.

white spirit A liquid hydrocarbon resembling kerosene obtained from petroleum, used as a solvent and in the manufacture of paints and varnishes.

Wilkinson, Sir Geoffrey (1921–96) British inorganic chemist. Wilkinson is noted for his studies of inorganic complexes. He shared the Nobel Prize for chemistry in 1973 with Ernst FISCHER for work on sandwich compounds. A theme of Wilkinson's work in the 1960s was the study and use of complexes containing a metal–hydrogen bond. Thus complexes of rhodium with triphenyl phosphine ($(C_6H_5)_3P$) can react with molecular hydrogen. The compound $RhCl(P(C_6H_5)_3)$, known as *Wilkinson's catalyst*, was the first such complex to be used as a homogeneous catalyst for adding hydrogen to the double bonds of alkenes (hydrogenation). This type of compound can also be used as a catalyst for the reaction of hydrogen and carbon monoxide with alkenes (hydroformylation). It is the basis of industrial low-pressure processes for making aldehydes from ethene and propene.

witherite *See* barium carbonate.

wolfram A former name for tungsten.

Wood's metal A fusible alloy containing 50% bismuth, 25% lead, and 12.5% tin and cadmium. Its low melting point (70°C) leads to its use in fire-protection devices.

Woodward–Hoffmann rules A set of rules concerning the course of chemical reactions proposed by the American chemists Robert Burns Woodward and Roald Hoffmann in the late 1960s. These rules can be stated in terms of molecular orbital theory (particularly FRONTIER ORBITAL THEORY). This gives rise to the Woodward–Hoffmann rules being referred to as the *conservation of orbital symmetry*. The rules were initially developed for organic reactions but can also be applied to inorganic reactions.

work function *See* photoelectric effect.

wrought iron A low-carbon steel obtained by refining the iron produced in a blast furnace. Wrought iron is processed by heating and hammering to reduce the slag content and to ensure its even distribution.

XYZ

xenon A colorless odorless monatomic gas, the fifth member of the rare-gases; i.e. group 18 (formerly VIIIA or 0) of the periodic table. It occurs in trace amounts in air from which it is recovered by fractional distillation. Xenon is used in electron tubes, strobe lighting, arc lamps, and lasers. Xenon compounds with fluorine and oxygen are known.

Symbol: Xe; m.p. –111.9°C; b.p. –107.1°C; mass density 5.8971 (0°C) kg m^{-3}; p.n. 54; most common isotope ^{132}Xe; r.a.m. 131.29.

x-radiation An energetic form of electromagnetic radiation. The wavelength range is 10^{-11} m to 10^{-8} meter. X-rays are normally produced when high-energy electrons are absorbed in matter. The radiation can pass through matter to some extent, hence its use in medicine and industry for investigating internal structures. It can be detected with photographic emulsions and devices like the Geiger-Müller tube.

X-ray photons result from electronic transitions between the inner energy levels of atoms. When high-energy electrons are absorbed by matter, an x-ray line spectrum results. The structure depends on the substance and is thus used in x-ray spectroscopy. The line spectrum is always formed in conjunction with a continuous background spectrum. The minimum (cutoff) wavelength λ_0 corresponds to the maximum x-ray energy, W_{max}. This equals the maximum energy of electrons in the beam producing the x-rays. Wavelengths in the continuous spectrum above λ_0 are caused when electrons decelerate and lose energy rapidly, such as when they collide with a nuclear target or pass through an electric field. X-radiation emitted in this process is called *bremsstrahlung* (German for braking radiation).

x-ray crystallography The study of the internal structure of crystals using the technique of x-ray diffraction.

x-ray diffraction A technique used to determine crystal structure by directing x-rays at the crystals and examining the diffraction patterns produced. At certain angles of incidence a series of spots are produced on a photographic plate; these spots are caused by interaction between the x-rays and the planes of the atoms, ions, or molecules in the crystal lattice. The positions of the spots are consistent with the Bragg equation $n\lambda = 2d\sin\theta$.

x-rays *See* x-radiation.

yellow phosphorus *See* phosphorus.

yocto- Symbol: y A prefix used with SI units, denoting 10^{-24}. For example, 1 yoctometer (ym) = 10^{-24} meter (m).

yotta- Symbol: Y A prefix used with SI units, denoting 10^{24}. For example, 1 yottameter (Ym) = 10^{24} meter (m).

ytterbium A soft malleable ductile silvery element belonging to the lanthanoid series of metals. It occurs in association with other lanthanoids in minerals such as gadolinite, monazite, and xenotime. Ytterbium has been used to improve the mechanical properties of steel. It is also used in ceramics.

Symbol: Yb; m.p. 824°C; b.p. 1193°C; r.d. 6.965 (20°C); p.n. 70; most common isotope ^{174}Yb; r.a.m. 173.04.

yttrium A silvery metallic element, the second element of group 3 (formerly IIIB) of the periodic table. It is found in almost every lanthanoid mineral, particularly monazite. Yttrium is used in various alloys, in yttrium–aluminum garnets used in the electronics industry and as gemstones, as a catalyst, and in superconductors. A mixture of yttrium and europium oxides is widely used as the red phosphor on television screens.

Symbol: Y; m.p. 1522°C; b.p. 3338°C; r.d. 4.469 (20°C); p.n. 39; r.a.m. 88.90585.

Zeeman effect The splitting of atomic spectral lines by a magnetic field. This effect was found by the Dutch physicist Pieter Zeeman (1865–1943) in 1896. Some of the patterns of line splitting that be explained both by classical electron theory and the BOHR THEORY of electrons in atoms. The Zeeman splitting that can be explained in these ways is known as the *normal Zeeman effect.* There exist more complicated Zeeman splitting patterns that cannot be explained either by classical electron theory or the Bohr theory. This more complicated type of Zeeman effect is known as the *anomalous Zeeman effect.* It was subsequently realized that the anomalous Zeeman effect occurs because of electron spin and that the normal Zeeman effect occurs only for transitions between singlet states.

At very high magnetic fields a splitting pattern known as the *Paschen–Back effect* occurs. In this pattern, named for the German spectroscopists Friedrich Paschen and Ernst Back in 1912, the basic pattern returns to that of the normal Zeeman effect, but with each line split up into a set of closely spaced lines. This occurs because the total orbital and spin angular momentum vectors of the atom, denoted L and S respectively, precess independently about the direction of the magnetic field.

Zeise's salt A complex of platinum and ethane (ethylene) first synthesized by W. C. Zeise in 1827. It has the formula $PtCl_3(CH_2CH_2)$ and was the first complex in which the metal ion was known to coordinate to a pi-electron system rather than to individual atoms.

zeolite A member of a group of hydrated aluminosilicate minerals, which occur in nature and are also manufactured for their ion-exchange and selective-absorption properties. They are used for water softening and for sugar refining. The zeolites have an open crystal structure and can be used as molecular sieves.

See also ion exchange; molecular sieve.

zepto- Symbol: z A prefix used with SI units, denoting 10^{-21}. For example, 1 zeptometer (zm) = 10^{-21} meter (m).

zero order Describing a chemical reaction in which the rate of reaction is independent of the concentration of a reactant; i.e.

$$\text{rate} = k[\text{X}]^0$$

The concentration of the reactant remains constant for a period of time although other reactants are being consumed. The hydrolysis of 2-bromo-2-methylpropane using aqueous alkali has a rate expression,

$$\text{rate} = k[\text{2-bromo-2-methylpropane}]$$

i.e. the reaction is zero order with respect to the concentration of the alkali. The rate constant for a zero reaction has the units mol dm^{-3} s^{-1}.

zero point energy The energy possessed by the atoms and molecules of a substance at absolute zero (0 K).

zetta- Symbol: Z A prefix used with SI units, denoting 10^{21}. For example, 1 zettameter (Zm) = 10^{21} meter (m).

Zewail, Ahmed H. (1946–) Egyptian-born American chemist. Zewail has pioneered the technique called *femtochemistry* for studying chemical reactions. Zewail and his colleagues have studied the detailed dynamics of many chemical reactions using this technique. This has justified the picture of chemical reactions given by Svante ARRHENIUS, Henry EYRING, and others as well as giving many surprising discoveries. Zewail won the 1999 Nobel Prize for chemistry.

zinc A bluish-white hard brittle reactive transition metal, the first element of group 12 (formerly IIB) of the periodic table. It occurs naturally chiefly as the sulfide (zinc blende) and carbonate (smithsonite). It is extracted by roasting the ore in air and then reducing the oxide to the metal using carbon. Zinc is used to galvanize iron, in alloys (e.g. brass), and in dry batteries. It reacts with acids and alkalis but only corrodes on the surface in air. Zinc ions are identified in solution by forming white precipitates with sodium hydroxide or ammonia solution, both precipitates being soluble in excess reagent. Zinc compounds are used in paints, cosmetics, and medicines. *See also* zinc group.

Symbol: Zn; m.p. 419.58°C; b.p. 907°C; r.d. 7.133 (20°C); p.n. 30; most common isotope ^{64}Zn; r.a.m. 65.39.

zincate A salt containing the ion ZnO_2^{2-}.

zinc-blende (sphalerite) structure A form of crystal structure that consists of a zinc atom surrounded by four sulfur atoms arranged tetrahedrally; each sulfur atom is similarly surrounded by four atoms of zinc. Zinc sulfide crystallizes in the cubic system. Covalent bonds of equal strength and length result in the formation of a giant molecular structure. If the zinc and sulfur atoms are replaced by carbon atoms the diamond structure is produced. Wurtzite (another form of zinc sulfide) is similar but belongs to the hexagonal crystal system.

zinc carbonate *See* calamine.

zinc chloride ($ZnCl_2$) A white crystalline solid prepared by the action of dry hydrogen chloride or chlorine on heated zinc. The anhydrous product undergoes sublimation. The hydrated salt may be prepared by the addition of excess zinc to dilute hydrochloric acid and crystallization of the solution. Hydrates with 4, 3, 5/2, 3/2, and 1 molecule of water exist. Zinc chloride is extremely deliquescent and dissolves easily in water. It is used in dentistry, as a flux, and as a dehydrating agent in organic reactions.

zinc group A group of metallic elements in the periodic table, consisting of zinc (Zn), cadmium (Cd), and mercury (Hg). The elements all occur at the ends of the three transition series and all have outer $d^{10}s^2$ configurations; Zn $[Ar]3d^{10}4s^2$, Cd $[Kr]4d^{10}5s^2$, $Hg[Xn]5d^{10}6s^2$. The elements all use only outer s-electrons in reaction and in combination with other elements unlike the coinage metals, which immediately precede them. Thus all form dipositive ions, and oxidation states higher than +2 are not known. In addition mercury forms the mercurous ion, sometimes written as Hg(I), but actually the ion $^{+}Hg\text{-}Hg^{+}$. Although the elements fall naturally at the end of the transition series their properties are more like those of main-group elements than of transition metals:

1. The melting points of the transition elements are generally high whereas those of the zinc group are low (Zn 419.5°C, Cd 329.9°C, Hg –38.87°C).

2. The zinc-group ions are diamagnetic and colorless whereas most transition elements have colored ions and many paramagnetic species.

3. The zinc group does not display the variable valence associated with transition metal ions. However the group does have the transition-like property of forming many complexes or coordination compounds, such as $[Zn(NH_3)_4]^{2+}$ and $[Hg(CN)_4]^{2-}$.

Within the group, mercury is anomalous because of the Hg_2^{2+} ion mentioned previously and its general resistance to reaction. Zinc and cadmium are electropositive and will react with dilute acids to release hydrogen whereas mercury will not. When zinc and cadmium are heated in air they burn to give the oxides, MO, which are stable to further strong heating. In contrast mercury does not react readily with oxygen (slow reaction at the boiling point), and mercuric oxide, HgO, decomposes to the metal and oxygen on further heating. All members of the group form dialkyls and diaryls, R_2M.

zincite (spartalite) A red-orange mineral form of zinc oxide, ZnO, often containing also some manganese. It is an important ore of zinc. *See* zinc oxide.

zinc oxide (ZnO) A compound prepared by the thermal decomposition of zinc nitrate or carbonate; it is a white powder when cold, yellow when hot. Zinc oxide is an amphoteric oxide, almost insoluble in water, but dissolves readily in both acids and alkalis. If mixed with powdered carbon and heated to red heat, it is reduced to the metal. Zinc oxide is used in the paint and ceramic industries. Medically it is used in zinc ointment.

zinc sulfate ($ZnSO_4$) A white crystalline solid prepared by the action of dilute sulfuric acid on either zinc oxide or zinc carbonate. On crystallization the hydrated salt is formed. The heptahydrate ($ZnSO_4.7H_2O$) is formed below 30°C, the hexahydrate ($ZnSO_4.6H_2O$) above 30°C, and the monohydrate ($ZnSO_4.H_2O$) at 100°C; at 450°C the salt is anhydrous. Zinc sulfate is extremely soluble in water. It is used in the textile industry.

zinc sulfide (ZnS) A compound that can be prepared by direct combination of zinc and sulfur. Alternatively it can be prepared as a white amorphous precipitate by bubbling hydrogen sulfide through an alkaline solution of a zinc salt or by the addition of ammonium sulfide to a soluble zinc salt solution. Zinc sulfide occurs naturally as the mineral zinc blende. It dissolves in dilute acids to yield hydrogen sulfide. Impure zinc sulfide is phosphorescent. It is used as a pigment and in the coatings on luminescent screens.

zircon *See* zirconium.

zirconium A hard lustrous silvery transition element the second element of group 4 (formerly IVB) of the periodic table. It occurs chiefly in the mineral zircon ($ZrSiO_4$) some deposits of which are of gemstone quality. It is extracted by chlorination, followed by reduction with magnesium. It is used in some strong alloy steels and as a protective coating. It is very corrosion-resistant.

Symbol: Zr; m.p. 1850°C; b.p. 4380°C; r.d. 6.506 (20°C); p.n. 40; most common isotope ^{90}Zr; r.a.m. 91.224.

zwitterion (ampholyte ion) An ion that has both a positive and negative charge on the same species. Zwitterions occur when a molecule contains both a basic group and an acidic group.

APPENDIXES

Appendix I

Periodic Table of the Elements - giving group, atomic number, and chemical symbol

Period	1	2	3	4	5	6	7	8	9	10	11	12	13	14	15	16	17	18
1	1 H																	2 He
2	3 Li	4 Be											5 B	6 C	7 N	8 O	9 F	10 Ne
3	11 Na	12 Mg											13 Al	14 Si	15 P	16 S	17 Cl	18 Ar
4	19 K	20 Ca	21 Sc	22 Ti	23 V	24 Cr	25 Mn	26 Fe	27 Co	28 Ni	29 Cu	30 Zn	31 Ga	32 Ge	33 As	34 Se	35 Br	36 Kr
5	37 Rb	38 Sr	39 Y	40 Zr	41 Nb	42 Mo	43 Tc	44 Ru	45 Rh	46 Pd	47 Ag	48 Cd	49 In	50 Sn	51 Sb	52 Te	53 I	54 Xe
6	55 Cs	56 Ba	57-71 La-Lu	72 Hf	73 Ta	74 W	75 Re	76 Os	77 Ir	78 Pt	79 Au	80 Hg	81 Tl	82 Pb	83 Bi	84 Po	85 At	86 Rn
7	87 Fr	88 Ra	89-103 Ac-Lr	104 Rf	105 Db	106 Sg	107 Bh	108 Hs	109 Mt	110 Ds	111 Uuu	112 Uub	113	114 Uuq	115	116 Uuh		

Period																
6	Lanthanides	57 La	58 Ce	59 Pr	60 Nd	61 Pm	62 Sm	63 Eu	64 Gd	65 Tb	66 Dy	67 Ho	68 Er	69 Tm	70 Yb	71 Lu
7	Actinides	89 Ac	90 Th	91 Pa	92 U	93 Np	94 Pu	95 Am	96 Cm	97 Bk	98 Cf	99 Es	100 Fm	101 Md	102 No	103 Lr

The above is the modern recommended form of the table using 18 groups. Older group designations are shown below.

Modern form	1	2	3	4	5	6	7	8	9	10	11	12	13	14	15	16	17	18
European convention	IA	IIA	IIIA	IVA	VA	VIA	VIIA	VIII (or VIIIA)			IB	IIB	IIIB	IVB	VB	VIB	VIIB	0 (or VIIIB)
N. American convention	IA	IIA	IIIB	IVB	VB	VIB	VIIB	VIII (or VIIIB)			IB	IIB	IIIA	IVA	VA	VIA	VIIA	VIIIA (or 0)

Appendix II

The Chemical Elements

(* *indicates the nucleon number of the most stable isotope*)

Element	*Symbol*	*p.n.*	*r.a.m*	*Element*	*Symbol*	*p.n.*	*r.a.m*
actinium	Ac	89	227*	europium	Eu	63	151.965
aluminum	Al	13	26.982	fermium	Fm	100	257*
americium	Am	95	243*	fluorine	F	9	18.9984
antimony	Sb	51	112.76	francium	Fr	87	223*
argon	Ar	18	39.948	gadolinium	Gd	64	157.25
arsenic	As	33	74.92	gallium	Ga	31	69.723
astatine	At	85	210	germanium	Ge	32	72.61
barium	Ba	56	137.327	gold	Au	79	196.967
berkelium	Bk	97	247*	hafnium	Hf	72	178.49
beryllium	Be	4	9.012	hassium	Hs	108	265*
bismuth	Bi	83	208.98	helium	He	2	4.0026
bohrium	Bh	107	262*	holmium	Ho	67	164.93
boron	B	5	10.811	hydrogen	H	1	1.008
bromine	Br	35	79.904	indium	In	49	114.82
cadmium	Cd	48	112.411	iodine	I	53	126.904
calcium	Ca	20	40.078	iridium	Ir	77	192.217
californium	Cf	98	251*	iron	Fe	26	55.845
carbon	C	6	12.011	krypton	Kr	36	83.80
cerium	Ce	58	140.115	lanthanum	La	57	138.91
cesium	Cs	55	132.905	lawrencium	Lr	103	262*
chlorine	Cl	17	35.453	lead	Pb	82	207.19
chromium	Cr	24	51.996	lithium	Li	3	6.941
cobalt	Co	27	58.933	lutetium	Lu	71	174.967
copper	Cu	29	63.546	magnesium	Mg	12	24.305
curium	Cm	96	247*	manganese	Mn	25	54.938
darmstadtium	Ds	110	269*	meitnerium	Mt	109	266*
dubnium	Db	105	262*	mendelevium	Md	101	258*
dysprosium	Dy	66	162.50	mercury	Hg	80	200.59
einsteinium	Es	99	252*	molybdenum	Mo	42	95.94
erbium	Er	68	167.26	neodymium	Nd	60	144.24

The Chemical Elements

Element	*Symbol*	*p.n.*	*r.a.m*	*Element*	*Symbol*	*p.n.*	*r.a.m*
neon	Ne	10	20.179	scandium	Sc	21	44.956
neptunium	Np	93	237.048	seaborgium	Sg	106	263*
nickel	Ni	28	58.69	selenium	Se	34	78.96
niobium	Nb	41	92.91	silicon	Si	14	28.086
nitrogen	N	7	14.0067	silver	Ag	47	107.868
nobelium	No	102	259*	sodium	Na	11	22.9898
osmium	Os	76	190.23	strontium	Sr	38	87.62
oxygen	O	8	15.9994	sulfur	S	16	32.066
palladium	Pd	46	106.42	tantalum	Ta	73	180.948
phosphorus	P	15	30.9738	technetium	Tc	43	99*
platinum	Pt	78	195.08	tellurium	Te	52	127.60
plutonium	Pu	94	244*	terbium	Tb	65	158.925
polonium	Po	84	209*	thallium	Tl	81	204.38
potassium	K	19	39.098	thorium	Th	90	232.038
praseodymium	Pr	59	140.91	thulium	Tm	69	168.934
promethium	Pm	61	145*	tin	Sn	50	118.71
protactinium	Pa	91	231.036	titanium	Ti	22	47.867
radium	Ra	88	226.025	tungsten	W	74	183.84
radon	Rn	86	222*	uranium	U	92	238.03
rhenium	Re	75	186.21	vanadium	V	23	50.94
rhodium	Rh	45	102.91	xenon	Xe	54	131.29
rubidium	Rb	37	85.47	ytterbium	Yb	70	173.04
ruthenium	Ru	44	101.07	yttrium	Y	39	88.906
rutherfordium	Rf	104	261*	zinc	Zn	30	65.39
samarium	Sm	62	150.36	zirconium	Zr	40	91.22

Appendix III

The Greek Alphabet

Α	α	alpha	Ν	ν	nu
Β	β	beta	Ξ	ξ	xi
Γ	γ	gamma	Ο	ο	omikron
Δ	δ	delta	Π	π	pi
Ε	ε	epsilon	Ρ	ρ	rho
Ζ	ζ	zeta	Σ	σ	sigma
Η	η	eta	Τ	τ	tau
Θ	θ	theta	Υ	υ	upsilon
Ι	ι	iota	Φ	φ	phi
Κ	κ	kappa	Χ	χ	chi
Λ	λ	lambda	Ψ	ψ	psi
Μ	μ	mu	Ω	ω	omega

Appendix IV

Fundamental Constants

speed of light	c	$2.997\ 924\ 58 \times 10^{8}$ m s^{-1}
permeability of free space	μ_o	$4\pi \times 10^{-7}$ = $1.256\ 637\ 0614 \times 10^{-6}$ H m^{-1}
permittivity of free space	$\varepsilon_0=\mu_0^{-1}c^{-2}$	$8.854\ 187\ 817 \times 10^{-12}$ F m^{-1}
charge of electron or proton	e	$\pm 1.602\ 177\ 33 \times 10^{-19}$ C
rest mass of electron	m_e	$9.109\ 39 \times 10^{-31}$ kg
rest mass of proton	m_p	$1.672\ 62 \times 10^{-27}$ kg
rest mass of neutron	m_n	$1.674\ 92 \times 10^{-27}$ kg
electron charge-to-mass ratio	e/m	$1.758\ 820 \times 10^{11}$ C kg^{-1}
electron radius	r_e	$2.817\ 939 \times 10^{-15}$ m
Planck constant	h	$6.626\ 075 \times 10^{-34}$ J s
Boltzmann constant	k	$1.380\ 658 \times 10^{-23}$ J K^{-1}
Faraday constant	F	$9.648\ 531 \times 10^{4}$ C mol^{-1}

Appendix V

Webpages

Chemical society webpages include:

American Chemical Society	www.chemistry.org
Royal Society of Chemistry	www.rsc.org
The International Union of Pure and Applied Chemistry	www.iupac.org

Information on nomenclature is available at:

Queen Mary College, London	www.chem.qmul.ac.uk/iupac
Advanced Chemistry Development, Inc	www.acdlabs.com/iupac/nomenclature

The definitive site for the chemical elements is:

WebElements Periodic Table	www.webelements.com

Bibliography

There are a number of comprehensive texts covering inorganic chemistry. These include:

Cotton, F. A.; Murillo, C.; Wilkinson, G.; Bochmann, M. & Grimes, R. *Advanced Inorganic Chemistry*. 6th ed. New York: Wiley, 1999

Greenwood, N. N. & Earnshaw, A. *Chemistry of the Elements*. Oxford, U.K.: Butterworth-Heinemann, 1997

Shriver, D. F. & Atkins, P. W. *Inorganic Chemistry*. 3rd ed. Oxford, U.K.: Oxford University Press, 1999

Additional useful sources are:

Emsley, J. *Nature's Building Blocks – An A-Z Guide to the Elements*. Oxford, U.K.: Oxford University Press, 2001

King, R.B. (ed.) *Encyclopedia of Inorganic Chemistry*. New York: Wiley, 1994